Niels Bohr and the
Philosophy of Physics

Also available from Bloomsbury

The Bloomsbury Companion to the Philosophy of Science,
edited by Steven French and Juha Saatsi
Philosophy of Science: Key Concepts, Steven French
Thomas Kuhn's Revolutions, James A. Marcum

Niels Bohr and the Philosophy of Physics

Twenty First-Century Perspectives

Edited by
Jan Faye and Henry J. Folse

BLOOMSBURY ACADEMIC
LONDON • NEW YORK • OXFORD • NEW DELHI • SYDNEY

BLOOMSBURY ACADEMIC
Bloomsbury Publishing Plc
50 Bedford Square, London, WC1B 3DP, UK
1385 Broadway, New York, NY 10018, USA

BLOOMSBURY, BLOOMSBURY ACADEMIC and the Diana logo
are trademarks of Bloomsbury Publishing Plc

First published 2017
Paperback edition first published 2019

A catalogue record for this book is available from the British Library.

A catalog record for this book is available from the Library of Congress.

ISBN: HB: 978-1-3500-3511-9
PB: 978-1-3501-0903-2
ePDF: 978-1-3500-3512-6
ePub: 978-1-3500-3513-3

Series: Social Theory and Methodology in Education Research

Typeset by Newgen Knowledge Works Pvt. Ltd., Chennai, India

To find out more about our authors and books visit
www.bloomsbury.com and sign up for our newsletters.

Contents

Introduction *Jan Faye and Henry J. Folse*   1

Part One   Understanding Bohr's Philosophical Background   17

1   Why Do We Find Bohr Obscure?: Reading Bohr as a Philosopher of
    Experiment *Kristian Camilleri*   19

2   On Bohr's Transcendental Research Program *Michel Bitbol*   47

3   Transcendental versus Quantitative Meanings of Bohr's Complementarity
    Principle *Patricia Kauark-Leite*   67

4   Complementarity and Pragmatic Epistemology: A Comparison of Bohr
    and C. I. Lewis *Henry J. Folse*   91

5   Complementarity and Human Nature *Jan Faye*   115

6   Bohr's Relational Holism and the Classical-Quantum Interaction
    Mauro Dorato   133

7   Complementarity as a Route to Inferentialism *Stefano Osnaghi*   155

8   Fragmentation, Multiplicity, and Technology in Quantum Physics:
    Bohr's Thought from the Twentieth to the Twenty-First Century
    Arkady Plotnitsky   179

Part Two   Bohr's Interpretation of Quantum Mechanics in
            Twenty-First-Century Physics   205

9   Complementarity and Quantum Tunneling *Slobodan Perović*   207

10 Bohr and the Problem of the Quantum-to-Classical Transition
    Maximillian Schlosshauer and Kristian Camilleri   223

11 On Bohr's Epistemological Contribution to the Quantum-Classical Cut
    Problems *Manuel Bächtold*   235

12 Individuality and Correspondence: An Exploration of the History and
    Possible Future of Bohrian Quantum Empiricism *Scott Tanona*   253

13 An Everett Perspective on Bohr and EPR *Guido Bacciagaluppi*   289

14 Niels Bohr and the Formalism of Quantum Mechanics *Dennis Dieks*   303

15 Bohrification: From Classical Concepts to Commutative Operator
    Algebras *Klaas Landsman*   335

16 Why QBism Is Not the Copenhagen Interpretation and What John Bell
    Might Have Thought of It *N. David Mermin*   367

Index   379

Introduction

Jan Faye and Henry J. Folse

Roughly a quarter century ago, in the years following the centenary of Niels Bohr's birth in 1885, there was a rather sudden surge in publications presenting somewhat more sophisticated and subtle interpretations of his thinking than—with a few exceptions—had been the general rule in earlier decades. We took advantage of this turn of events to collect a set of seventeen chapters by some of the scholars who had contributed to this new literature. It was published in 1994 in the series *Boston Studies in Philosophy of Science* under the title of *Niels Bohr and Contemporary Philosophy*

Now we are more than a decade into the twenty-first century and approaching the centenary of Niels Bohr's interpretation of quantum mechanics. Meanwhile a whole new generation of scholars writing on Bohr has come to the fore. The issues that were most conspicuous in our previous collection were very different from those that are prominent in the present volume. In the 1990s many of our contributors considered questions related to the then contemporary debates over scientific realism. Now those issues are largely in the background, and our present contributors are much more concerned with Bohr's views on the use of classical concepts, how to draw the distinction between classical and quantum descriptions, and how Bohr's interpretation relates to de-coherence, QBism, and other non-collapse views. Furthermore many new developments in quantum physics naturally lead us to ask how Bohr's views would stand with respect to them. Thus it is appropriate to pause and reflect upon the philosophical traditions that may have nurtured his interpretation as well as its current status in the physics of today. Over the years an astounding number of books and articles have discussed manifold aspects of Bohr's thinking. Nevertheless, like the interpretation of quantum theory itself, the interpretation of Bohr's philosophical viewpoint still remains a matter of lively dispute.

1.

Even though complete consensus concerning the whole of Bohr's philosophy is likely to be unattainable, clarification of his philosophical intentions is a rational goal, for he stood squarely at the center of that iconoclastic transformation in physics that we know as the "quantum revolution." In this context it is encouraging that contemporary scholarship has converged on at least some prominent themes that present great promise for a better understanding of Bohr. For example, more and more scholars point to Kant's philosophy as a possible source for understanding the "viewpoint" Bohr called "complementarity." The contributors to this volume are no exception; among them there are several who see strong similarities between Bohr's philosophy and the Kantian way of thinking. However, other scholars recognize elements of epistemological naturalism in general and of pragmatism in particular in Bohr's approach to quantum mechanics.

These different roots of complementarity need not be regarded as exclusive of each other. Although the naturalistic attitude of most pragmatists is starkly opposed to Kant's transcendental method, and the pragmatists saw themselves as turning the page on epistemology as it was conceived during the Enlightenment, there is undeniable evidence that in his philosophical upbringing Bohr was exposed to both of these influences. Bohr had a long and close relationship to the Danish philosopher Harald Høffding, who not only taught Kant's philosophy but also exhibited Kant's influence in his own thought. Moreover, Høffding corresponded with William James and visited him while Bohr was a student. In the year after Bohr enrolled at the University of Copenhagen, Høffding held a series of seminars about modern philosophical theories that Bohr seems to have attended. Here American pragmatism, focusing on the thought of William James, was one of the topics.

Although they had different goals, pragmatists broadly agreed with Kant that the mind provides a conceptual element that it imposes on a sensory input in generating the empirical world of phenomena with which natural science is concerned. And this is an important theme that appears repeatedly in Bohr's references to the "epistemological lesson" that the quantum revolution has taught. Nevertheless, Kant is working with a conception of knowledge that requires certainty, and this is what leads him to embark on his transcendental approach, which aspires to prove the scheme of categories that Kant claimed to deduce from the nature of Reason are universal and necessary for all human experience. In contrast the pragmatists appealed to a naturalistic account of knowledge, which relinquishes the claim of certainty in favor of a much less stringent criterion of pragmatic justification in terms of successful expectations and actions. In accepting a fallibilist approach to knowledge, pragmatists allow the possibility that what it is reasonable to say we know today may not be reasonable to say we know tomorrow. Thus pragmatism points the way to a dynamic theory of knowledge deeply intertwined with its growth over time. This again is important for appreciating Bohr's thought, for standing at the epicenter of the quantum revolution as he did, he was deeply personally invested in making the case for conceptual change in atomic physics. Kantians and pragmatists agree that we are not merely spectators of an objectively existing world, but that we actively interpret and interact with the world. However, only pragmatists are open to the possibility that the conceptual element in knowledge changes over time as new cognitive standards and goals replace old ones with the expansion of human experience into new domains of phenomena. The pragmatist grounds these changing criteria in human nature as biological, physical beings immersed in the empirical world. While William James was not himself a defender of epistemological naturalism, from its inception with C. S. Peirce pragmatism strongly emphasized scientific knowledge as the paradigm of empirical knowledge and disdained the characteristic nineteenth-century interest in matters transcendent. The methodology of natural science was understood as embodying the key tenets of pragmatism.

Another theme that connects Bohr with Kant and the pragmatists is their rejection of our ability to have knowledge of the world as it is in itself. Both of these traditions accept that our scheme of categories provides non-analytic knowledge only as applied to sense impressions and does not justify any application beyond our sensory experience. We can very well understand the world as it is given empirically, but as soon as we attempt to claim knowledge of the world as it really is independently of our experience of it, we speculate beyond the proper limits of our cognitive faculties. No scientific knowledge is possible about the things-in-themselves, which are empirically inaccessible hidden behind the veil of phenomena. While Kant himself spoke of things-in-themselves as a limiting notion, the name for an unknown realm that the Understanding could not survey, the ink was hardly dry on his first *Critique* before philosophers began to try to find ways around this limit. However, pragmatists, who see the whole of human cognitive apparatus and its products in terms of the needs of natural beings to cope with the world of experience, a realm of objects beyond human experience, transcendent beings, are, quite literally, of no interest. And this also seems to be a view with which Bohr's whole philosophical outlook was in warm accord.

Another theme on which many interpreters have focused is Bohr's insistence on the retention of classical concepts within quantum mechanics. This can be understood as parallel to Kant's claim to have proved that space and time are the forms of intuition necessary for any possible experience. Our sense impressions provide the content for our experience of the world only after they are synthesized in representations with the spatial and temporal forms of intuition and the categories of judgment such as substance, unity, plurality, and causation. In contrast pragmatists seek no such transcendental defense for the unavoidability of the classical concepts, but anchor the indispensable nature of their use in the claim that like all of our faculties, our sensory perceptual faculties are the way they are due to the evolutionary pressure of survival. Thus the classical concepts have evolved for unambiguous communication about the objects of everyday human experience: the objects with which human beings daily interact. According to Bohr, classical physics has been able to define and operationalize these categories to such a degree that their use can be extended beyond the objects of common-sense experience to the physical interpretation of mathematical theories. Our natural language embodies the categories in terms of which we have learned to form a world of causally interacting physical objects in space and time, and it is this use of natural language which allows us to give an empirical meaning to the mathematical symbolism, and thus to use the results of experiments and observation to justify that theoretical structure.

Therefore Bohr held that the classical concepts could not be replaced by other concepts since they are the only concepts by which we can understand and communicate "what we have seen and learned." But at the same time, he was also acutely conscious of the one big obstacle for applying the classical concepts to the results of quantum mechanics. The discovery of the quantum of action, symbolized by Planck's constant, made it impossible to give a classical description of atomic objects. Classical concepts like position, duration, energy, and momentum cannot apply unambiguously to the trajectories of free particles, isolated from any interaction, because in quantum mechanics such trajectories are in principle unobservable since observation requires that the object be interacting with the observing instruments in a way that cannot be ignored or compensated for. But Bohr clearly maintained that the classical concepts could—and indeed must—still be used to describe the outcome of all experiments. There would never be inconsistency in the ascription of incompatible properties to the atomic object based on classical concepts as long as this application was understood to be only in relation to a particular experimental phenomenon. This insistence on the classical concepts while at the same time limiting their proper use gave rise to the viewpoint he called "complementarity." Thus, he argued that the application of the classical concepts is well defined only in relation to a specific phenomenon observed in a particular experimental context and has no physical meaning outside of an experimental context. The consequence is that a quantum system does not have well-defined kinematic or dynamical properties independently of any measurement.

A related theme essential to Bohr's understanding of quantum mechanics is his denial of the representational function of the mathematical formalism. Much of what has been proposed concerning the interpretation of quantum theory fails to follow Bohr's lead in this respect, thus becoming lost in a maze of branching universes and the mystery of what is represented by the nonphysical "collapse" of the wave function. Like all empiricists, Bohr held that the acceptability of a theory is a function of its utility in making successful predictions, but he seems to add to that the pragmatic position that scientific theories are tools or instruments for handling and describing reality as it appears to us, rather than attempts at a representation of the world as it is in itself. So Bohr refused to regard the so-called collapse of Schrödinger's wave function as an issue calling out for interpretation, and he never addressed the "paradox" of Schrödinger's infamous cat. The so-called collapse was beyond any possible experimental inquiry. Where Schrödinger and the anti-Copenhagen physicists saw a paradox, Bohr saw a misunderstanding of the significance of physical theory.

Although Bohr never publically distanced himself from any of the various realist interpretations of the wave function, he repeatedly referred to it as a symbol for calculating probabilities. He seems to have thought that the Born rule that interprets the wave function as a probability amplitude is the only role for the wave function. He certainly never refers to the "collapse" of the wave function as representing a physical process of some abstract, non-empirical form. Moreover, Bohr strongly believed that any attempt to use the quantum formalism to try to characterize a world "behind the phenomena" would be mere metaphysical speculation, and an abuse of the concepts which were well defined for describing the objects of human experience. To him the formalism was a tool for predictions that was "symbolic" and should not be treated as if it gave us a "picture" of what the world looks like when no one is looking at it.

In earlier decades when the influence of Karl Popper was at its peak it was common to treat Bohr as advocating a subjectivistic interpretation, but today no recent scholars or philosophers of science would assume that Bohr understood complementarity to imply that the mind or consciousness of the observer has any influence on

experimental outcomes in quantum mechanics. Bohr's philosophical roots in a naturalistic pragmatism make it evident that earlier accusations of subjectivism are without
any historical support. While it is easy to find such misinterpretations in the popular
press, anyone who reads Bohr carefully can see that his references to "the observer"
refer to the observer qua physical system, not qua consciousness. It would appear that
this view was often imputed to Bohr by those anxious to discredit the "Copenhagen
interpretation," and it is true that at least some physicists—who may or may not have
imagined that they were upholding Bohr's position—have indeed held this view. In
fact Bohr did emphasize that the physicist makes a free conscious choice in selecting
the physical context in which a particular dynamical variable is well-defined, but once
that context has been chosen no conscious mind is regarded as having any influence
on the result.

At one point he expressed his dismay with what he considered to be Heisenberg's
more subjectivist view. Heisenberg had once suggested that the observer partly determines the outcome by his reading of the measuring instrument. Referring to the
debate that took place at the Solvay conference in 1927 in "Discussion with Einstein
about Epistemological Problems in Atomic Physics" Bohr (1958, 51) writes about this
proposal:

> On that occasion an interesting discussion arose also about how to speak of the
> appearance of phenomena for which only predictions of statistical character can
> be made. The question was whether, as to the occurrence of individual effects,
> we should adopt a terminology proposed by Dirac, that we were concerned with
> a choice on the part of "nature" or, as suggested by Heisenberg, we should say
> that we have to do with a choice on the part of the "observer" constructing the
> measuring instruments and reading their recording. Any such terminology would,
> however, appear dubious since, on the one hand, it is hardly reasonable to endow
> nature with volition in the ordinary sense, while, on the other hand, it is certainly
> not possible for the observer to influence the events which may appear under the
> conditions he has arranged.

Bohr (1963, 6–7) certainly appreciated the fact that in its emphasis on motion as
relative to a frame of reference the theory of relativity was misunderstood by some
as injecting subjectivity into the description of motion, and an analogous misunderstanding was common in the case of quantum theory because of its emphasis on the
role of the observing instruments:

> Notwithstanding all difference in the typical situations to which the notions of
> relativity and complementarity apply, they present in epistemological respects far-
> reaching similarities. Indeed, in both cases we are concerned with the exploration
> of harmonies which cannot be comprehended in the pictorial conceptions adapted
> to the account of more limited fields of physical experience. Still, the decisive point
> is that in neither case does the appropriate widening of our conceptual framework
> imply any appeal to the observing subject, which would hinder unambiguous
> communication of experience. In relativistic argumentation, such objectivity is

secured by due regard to the dependence of the phenomena on the reference frame of the observer, while in complementary description all subjectivity is avoided by proper attention to the circumstances required for the well-defined use of elementary physical concepts.

Thus, although Bohr accepted that the experimenter makes a conscious choice in deciding what experiment to perform, he rejected the observer's influence on the *outcome* without any doubt. What happens in an experiment depends entirely on the physical interaction of the apparatus with the atomic object being investigated.

Nevertheless, at least one reason for mistakenly attributing subjectivism to Bohr's view of complementarity is due to the fact that the label "the Copenhagen interpretation" is often used to cover a broad spectrum of quite different interpretations. Thus when some physicists, who were associated with Bohr and the Copenhagen Institute (and that number included a large portion of the first generation of quantum physicists), proposed that the observer as a conscious mind seems to play an active role in getting a certain measuring result, it came to be mistakenly believed that Bohr himself held this view. It is worth remembering that Bohr never used the phrase "Copenhagen interpretation," nor "orthodox interpretation," although both have now become too entrenched to be in danger of extinction. However, recent scholarship has gone a long way toward demonstrating that the many various versions of the so-called orthodox interpretation or Copenhagen interpretation were often inconsistent with each other and that there was no uniform agreement among all the diverse physicists (and philosophers) who considered themselves to be supporting the Copenhagen point of view.

In this connection, an important point was made at the beginning of our current century by Don Howard (2004) who argued that Heisenberg invented the label "Copenhagen interpretation" as part of his campaign to reconcile himself with Bohr after their break through World War II. He points out that "until Heisenberg coined the term in 1955, there was no unitary Copenhagen interpretation of quantum mechanics" (680). Howard also notes that at the center of Heisenberg's picture of the Copenhagen interpretation was his own "distinctively subjectivist view of the role of the observer." In particular, it seems this came to be the picture of the thus named "Copenhagen interpretation" that a new generation of physicists and philosophers of science began to question. And not uncommonly these critics missed the differences between Bohr and Heisenberg.

Recently Kristian Camilleri (2006, 2007) has cogently argued that sometimes Heisenberg completely misunderstood Bohr's view in his attempt to show that Bohr's and his own understanding were very similar. Camilleri observes that in the Como paper, Bohr defines complementarity as holding between space-time description and the causal description of the stationary states of the atom in terms of well-defined energy. In 1958 Heisenberg claimed that

> Bohr uses the concept of "complementarity" at several places in the interpretation of quantum theory … The space-time description of the atomic events is complementary to their deterministic description. The probability function obeys an equation of motion as did the co-ordinates in Newtonian mechanics; its change in

the course of time is completely determined by the quantum mechanical equation; it does not allow a description in space and time but breaks the determined continuity of the probability function by changing our knowledge of the system. (50)

But this interpretation completely misreads Bohr by identifying the continuous evolution of the time-dependent wave function with a "deterministic" causal description. So it was Heisenberg, not Bohr, who identified the causal description with the deterministic evolution of the wave function, since it gave a probabilistic description of a free particle in configuration space. In contrast, Bohr identified a causal description with the dynamical conservation principles. Thus, Heisenberg's misinterpretation of Bohr has generated some important distortions in the general understanding of Bohr's point of view.

The Copenhagen interpretation, often branded as the "orthodox" interpretation to put it in a pejorative light, came to be associated with the "collapse of the wave function" when the measuring process reduces the superposition of possible state values of a system into the actual detected value. This way of speaking—and thinking—would have been common only after J. von Neumann's version of Dirac's formalism in 1932, well after Bohr had already arrived at complementarity. This makes sense if one follows Heisenberg's view that the so-called collapse is an objective physical process. But Bohr never once mentions the "collapse of the wave function." To him the wave function was symbolic because it is not a function of the classical dimensions of space and time, by which we represent to ourselves the world of everyday experience and classical physics. It cannot represent anything visualizable in space and time, so Bohr concluded that at least in the atomic domain it is a mistake to try to demand that a theory represent the world as it is in-itself. And this led him to view quantum theory in a way highly compatible with the current of pragmatism which, as we have noted above, was one of the influences that nurtured his philosophical thinking.

Therefore Bohr held that the wave function, which serves as an invaluable symbolic tool for making statistical predictions concerning all possible observational outcomes, cannot be given an ontological reading by taking it to represent a world existing independently of our interactions with it. Because the unavoidable interaction with the observing instruments in quantum mechanics is an essential condition for the occurrence of the phenomena we seek to describe, the criterion of objective description cannot lie in picturing the world apart from our interactions with it; thus the theory's acceptability must rely on the statistical predictions it makes: "The ingenious formalism of quantum mechanics ... abandons pictorial representation and aims directly at a statistical account of quantum processes" (Bohr 1998, 152), and "the formalism thus defies pictorial representation and aims directly at prediction of observations appearing under well-defined conditions" (172). Moreover, as a complex-valued function the wave function is mathematically defined in an abstract vector space, and not in a real physical space. So to make the collapse of the wave function a real physical process, an abstract vector space would have to transform into real physical space a transmutation that lies outside the scope of physical explanation.

Bohr might have argued instead that the wave function must be given an epistemological reading in the sense that the function represented our total knowledge of the quantum system. In this case the so-called reduction of the wave function does not refer to a *physical* process, but to the *epistemic* event of coming to know the result of a measurement; it is the reduction of a range of possible outcomes we *might* know to what we actually *do* know when we make a measurement. While such a view may have corresponded to Heisenberg's earliest view, Bohr abstained from making any such epistemic claim. Bohr frequently refers to an "epistemological lesson" concerning the recognition that observation requires a physical interaction that produces the phenomena we seek to describe, and that therefore the description of the phenomena is "objective" only when the physical conditions of that interaction are included. But nowhere does he identify this lesson with the symbolism of the "reduction of the wave packet." The predictive utility of the wave function gives us no grounds for holding that it is an ontological representation of reality independently of our experience of it; nor that it represents epistemologically all possible empirical knowledge.

There is an abundance of textual evidence for Bohr's refusal to regard the wave function as *representing* anything and his regarding it as only an abstract symbolic tool. The statistical interpretation of the wave function is stipulated already in Como:

> Another application of the method of Schrödinger, important for the further development, has been made by Born in his investigation of the problem of collisions between atoms and free electric particles. In this connection he succeeded in obtaining a statistical interpretation of the wave functions, allowing a calculation of the probability of the individual transition processes required by the quantum postulate. (Bohr 1934, 75)

Not only is it part of his outlook in Como, it's there from the beginning to the end of his life. Here is an example from his very last year, 1962, recalling the 1927 Solvay meeting:

> A main theme for the discussion was the renunciation of pictorial deterministic description implied in the new methods. A particular point was the question, as to what extent the wave mechanics indicated possibilities of a less radical departure from ordinary physical description than hitherto envisaged in all attempts at solving the paradoxes to which the discovery of the quantum of action had from the beginning given rise. Still, the essentially statistical character of the interpretation of physical experience by wave pictures was not only evident from Born's successful treatment of collision problems, but the symbolic character of the whole conception appeared perhaps most strikingly in the necessity of replacing ordinary three-dimensional space coordination by a representation of the state of a system containing several particles as a wave function in a configuration space with as many coordinates as the total number of degrees of freedom of the system. (Bohr 1963, 89)

In the very next paragraph he clearly connects this symbolic character of the formalism with the claim that not only is the "wave picture" to be considered symbolic, but equally so the "particle picture" is not to be taken as a representation of a real object, as it was used in the classical framework. Thus he continues:

In the course of the discussions the last point was in particular stressed in con-nection with the great progress already achieved as regards the treatment of sys-tems involving particles of the same mass, charge and spin, revealing in the case of such "identical" particles a limitation of the individuality implied in classical corpuscular concepts. Indications of such novel features as regards electrons were already contained in Pauli's formulation of the exclusion principle, and in connec-tion with the particle concept of radiation quanta Bose had at an even earlier stage called attention to a simple possibility of deriving Planck's formula for temperature radiation by the application of a statistics involving a departure from the way fol-lowed by Boltzmann in the counting of complexions of a many-particle system, which had proved so adequate for numerous applications of classical statistical mechanics. (89–90)

Sometimes Bohr speaks about the wave function as a symbolic representation in con-trast to a pictorial representation as when he says: "Even if the momentum of this particle is completely known before it impinges on the diaphragm, the diffraction by the slit of the plane wave giving the symbolic representation of its state will imply an uncertainty in the momentum of the particle, after it has passed the diaphragm, which is the greater the narrower the slit" (Bohr 1998, 75). The difference between a pictorial and a symbolic representation is that a symbolic representation is not a representation in the sense of establishing an isomorphic relation between the representation and its object. We have a pictorial representation when there exist some visual or structural similarities between the representation and what is represented. However, a symbolic representation is an abstract representation; it is *conventionally* established and does not show, even only partially, how the world really is. Nevertheless, as a symbolic tool the formalism has proved its utility in successfully predicting the statistical distribu-tion of observational results with all known quantum phenomena, but it does not pro-vide us with any grounds for a realist interpretation of *the wave function*. Instead, Bohr held that the wave formalism is a mathematical tool by which we can calculate in terms of probabilities what can happen in between preparation and measurement.

Other authors have also focused on the contrasting differences between vari-ous versions of the Copenhagen interpretation. Henderson (2010) discusses Werner Heisenberg's, Johann von Neumann's, and Eugene Wigner's contributions to their interpretations of quantum mechanics. He shows how the latter two physicists explicitly refer to the human mind as having an active part in the collapse of the wave function. They believed that the mind could not be given a quantum mechanical description, and indeed that this psycho-physical interaction cannot be explained by physics at all. The mind was taken to exist outside the physical world but could nevertheless have a causal influence on what happened in the physical world by virtue of its power to col-lapse the superposition of quantum states. No doubt these thinkers tended to see the quantum revolution as in some way addressing what they took to be the imbalance of classical mechanism, which had banned mind or consciousness from any role in natural science. Therefore, von Neumann and Wigner saw the problem of observation in quantum mechanics as allowing one to make a scientific case for human free will. Although Bohr did make comments favorable toward the complementarity between mechanistic and volitional descriptions of behavior as indispensable in psychology, the subjectivistic view that mind has a causal role in nature was far from anything he ever considered possible.

Quite apart from the question of whether or not their advocates imagined themselves as agreeing with Bohr, many other interpretations of quantum mechanics have surfaced since Bohr first proposed his interpretation. Nearly all of them are based on representationalist approaches to the quantum formalism and a classical realist, spectator view of knowledge. Those searching for a way to interpret quantum mechanics realistically may be more attracted to one of these interpretations, such as the many worlds-interpretations. Obviously anyone who takes this kind of approach must assume that it is meaningful to think about reality apart from human experience, and that giving an explanatory role to "worlds" which remain in principle beyond any possibility of empirical access is methodologically acceptable.

No doubt these kinds of disagreements among philosophers and physicists over quantum mechanics are at least partially due to differences in the philosophical expectations and temperaments of the individuals involved, as well as different disciplinary goals that distinguish philosophy from physics. However, many physicists and philosophers supporting one of these other interpretations have also argued that Bohr's interpretation (even if correctly understood) cannot account for some of the most recent developments in quantum physics. In this context the primary criterion in evaluating any of these interpretations rests on how well it can be argued that any one of them can handle current developments in physics better than the interpretation justifiably attributed to Bohr. And of course there will always remain future unanticipated developments against which to judge rival interpretations in the years to come. What the future may bring nobody knows. But the contemporary development of quantum theory has produced some new theoretical concepts and empirical phenomena that were not available in Bohr's lifetime. In particular since the 1970s quantum decoherence has been on the physicists' table. This theoretical suggestion assumes that a quantum system is loosely coupled with its environment, and that decoherence occurs when the system interacts with its environment in such a way that its state of superposition is suppressed.

2.

Among our contributors there is a reasonable degree of consensus regarding the matters discussed above; nevertheless, there is still ample room for disagreement and diversity of opinion on many questions. Today the issues in understanding Bohr's view and its implications for physics and the philosophy of science are not the same as those that concerned most of the contributors to our earlier collection more than twenty years ago. The chapters in this collection reveal that there are a new set of concerns for those who endeavor to interpret Bohr's message and what it has to tell us about contemporary discussions in physics and philosophy.

The contributions collected here have been divided into two parts. The first is essentially historical, examining influences that shaped Bohr's thinking, and the second is concerned with how Bohr's interpretation fares with respect to a variety of issues that have arisen only since his lifetime. Part I, which we have titled "Understanding Bohr's Philosophical Background," includes eight contributions focusing on Kantian, pragmatic, and naturalistic themes. Part II, "Bohr's Interpretation of Quantum Mechanics in Twenty-First-Century Physics," includes eight chapters concentrating on issues concerning such topics as decoherence, the classical-quantum partition, and the relation of the quantum formalism to Bohr's view, in addition to the question of the relation of Bohr's interpretation to other interpretations normally seen as rivals to complementarity, including QBism.

Complaints about Bohr's "obscurity" are more numerous than his uses of "notwithstanding." Kristian Camilleri opens this collection by asking "Why Do We Find Bohr Obscure?" Surely not all of this obscurity is in the eye of the beholders, but part of the problem also lies in the tendency of earlier generations to attempt to shoehorn Bohr into established philosophical categories, especially positivism. Camilleri seizes on Bohr's much discussed insistence on classical concepts for describing measurement phenomena to make a case for another reason: we have tended to interpret Bohr from the point of view of the finished formalism rather than the empirical phenomena which formed the foundation of his own approach to atomic physics. Camilleri proposes reading Bohr as a philosopher of experiment and calls attention to how we have often failed to recognize the extent to which Bohr's thinking is grounded in the experimental foundation of quantum theory.

Michel Bitbol turns to the Kantian heritage that surely influenced Bohr's understanding of the problems of scientific knowledge as another avenue of approach to the "central enigma" of Bohr's views regarding the "classical concepts." Borrowing Imre Lakatos's well-known notion of a "research programme," Bitbol envisions Kant as the progenitor of a "transcendental research programme" that has continued to evolve in epistemology since Kant's time. Bohr's philosophy, he concludes, can be understood as a recent development in the evolution of that epistemological "research programme."

Einstein was the first in a long line of commentators to express dismay over Bohr's failure to enunciate a clear statement of his "principle of complementarity." Bohr's use of "complementarity" is hardly univocal, and this fact motivates Patricia Kauark-Leite to distill three distinct uses from his writings. Bohr's first use of the term in Como referred to the relation between space-time description and "the claims of causality," and unlike the other two uses, wave/particle complementarity and the complementarity of phenomena measuring conjugate observables, this form is not a direct consequence of the formal theory, nor is it reducible to the other complementary relations. Kauark-Leite agrees with Bitbol in situating Bohr in a Kantian framework in which space and time are treated as a priori forms of intuition, embodying a transcendental approach to the analysis of human knowledge. Thus armed, she analyzes the Afshar experiment claiming to refute complementarity, and finds such a claim simply wrong.

There is no doubt that a great deal of the stage-set on which the story of epistemology was told, especially during the period of Bohr's education, was designed by Kant, but there is also evidence that Bohr's thinking resonated with many of the chords struck by American pragmatists. Quite apart from any question of historical influences, Henry Folse considers the ways in which Bohr's thinking displays pragmatic modes of thought as typified by the American pragmatist C. I. Lewis, whose "conceptual pragmatism" is presented as a pragmatized version of Kantianism. Folse suggests that Bohr the physicist and Lewis the epistemologist have complementary talents, which might help us to understand Bohr's views about the unavoidability of "the classical concepts."

Jan Faye continues in the naturalistic vein opened up by the discussion of pragmatism, also looking at Bohr's arguments for the role of classical concepts. Faye finds inspiration in Darwinian evolutionary principles for understanding how human perceptual faculties are adaptations formed by natural selection. Faye proposes that Bohr's argument that only the classical concepts can provide the basis for unambiguously describing measurements is a natural outgrowth of evolutionary principles that have shaped our empirical knowledge. He also points out that a consequence of this indispensability of classical concepts is Bohr's nonrepresentational view about the interpretation of the formal language of quantum mechanics, a view which he then compares with Wittgenstein's philosophy of language.

Mauro Dorato also discusses Bohr's claim regarding the indispensability of classical concepts and connects this to another characteristic claim of Bohr's point of view, namely, the indivisibility or wholeness of interaction at the atomic level. Dorato calls attention to the tension between the contextual nature of Bohr's claims regarding the need to make an unambiguous partition between what is described quantum mechanically and what is described classically in a measurement phenomenon and his simultaneous insistence on the wholeness or indivisibility of the interaction. While Bohr's most severe critics have found here an outright contradiction, Dorato concludes that by making appropriate distinctions Bohr's position can be made consistent.

Stefano Osnaghi recognizes that Bohr viewed attempts to solve the measurement problem as "fundamentally misguided" and "irrelevant," because he implicitly rejected the semantic approach of representationalism that this "problem" assumes. Instead, he proposes that the pragmatist view of semantic inferentialism provides a better way to understand Bohr's doctrine of classical concepts, thus connecting his contribution with the pragmatist theme sounded in several of the other chapters. If one associates the meaning of a proposition not with its assumed representational role, but with the practical consequences of accepting it, then Osnaghi concludes we may say that the statistical predictions of the quantum formalism are the *explanans* of the connection between theory and empirical reality, rather than a mysterious *explanandum* begging for an explanation.

Arkady Plotnitsky turns the spotlight on quantum electrodynamics and quantum field theory as embodying the "spirit of Copenhagen" in a different experimental context: the recent detection of the Higgs boson by the Large Hadron Collider. He sees this discovery as manifesting two forms of *techne*, in the broad Heideggerian sense, involving both experimental technique and mathematical manipulation of the formalism (as well as the *techne* of digital computing). Plotnitsky concludes that the role of Bohr's interpretation of quantum theory in this achievement testifies to the enduring power and significance of his outlook.

Slobodan Perovic calls our attention to a different experimental phenomenon already well known in Bohr's own lifetime but rather neglected in discussions of his views: quantum tunneling. He seeks to demonstrate how Bohr's approach grounded in experimental phenomena—a continuation of the tradition initiated long ago by Francis Bacon—was central to the work of both F. Hund and L. Nordheim in their distinct analyses of this phenomenon. Perovic concludes that far from having run its course and being only of historical interest, Bohr's complementarity framework continues to provides "a useful methodological guideline" that has proved to be "invaluable" in ongoing experimental investigations of tunneling times.

The experimental roots of Bohr's thinking, emphasized by quite a few of our contributors, continue to keep the quantum to classical transition in the forefront of discussion. While Bohr himself may very well have regarded his "correspondence principle" as an adequate bridge to make this transition, most view it as a pragmatically justified heuristic aide, but not an acceptable explanation. Many commentators have expressed their frustration with Bohr's refusal to provide a dynamical account of the transition from the quantum mechanical description of a microsystem to the classical description of the measuring instruments, and Maximilian Schlosshauer and Kristian Camilleri address this frustration. In spite of common misunderstandings otherwise, there can be no doubt that Bohr held that in principle any physical system *could* be described quantum mechanically, but if a measurement is to serve its purpose, it *must* be described classically. This doctrine can be seen as the consequence of what they term his "epistemology of experiment." Schlosshauer and Camilleri argue that the current actively discussed work on "decoherence" is best understood as an attempt to address this frustration. They contend that rather than being understood as an attempt to "refute" the spirit of Copenhagen, it is a natural outgrowth of the questions bequeathed by Bohr's attitude. Thus not conflict, but "peaceful coexistence" should rule between those who see themselves as continuing in the line of thinking initiated by Bohr and the advocates of decoherence.

Manuel Bächtold continues the focus on the quantum to classical transition and its relation to the role of decoherence. Although some assuming the premise of realism have understood this transition ontologically as a move from one "world" to another, each ruled by different laws, Bohr certainly did not take this stance. For him the quantum/classical relation is an epistemological distinction between two types of description. Decoherence addresses the *problem in physics* of how the classical behavior of the objects of human experience arises from the quantum behavior of the micro-level, but this was not Bohr's primary concern, which was instead the *epistemological problem* of why we must use different conceptual schemes for atomic objects and measuring instruments and the results they record. Bächtold concludes that Bohr did not hold that a quantum mechanical account of a measurement was impossible, but that objectivity in Bohr's sense implies that we must make a clear cut between observing instruments and observed object, and inasmuch as in the quantum description the systems are entangled, we must transition to the classical concepts to describe the measurement outcome and the apparatus which records it.

Scott Tanona reiterates the theme of many of our contributors that Bohr's thinking is grounded in empirical phenomena, not the mathematical formalism, an approach largely conditioned by his work with his 1913 atomic model. However, he regards Bohr's interpretation as incomplete and outlines an empiricist, but not anti-realist, interpretation, partially informed by decoherence to "update" it. Tanona's proposal starts with what he regards as Bohr's central insight: the inescapable empirically given fact that measurements do have definite results, rather than the assumption that the state vector is a representation of some reality behind the phenomena. From this perspective the so-called measurement problem with its mysterious collapse does not need to be solved; it disappears. Decoherence cannot "explain" the physical fact of definite outcomes, because it starts with this very assumption. Tanona concludes that if we start with the empirically given, Bohr's attitude does not deny realism and embrace a purely instrumentalist understanding of the formalism. Instead what the theory provides is a way to define and describe the physical circumstance under which one may attribute properties to quantum mechanical systems.

If Bohr is seen through the lens of positivism, his views must appear as the very antithesis of the interpretation that was first put forward by Hugh Everett during Bohr's lifetime and that has experienced a considerable evolution in the ensuing years. As we have urged above, most commentators today regard this positivistic reading of Bohr as a misinterpretation to be avoided, and Guido Bacciagaluppi argues that it is revealing to explore the affinities between Bohr's outlook and Everett's. Although Everett himself believed his theory offered a "bridge" between the outlooks of Bohr and Einstein, the historical records reveal a "rather tragic" case of misunderstanding. Analyzing Bohr's reply to Einstein, Podolsky, and Rosen's famous thought-experiment (EPR) from Everett's perspective, Bacciagaluppi suggests that Everett's theory not only recaptures Bohr's basic ideas regarding complementarity, but also indicates where Bohr's views continue to be problematic.

While recognizing that Bohr's writings are rather distinctive in their lack of reliance on the formalism of quantum theory, Dennis Dieks argues that the link between Bohr's thinking and the quantum formalism is much stronger than "usually acknowledged," and not so much the expression of a "preconceived philosophical doctrine." He shows that, despite the notorious obscurity and vagueness for which Bohr has been excoriated, it is possible to construct a coherent account of Bohr's interpretation of quantum theory that is supported by the textual evidence and the quantum formalism. This result also leads Dieks to conclude that Bohr's understanding of quantum theory is "not so far removed" from the non-collapse interpretations as has been commonly thought.

Klaas Landsman continues with the interest in Bohr's partition between classical and quantum descriptions. He focuses on the initial tension between Bohr's doctrine of classical concepts and the correspondence principle which he then argues can be overcome with the correct understanding of the relationship between Bohr's conceptual analyses and the mathematical extension of von Neumann's operator-algebraic formulation of quantum theory. Naming his program "Bohrification" Landsman operates with what he calls an exact Bohrification and an asymptotic Bohrification. The first approach is the appropriate framework for the Kochen-Specker Theorem and for (intuitionistic) quantum logic, whereas the asymptotic Bohrification, inspired by the correspondence principle, is the right conceptual and mathematical framework for the explanation of the emergence of the classical world from quantum theory.

At the dawn of the twenty-first century a new interpretation of quantum mechanics has surfaced under the name "QBism," an abbreviation for "Quantum Bayesianism" and partly inspired by quantum information theory. This interpretation is sometimes compared to Bohr's view, and some physicists have claimed that the Copenhagen interpretation and QBism are not that different. Indeed several similarities between Bohr's view and QBism are noticeable. They both are rooted in a pragmatist approach to knowledge; they both deny that the reduction of the wave function is a physical process; and they both hold that our knowledge of quantum world is contextualized by the experimental agent (Faye 2016). However there are also some significant differences. Where Bohr considered the wave function to be a mathematical tool for statistical prediction without any representational function, QBism understands the wave function as a representation reflecting the agent's personal degrees of belief about the future content of his or her experience. In the final contribution, the only chapter reprinted from an earlier publication, David Mermin agrees that the QBist vision of science, which he stresses is not limited to quantum mechanical matters, shares with Bohr a pragmatist epistemology that rejects the traditional representational approach in favor of a view that beliefs originate in interactions between the epistemic agent and the world. But neither Bohr's view, nor any claiming loyalty to the banner of Copenhagen, permits admitting "the scientist" in any sense beyond "as proprietors of a single large, classical measurement apparatus"; QBism rejects this essential Copenhagen tenet because it holds "the observer" refers to the individual epistemic agent and his or her beliefs about future experiences.

The present chapters cover a variety of topics connected with Bohr's interpretation of quantum mechanics, but it seems fair to say that all of these papers build upon insights into Bohr's thinking gained by earlier scholars and scientists. Thus here well into the twenty-first century we have a more substantial and a less dogmatic grasp of his thinking than ever before. By virtue of historical hindsight today we are in a much better position than earlier generations to spot Bohr's perennial contribution to philosophy and the interpretation of quantum mechanics.

References

Bohr, N. 1934 [1987]. *Atomic Theory and the Description of Nature*, reprinted as *The Philosophical Writings of Niels Bohr, Vol. I*. Woodbridge, CT: Ox Bow Press.

Bohr, N. 1958 [1987]. *Essays 1932–1957 on Atomic Physics and Human Knowledge*, reprinted as *The Philosophical Writings of Niels Bohr, Vol. II*. Woodbridge, CT: Ox Bow Press.

Bohr, N. 1963 [1987]. *Essays 1958–1962 on Atomic Physics and Human Knowledge*, reprinted as *The Philosophical Writings of Niels Bohr, Vol. III*. Woodbridge, CT: Ox Bow Press.

Bohr, N. 1998. *Causality and Complementarity*, supplementary papers edited by Jan Faye and Henry J. Folse as *The Philosophical Writings of Niels Bohr, Vol. IV*. Woodbridge, CT: Ox Bow Press.

Camilleri, K. 2006. "Heisenberg and the wave-particle duality." *Studies in History and Philosophy of Modern Physics* 37: 298–315.

Camilleri, K. 2007. "Bohr, Heisenberg and the divergent views of complementarity." *Studies in History and Philosophy of Modern Physics* 38: 514–28.

Faye, J. 2016. "Darwinism in disguise? A comparison between Bohr's view on quantum mechanics and QBism." *Philosophical Transactions of the Royal Society* A 374 (2068).

Faye, J., and H. J. Folse (eds). 1994. *Niels Bohr and Contemporary Philosophy*. In *Boston Studies in the Philosophy of Science*, Vol. 158. Dordrecht: Kluwer Academic Publisher.

Heisenberg, W. 1958. *Physics and Philosophy: The Revolution in Modern Science*. London: George Allen & Unwin.

Henderson, J. R. 2010. "Classes of Copenhagen interpretations: Mechanisms of collapse as a typologically determinative." *Studies in History and Philosophy of Modern Physics* 41: 1–8.

Howard, D. 2004. "Who invented the 'Copenhagen interpretation?' A study in mythology." *Philosophy of Science* 71: 669–82.

Part One

Understanding Bohr's Philosophical Background

Why Do We Find Bohr Obscure?

Reading Bohr as a Philosopher of Experiment

Kristian Camilleri

It might be difficult to leave Bohr behind, but we continue to discover new ways of reading Bohr.

(Plotnitsky 2006, 142)

1. Introduction

In a letter to Imre Lakatos in 1968, Paul Feyerabend declared "the idea of Bohr's obscurity" to be "nothing but a myth," perpetuated by critics like Popper, who had simply never bothered to read him properly (Feyerabend to Lakatos, January 28, 1968, in Feyerabend and Lakatos 1986, 126–7). Feyerabend had a point. Many of Bohr's critics in those years tended to rely on second-hand sources, and seldom made a serious study of Bohr's philosophical writings for themselves. Yet Feyerabend's efforts to render Bohr's notion of complementarity more intelligible, through a detailed analysis, did little to change the perception in the years that followed (Feyerabend 1968 and 1969). When a later generation of scholars did devote themselves to a close reading Bohr's texts, any illusions that Bohr's meaning would become transparent if only one read him carefully enough were quickly dispelled. In 1985 Abner Shimony expressed the frustration of many, confessing "that after 25 years of attentive—and even reverent—reading of Bohr, I have not found a consistent and comprehensive framework for the interpretation of quantum mechanics" (109). In the three decades since then, Bohr's philosophy of physics has been the subject of numerous books and articles. Thanks to this work, much of which began to emerge in the 1980s and 1990s, we now have a far more nuanced view of his philosophy than we did in the 1960s, when the positivist image of Bohr prevailed. Yet, scholarly opinion on how we should understand his viewpoint remains deeply divided. Bohr, it seems, remains as obscure as ever.

In this chapter I want to revisit the question, "why do we find Bohr obscure?" One obvious response is of course that Bohr *was* obscure. While there is undoubtedly some truth in this, I would suggest that familiar complaints about Bohr's impenetrable writing style miss the point.[1] Undeniably part of the problem, as Makoto Katsumori (2011, 61) rightly points out, is that Bohr's central notion of complementarity is, by its very nature, not easy to express in the traditional categories philosophers often work with. Catherine Chevalley (1994 and 1999), whose work represents the most systematic treatment of this question to date, suggests that the distortion of Bohr's original line of thought and the failure to locate his thinking in the appropriate cultural and intellectual context are largely to blame. Although I find much to admire in her analysis, my historiographical approach differs from hers in a number of crucial respects. Whereas Chevalley argues that Bohr's words only make sense when read against the background of the post-Kantian philosophical tradition, I take my inspiration from R. G. Collingwood's insight that "we can understand a text only when we have understood the question to which it is an answer."

In addressing the question of Bohr's obscurity, I begin by providing a brief historical overview of the various philosophical interpretations, and historiographical approaches, which have attempted to make sense of Bohr's views. Beginning in the 1930s, many of Bohr's contemporaries attempted to read complementarity through the lens of a particular philosophical tradition, such as logical empiricism, neo-Kantianism, or dialectical materialism, in spite of the fact that Bohr never aligned himself with any school of thought. More sophisticated historiographical approaches began to emerge by the 1970s, as scholars attempted to unravel the web of meanings associated with Bohr's writings. One of the important directions in Bohr scholarship over the past few decades has been the attempt to make sense of Bohr by exploring the relationship between his epistemological viewpoint and one or another philosophical tradition, position, or school of thought. A somewhat different, though not necessarily unrelated, approach seeks to make sense of Bohr's thought by situating him in the right intellectual tradition, through an examination of the historical or philological context in which his views took shape. Much of this work is focused on providing a "philosophical reconstruction" of Bohr's view by drawing on critical methods of textual analysis. While many new insights emerged from these efforts, Bohr has remained something of a philosophical enigma.

Here I add a new layer to this discussion, by turning my attention to one of Bohr's key doctrines—the indispensability of classical concepts. I have deliberately chosen to focus specifically on this aspect of Bohr's thought, and not others, because this has undoubtedly been one of the most intriguing and puzzling aspects of Bohr's philosophy, and as Don Howard (1994, 202) has rightly pointed out, it is "more fundamental to Bohr's philosophy of physics than are better-known doctrines, like complementarity." As Bohr ([1929] 1987a, 16) explained:

> It would be a misconception to believe that the difficulties of atomic theory may be evaded by eventually replacing the concepts of classical physics by new conceptual forms. Indeed, as already recognized, the recognition of the limitation of our forms of perception by no means implies that we can dispense with our customary ideas or their direct verbal expressions when reducing our sense impressions to order. Nor is it likely that the fundamental concepts of the

classical theories will ever become superfluous for the description of physical experience.

Here I argue that Bohr's meaning has remained elusive because his central preoccupations lay not so much with an interpretation of the quantum-mechanical formalism, which many of his contemporaries saw as *the* problem of quantum theory, but rather with the question of what kind of knowledge of objects can be obtained by means of experiment. At the heart of this view is a functional, as opposed to a structural, conception of the *experimental apparatus*. To this end, I suggest that we should read Bohr first and foremost as a philosopher of experiment.

In the final section, I explore Bohr's doctrine of classical concepts from another angle, this time in the description of quantum *objects*. Bohr's emphasis on the primacy of classical language has often been misconstrued as a doctrine about the *theoretical* language we must use in representing quantum objects, rather than about how concepts function in experimental *practice*. Appreciating this point requires a shift of perspective away from the traditional preoccupation with meaning, characteristic of both positivist and post-positivist philosophy of language, to an understanding of how concepts are *used* in particular contexts of inquiry. Here, I take my cues from some recent work, which has typically been associated with the "turn to practice" in history and philosophy of science. While I am not suggesting that Bohr should be read as anticipating these later developments, I do think that by paying careful attention to the way concepts function in experimental practice, we can open up a different perspective on Bohr's epistemology that serves to correct the persistent failure to read him on his own terms.

2. A historiographical overview: The challenges of reading Bohr

Bohr's obscurity: Early philosophical interpretations

While Bohr's view of complementarity is still commonly seen as having formed the central plank in a unified and widely shared orthodox view that emerged in the late 1920s, commonly known as the "Copenhagen interpretation," extensive historical scholarship over the past thirty years has challenged, if not seriously undermined, the view that Bohr's views were well understood and widely accepted, even among his closest collaborators.[2] Bohr's notion of complementarity was hailed by many of his followers as "the most significant result for philosophy that crystallized out of modern physics," but it was given a variety of different philosophical interpretations by his various spokesmen, and was appropriated by various philosophical camps during the 1930s who sought to enlist him as an ally in pursuit of the their own agendas (Jordan 1944, 131). By the 1960s it was no longer possible to easily distinguish Bohr's own views from those who professed to speak on his behalf (Chevalley 1999; Howard 1994).[3]

The rise of logical positivism in the 1930s undoubtedly played a significant role in shaping the early reception of Bohr's views. In the preface to his textbook on quantum theory in 1936 and an earlier article, Pascual Jordan (1936, vii; 1934) argued that the

development of quantum mechanics and Bohr's complementarity interpretation signi-
fied a resounding victory for positivism over realism. As Jan Faye and Ulrich Röseberg
have argued, during the 1930s, leading figures of the Vienna Circle, Otto Neurath,
Philip Frank, and the Danish philosopher Jørgen Jørgensen, engaged in an ongoing
dialogue with Bohr, and formed the view that while he was occasionally prone to lapse
into metaphysical language, his philosophical outlook was basically favorable to positivism. In a letter to Carnap in 1934, Neurath reported that Bohr's writings were "full of
crass metaphysics," but he was convinced from discussions with Bohr in Copenhagen
that his "basic attitude" was in accordance with logical positivism (Neurath to Carnap,
November 14, 1934, cited in Röseberg 1995, 112). Writing to Bohr in 1935–36, Frank
complained that Bohr's papers were not always expressed as clearly as they might have
been, and to this extent, Frank saw it as his task to provide a clarification of his views—
to "explain what Bohr really meant" (Röseberg 1995, 109–10; Jacobsen 2012, 128–9;
Faye 2007, 37–8). Complementarity, Frank (1975, 179) would confidently declare in
1938, was "fully compatible with the formulations of logical empiricism."

Yet the Vienna Circle was not the only school of the interwar period to lay claim to
Bohr as a philosophical ally. In Germany, where Kant's critical idealism continued to
exert a significant influence, Bohr was often read through the lens of one or another
of the various schools of neo-Kantianism (Weizsäcker 1936, 1941a,b, 1952; Hermann
1935, 1937a,b). In the 1930s, Heisenberg (1934, 700) called on philosophers to reex-
amine "the question raised by Kant, and much discussed ever since, concerning the
a priori forms of intuition and categories" in the light of quantum mechanics, for "as
Bohr particularly has stressed, the applicability of these forms of intuition, and of the
law of causality is the premise of every scientific experience even in modern physics."
Whereas Frank saw Bohr's viewpoint as signifying the triumph of positivism, philosophical discussions in Leipzig during the 1930s between Heisenberg, Weizsäcker, and
the visiting Kantian scholar Grete Hermann tended to revolve around Bohr's relationship to Kant. As Weizsäcker ([1966] 1994, 185) would put it some years later: "The
alliance between Kantians and physicists was premature in Kant's time, and still is; in
Bohr, we begin to perceive its possibility."

Further divisions between Bohr's followers would become apparent during the
1930s. Léon Rosenfeld, who served as Bohr's assistant in Copenhagen and became
increasingly interested in Marxism during this time, was deeply critical of physicists
who had succumbed to the "scourge" of neo-positivism or Kantianism, and in doing
so, mistakenly labeled Bohr a positivist or an idealist. After the Second World War,
Rosenfeld ([1953] 1979, 465; [1957] 1979) launched a vigorous defense of complementarity as a striking example of the dialectical method in epistemology. As Anja
Jacobsen (2007) has made clear, Rosenfeld's understanding of complementarity was
to a large extent shaped by his commitment to Marxism, which drew sharp criticism
from Max Born and Wolfgang Pauli during the 1950s.[4] While Rosenfeld later insisted
that he had "never played the game of putting a materialist label" on Bohr, he did argue
that complementarity constituted "the first example of a precise dialectical scheme"
(Rosenfeld to Bohm, December 6, 1966, LRP "Epistemology"; Rosenfeld [1953] 1979,
481). Indeed in private correspondence, Rosenfeld went so far as to claim that Bohr
was a Marxist who simply wasn't aware of it (Jacobsen 2007, 17).

The ideological context was in some cases crucial to the reception and interpretation of Bohr's views. This was particularly evident in the Soviet Union, where Bohr was often criticized as endorsing some form of subjectivism or positivism. Such criticisms intensified during the early years of the Cold War, resulting in a decision at the 1947 meeting of the Soviet Academy of Sciences to effectively ban complementarity. By the late 1950s, however, a more relaxed attitude began to emerge, resulting in a reconciliation of Bohr's views with Soviet dialectical materialism (Graham 1988, 311–13). After discussions with Bohr in Copenhagen in 1957, the Soviet physicist Vladimir Fock (1957, 646) declared that Bohr's view of quantum mechanics was essentially correct, but that his use of language had unfortunately given "rise to many misunderstandings and to an incorrect [positivist] interpretation." Here Fock (1958, 210) argued "it is not to be supposed that the use of such expressions reflects any subjectivity on Bohr's point of view; without question this is simply carelessness, and there is no real need to comment on such imprecise expressions." This appears to have been the shared experience of many of Bohr's interpreters, regardless of their philosophical orientation.

By the 1960s, new interpretations began to appear. Klauss Michael Myer-Abich's *Korrespondenz, Individualität und Komplementarität* published 1965 was one of the earliest book-length studies of Bohr's philosophy, and undoubtedly set the tone for much of the later work (Feyerabend 1968 and 1969; Scheibe 1973). Perhaps the most influential contribution in English appeared in two important works published in the 1960s by Aage Petersen (1963; 1968), who served as Bohr's assistant in Copenhagen from 1952 until his death in 1962. Petersen's philosophical reading of Bohr, much like Rosenfeld's, was informed less by careful textual analysis, and more by private recollections of his philosophical discussions with Bohr. Here we encounter a rather different image of Bohr—as a thinker preoccupied with the limits and the constitutive dimension of human language. Indeed it is from Petersen (1963, 11) that we have inherited the adage, frequently attributed to Bohr, that "we are suspended in language." Petersen's shift of focus to the primacy of language found a receptive audience among physicists and philosophers, who had otherwise struggled to make sense of Bohr's own writings.

Given the diversity of readings of Bohr, clashes were inevitable.[5] Yet Bohr never publicly endorsed any particular reading of his work, nor did he ever acknowledge an intellectual debt to particular philosophical tradition or align himself with a school. This was undoubtedly intentional. As Anja Jacobsen (2012, 124) has pointed out, "Bohr had a rather skeptical attitude to professional philosophers" and he frequently lamented the tendency to interpret quantum mechanics through the lens of one or another philosophical "-ism." As Rosenfeld explained in a letter to David Bohm, those who were well acquainted with Bohr did "not take very serious this game of putting labels with various 'isms' upon Bohr. No more seriously, in fact than he himself took it" (Rosenfeld to Bohm, December 6, 1966, LRP "Epistemology"). This attitude is evident in Bohr's (2005, 200) remarks in an interview for *Izvestia* in 1934:

> When one raises the question of which philosophical consequences arise from modern physics, one may not thereby understand the question to mean which old philosophical schools comply with modern physics ... Although some

consequences of modern physics have something in common with the view of many great philosophers, yet it seems to me that if men such as Spinoza or Marx were alive today, they would probably, together with the rest of us, enjoy learning new things from modern physics for the relevance of general philosophy.

Bohr often became disillusioned with philosophers, who appeared more interested in how complementarity related to the philosophical systems of the past than in what was genuinely new about it.[6] In a letter to Martin Strauss in 1935, Rosenfeld reported that Bohr had expressed his disappointment about Heisenberg's willingness to engage in discussions with philosophers about the relationship of quantum physics to Kantian philosophy. Bohr took the opportunity to remind his colleagues in Copenhagen that it was "psychologically (and of course in the first instance substantively!) important that *one does not enter into any compromises with philosophers*" (Rosenfeld to Strauss, November 16, 1935, in LRP "Correspondence particulière"; emphasis mine). Given his suspicion of the corrupting influence of professional philosophers, it is therefore unsurprising that in the famous "Last interview" with Kuhn, Petersen, and Rüdinger, Bohr boldly declared that "I think it would be reasonable to say that no man who is called a philosopher really understands what one means by the complementary description" (November 17, 1962, AHQP).

There is a certain irony here. Although Bohr never aligned himself with any philosophical school, he seldom made public his frustration with physicists who sought to interpret complementarity through the lens of a particular philosophical tradition. In many cases, important differences of opinion between Bohr and the leading physicists with whom he worked were confined to private correspondence (Bohr to Pauli, March 2, 1956, in Pauli 1996, 137). Bohr was often inclined to dismiss such disagreements as reflecting nothing more than matters of emphasis or choices of terminology, though this was not always the view of his interlocutors. While such disagreements rarely made it to print, his correspondence with physicists such as Heisenberg, Born, Pauli, Frank, Fock, and Weizsäcker, spanning more than two decades, reveals a series of hidden tensions and misunderstandings that were often never fully resolved (Beller 1999; Camilleri 2009a).

How should we read Bohr? Historiographical approaches

Given the disagreement among even Bohr's closest collaborators, it has been one of the important tasks of Bohr scholarship over the past thirty years to disentangle his views from those who professed to speak on his behalf. As Don Howard (1994, 204). has put it, the "history of misreadings of Bohr has so obscured his intentions, that one must first deconstruct the misreadings, so that one can reconstruct Bohr's words and their meanings." Yet here it is important to realize that the divergent, and in some cases misleading, interpretations that we have inherited from an earlier generation of commentators were as much a *response* to the difficulties they encountered in reading Bohr himself, as they were a *contributing factor* to those difficulties. The proliferation of interpretations can be seen as both a symptom and a cause of the hermeneutic challenges of reading Bohr.

Here I turn my attention from history to historiography. Over the past few decades, a number of distinctive historiographical approaches have emerged, which draw on different methods of analysis. While these approaches differ, they are not necessarily mutually exclusive, and have often been effectively combined in an effort to gain a deeper insight into Bohr's writings. One such approach signifies a continuation of the attempts to make sense of Bohr by reading him through the lens of a particular philosophical school or tradition, such as Kantianism or positivism, or to explore the relationship between his thought and modern philosophical positions such as realism and anti-realism. Michael Cuffaro (2010, 309), for example, has recently argued that "Bohr's complementarity interpretation of quantum mechanics" can only be properly understood once it is recognized that it "follows naturally from a broadly Kantian epistemological framework." While acknowledging that Bohr's views differ in certain crucial respects from those of Kant, he maintains that any proper "interpretation of Bohr should *start* with Kant" (310; emphasis in the original). Yet, even among those scholars who see certain parallels between Bohr and Kant, there is no agreement on how to understand this relationship.[7] Bohr's views do not appear to fit neatly within traditional philosophical categories, and there remains considerable debate and disagreement about how best to locate his philosophical position with respect to the contemporary realism/anti-realism debates.[8]

In an effort to locate Bohr's views with respect to familiar philosophical reference points, some scholars have sought to explore connections with twentieth-century analytic and continental philosophical traditions. Edward Mackinnon has suggested that Bohr's conception of language shows some similarities with certain views we find in Wittgenstein's *Philosophical Investigations*, while Roger Fjelland suggests that "there are interesting parallels between Bohr's philosophy and [Husserlian] phenomenology" (MacKinnon 1985, 115; Fjelland 2002, 58). More recently, Makoto Katsumori (2011, 89–151) has examined the connections between Bohr's idea of complementarity and post-Heideggerian hermeneutic philosophy and the deconstruction of Jacques Derrida. Katsumori's work is indicative of a recent trend, exemplified in the work of Arkady Plotnitsky (1994), John Honner (1994), and Karen Barad (2007), to explore certain parallels between Bohr's thought and some aspects of postmodernism, poststructuralism, or deconstruction.[9] Yet, as Folse (1994, 119) noted more than twenty years ago, these efforts leave one with the strong feeling that "the philosopher's categories cannot characterize complementarity without distortion."

A somewhat different historiographical approach can be seen in the work of scholars such as Jan Faye and Catherine Chevalley, both of whom have sought to gain a deeper insight into Bohr by locating him within his appropriate intellectual context. Faye's (1991) work focuses largely on Bohr's intellectual background in Denmark, in tracing the philosophical impulses that shaped the development of Bohr's thinking about atomic physics. Drawing on methods of intellectual biography, Faye (1991) shows that Bohr's early forays into philosophy were shaped by his close relationship with the Danish philosopher Harald Høffding, as well as his participation in the philosophical seminars and discussions of the Ekliptika circle in Copenhagen. More contentious perhaps is Faye's claim that Høffding exerted an important influence on Bohr's notion of complementarity, as is evident in the various strands of Kantian and pragmatist thought that can be traced in Bohr's philosophical writings of the 1920s and 1930s. Notwithstanding the criticisms of David Favrholdt, there is considerable evidence to suggest that Høffding was an important formative influence on Bohr's philosophical development. However, precisely what that influence was and to what extent Bohr's view of complementarity owes something to this influence remain difficult questions to answer.[10]

Like Faye, Catherine Chevalley (1991; 1994; 1999) adopts a historical contextualist approach, but her work focuses on tracing the historical sources of "Bohr's epistemological vocabulary." Chevalley (1994, 51) contends that Bohr's language has proved difficult to decipher "because we have become blind to the specific tradition in which it makes sense." This tradition, as she explains, emerged in the German-speaking world in the second half of the nineteenth century, in the work of such figures as Wilhelm von Humboldt, Herman von Helmholtz, and Heinrich Herz. Bohr's attempt to come to grips with the paradoxes of quantum theory in the 1920s, which culminated in the idea of complementarity, was "formulated within the philosophical language that developed in the German culture starting with Kant." To this extent, Chevalley (1999, 65–6) suggests that we need to read Bohr "in a philological way" because "Bohr's interpretation makes sense only if one takes into account both the detailed genesis of quantum theory and the very precise lexicon of post-Kantian epistemology."[11]

Chevalley's contextualist approach throws up many valuable insights. Yet language can be deceptive. As Folse (1985, 219) points out, while it is true that Bohr's "description of phenomenal objects has a certain Kant-like appearance," such an appearance is misleading, given that complementarity has nothing to do with "how phenomena arise in the subject's consciousness." Here we can see a certain parallel with the difficulties that emerged in the nineteenth-century reception of Kant. As Ernst Cassirer noted in 1929, much of the confusion surrounding Kant's critical turn in philosophy was a result of the fact that it was expounded in the language of eighteenth-century faculty psychology. Thus, while Kant sought to give new meaning to terms like "receptivity," "spontaneity," "sensation," and "understanding," the use of these terms tended to evoke the very metaphysical and psychologistic interpretations that Kant had explicitly set out to overcome (Cassirer [1929] 1957, 194–5). My suspicion is that we encounter something similar in Bohr. Recognition of this point raises the possibility that tracing the linguistic similarities between Bohr's vocabulary and the "lexicon of post-Kantian epistemology" might prove to be more of a hindrance than a help in making sense of his views.

Complementing the forms of contextualist analysis of scholars like Faye and Chevalley are the methods of textual analysis and reconstruction exemplified in the work of philosophers such as Clifford Hooker (1972 and 1994), Erhard Scheibe (1973), Henry Folse (1985), Dugald Murdoch (1987), and Don Howard (1994). While each of these authors differ on specific points of interpretation, they each attempt to reconstruct Bohr's meaning through a rigorous critical analysis of his texts and modes of argumentation, often paying careful attention to subtle shifts of terminology and the historicity of Bohr's thought, as it unfolded during the 1920s and 1930s (Röseberg 1994). Much of this scholarship builds on the earlier work of Feyerabend (1968, 309; and 1969, 103), who in the late 1960s urged philosophers to disregard the muddled and distorted versions of Bohr's views handed down to us by both his supporters and his critics, and to instead go "back to Bohr." In a similar vein, Howard advocates the "need to return to Bohr's own words, filtered through no preconceived dogmas," under the provisional assumption that "his words make sense." Here Howard (1994, 201) emphasizes the need to "apply the critical tools of the historian" and the "synthetic tools of the philosopher" to the difficult task of reconstructing from Bohr's writings a coherent philosophy of physics.

Yet, as Howard (1994) makes clear, we should not fall prey to the illusion that we can simply let Bohr's words "speak for themselves." One can no longer simply assume that we can "interpret Bohr's words as if they stand there unadorned, waiting for an informed and sympathetic eye to read their author's intentions" (204). At certain places in the text, it is necessary to go beyond Bohr's words, "in order to clarify the direction in which they were tending" (225). A philosophical reading of Bohr must therefore remain faithful to Bohr's words, but it can and should go beyond those words, "to ask what Bohr would have said, in certain contexts, consistent with what he says elsewhere" in order to "bring out better their intended meaning." At such points, "interpretation passes over into reconstruction" (204). I agree with Howard that any philosophical reading of Bohr will inevitably involve some degree of reconstruction. Yet here there is a danger of straying too far from the text. The plausibility of any reconstruction will not only depend on whether it makes good sense of Bohr, but also on what kind of textual (or other forms of) evidence can be marshaled in support of it. Where such evidence is lacking, the reconstruction may look suspiciously like a distortion of "what Bohr really meant."

The problems briefly touched on here are, to a large extent, reflective of more general methodological difficulties that will be familiar to intellectual historians of different stripes. Yet, Bohr's writings present a particularly acute example of the challenges of intellectual history. Reading Bohr therefore demands we remain attentive to these general historiographical issues, as well as the specific aspects of his philosophical viewpoint that have persistently remained opaque. As Henry Folse (1994, 119) has astutely observed, we typically find Bohr at his "most elusive" when we endeavor to discern in his writings answers to "questions which our philosophical perspective seems to make inevitable." Here I think Folse puts his finger on the crux of the problem. It is our "philosophical perspective," not so much in the doctrines we bring to the reading of Bohr's texts, but in the questions we pose, that has, more often than not, led us astray.

My contention is that Bohr's philosophical viewpoint has remained elusive in large part because we have persistently asked the wrong questions in interrogating his texts. There is still much to be said for R. G. Collingwood's (1939, 36–7) insight that "we can understand a text only when we have understood the question to which it is an answer." Much of the literature has simply uncritically assumed that Bohr's intention was to provide an "interpretation of quantum mechanics." Yet, Bohr never set out to defend a particular interpretation of quantum mechanics, at least in the sense in which that term is usually understood. Much to the frustration of many contemporary philosophers, Bohr resolutely refused to be drawn into what he saw as unhelpful philosophical debates about realism and idealism (Honner 1994). Questions about the physical meaning of the wave function, or the ontological status of the interphenomenal object, or the measurement problem were at best peripheral concerns, which only served to obscure the crucial epistemological problem that Bohr sought to make the central focus of his writings.[12] It is therefore no surprise that Bohr devoted relatively little attention to these questions, leaving it to others to reconstruct a coherent position from his cursory remarks on these subjects.

3. Niels Bohr as philosopher of experiment

The doctrine of classical concepts and the epistemology of experiment

By the 1930s Bohr's writings reflect his primary interest in the conditions of possibility of *experimental knowledge* of quantum objects. Putting experiment at the center of Bohr's epistemological concerns may seem trivial, but it provides a critical point of departure for overcoming many of the persistent difficulties that have plagued Bohr scholarship. Our preoccupations with the paradoxes of quantum *theory* have, to some extent at least, blinded us to what by the mid-1930s had assumed central importance for Bohr. While I can only sketch my account in the pages that follow, it is noteworthy that Bohr's foray into quantum electrodynamics in the early 1930s, culminating in his classic paper with Rosenfeld in 1933, was not only an attempt to make sense of the newly emerging quantum field theory, but was intended to provide a deeper insight into the measurability of field and charge quantities (Bohr and Rosenfeld [1933] 1979). The importance of this work for the development of complementarity has, to my knowledge, received surprisingly little attention, with the notable exception of Arkady Plotnitsky (2006, 119–42; 2012, 89–106), though it marks an important step in the unfolding of Bohr's epistemological views on quantum theory in the 1930s.[13]

Given the restrictions of space, I will not deal here with the development of Bohr's early views on complementarity prior to 1935. The period from 1926 until the EPR paper, as many scholars have noted, saw a gradual refinement of Bohr's thought. Whereas Bohr's original formulation of complementarity at the Como conference in 1927 had focused primarily on stationary states, by the time he presented his paper at the Unity of Science Congress in 1936, we find a somewhat different account in which the mutually exclusive experimental arrangements now assumed central importance (Camilleri 2009b, 111–19). This mature position was developed further in lectures in the late 1930s, which reveal important terminological shifts from his earlier writings.[14] Yet, these shifts in the formulation of complementarity did not shake his conviction in the indispensability of classical concepts. In striking contrast to the views expressed by Ernst Cassirer, Bohr remained convinced that classical concepts would continue to play an ineliminable role in modern physics.[15] In perhaps his most widely quoted passage on the subject, Bohr (1949, 209) wrote:

> It is decisive to recognize that, *however far the phenomena transcend the scope of classical physical explanation, the account of all evidence must be expressed in classical terms.* The argument is simply that by the word "experiment" we refer to a situation where we can tell others what we have done and what we have learned and

that, therefore, the account of the experimental arrangement and of the results of the observations must be expressed in unambiguous language with suitable application of the terminology of classical physics. (Emphasis in the original)

What should we make of this passage? In order to reconstruct the steps that led Bohr to this view, we must first understand that Bohr had long recognized that the entanglement of the object and the instrument posed difficulties for the ordinary understanding of experimental observation. A quantum-mechanical treatment of the observational interaction would paradoxically make the very distinction between object and instrument ambiguous. The impossibility of "separating the behaviour of the objects from their interaction with the measuring instruments" in quantum mechanics "implies an ambiguity in assigning conventional attributes to atomic objects" (Bohr 1948, 317). At the same time, such a distinction is a necessary condition for empirical inquiry. After all, an experiment is carried out precisely to reveal information about the "autonomous behavior of a physical object" (Bohr 1937, 290). To speak of an interaction between two separate systems—an object and measuring instrument—is to speak in terms of classical physics.

Bohr regarded this condition of isolation to be a simple logical demand, because, without such a presupposition, an electron cannot be an object of experimental knowledge at all. The crucial point, as Bohr explained at the 1936 Unity of Science Congress, is that in contrast with the situation in classical physics, in quantum mechanics "it is no longer possible sharply to distinguish between the autonomous behaviour of a physical object and its inevitable interaction with other bodies serving as measuring instruments." Yet it lies in "the very nature of the concept of observation itself" that we can draw such a distinction (Bohr 1937, 290). Bohr ([1938] 1987c, 25–6) elaborated on this point at some length from his 1937 lecture "Natural Philosophy and Human Cultures":

> We are faced here with an epistemological problem quite new in natural philosophy, where all description of experiences so far has been based on the assumption, already inherent in the ordinary conventions of language, that it is possible to distinguish sharply between the behaviour of objects and the means of observation. This assumption is not only fully justified by everyday experience, but even constitutes the whole basis of classical physics . . . [In light of this situation] *we are, therefore, forced to examine more closely the question of what kind of knowledge can be obtained concerning objects.* In this respect, we must . . . realize that *the aim of every physical experiment*—to gain knowledge under reproducible and communicable conditions—leaves us no choice but to use everyday concepts, perhaps refined by the terminology of classical physics, not only in accounts of the construction and manipulation of measuring instruments but also in the description of actual experimental results. (Emphasis mine)

The real question for Bohr was not what kind of reality is described by quantum mechanics, or how we should interpret the quantum formalism, but rather "what kind of knowledge can be obtained concerning objects" by means of experiment. Thus, for Bohr, the epistemological problem was how experimental knowledge of quantum objects is possible. To this extent, Bohr's philosophical preoccupations were fundamentally at odds with, or at least rather different from, what many physicists and philosophers of physics see as the problem of quantum mechanics, namely, the interpretation of its formalism.[16] To put it simply, Bohr's doctrine of classical concepts is not primarily an interpretation of quantum mechanics (although it certainly bears on it), but rather is an attempt to elaborate an epistemology of experiment.

In order to bring out this point clearly, it is instructive to borrow the terminology employed by Peter Kroes (2003, 76) in his distinction between a structural and a functional description of an experimental apparatus: "The structural description represents the object [serving as the measuring instrument] as a physical system, whereas the functional description represents the object as a technological artefact." While it is always possible to conceptualize a measuring instrument, such as a mercury thermometer, "from a purely physical (structural) point of view" as an object with certain dynamical properties and obeying physical laws, the same object can also be described "from an intentional (functional) point of view" as an instrument designed "to measure a particular physical quantity" such as temperature. From the epistemological point of view, the functional description is more fundamental (75). As Kroes explains:

> Every experiment has a goal (to measure x or to detect y, or to show phenomenon z, etc.) and it is in relation to this goal that every part of the experimental setup is attributed a function, as well as actions performed during the experiment. *For describing and understanding an experiment, reference to functions is unavoidable.* In contrast the description of the results of the outcome of an experiment (the observations, data, measurements) is free of any reference to functions at all . . . Thus whereas experiments are described in a functional way, the description of the results of experiments and of physical reality as constructed on the basis of those results is of a structural kind. *This means that the structural description of physical reality rests implicitly on a functional description of at least part of the world.* (74; emphasis added)

This passage provides a valuable insight through which we can make good sense of why Bohr saw it as necessary to draw an epistemological distinction between the functioning of the instrument and the object under investigation. Indeed, the language Kroes employs here bears a striking similarity with the way Bohr often expressed himself. In his discussion of the interaction of object and instrument, Bohr always refers to the aims and functions of the experiment. Here it is worth noting that for Kroes, a "functional description also makes use of structural concepts; it makes reference to the structural properties of the object [serving as the measuring instrument], not only in describing but also in explaining the design features of the object" (76).

Bohr's central insight was that if a measuring instrument is to serve its *purpose of furnishing us with knowledge of an object*—that is to say, if it is to be described functionally—it must be described classically. Of course, it is always possible to represent the experimental apparatus from a purely structural point of view as a quantum-mechanical system without any reference to its function. However, any functional description of the experimental apparatus, in which it is treated as a means to an end, and not merely as a dynamical system, must make use of the concepts of classical physics. This is true even when we measure quantum properties, such as spin. In the Stern–Gerlach experiment, for example, the experimental apparatus and the magnetic field must be treated classically for the purpose of performing the experiment. Put simply, any functional description of the experimental apparatus must be couched in the language of classical physics. In Bohr's (1958b, 310) view, "All unambiguous information concerning atomic objects is derived from permanent marks—such as a spot on a photographic plate, caused by the impact of an electron—left on the bodies which define the experimental conditions." In his lecture "On Atoms and Human Knowledge," Bohr (1958a, 169–70) expanded on this point:

> In the analysis of single atomic particles, this is made possible by irreversible amplification effects—such as a spot on a photographic plate left by the impact of an electron, or an electric discharge created in a counter device—and the observations concern only where and when the particle is registered on the plate or its energy on arrival with the counter. Of course, *this information presupposes knowledge of the position of the photographic plate relative to other parts of the experimental arrangement*, such as regulating diaphragms and shutters defining space–time coordination or electrified and magnetized bodies which determine the external force fields acting on the particle and permit energy measurements. The experimental conditions can be varied in many ways, but the point is that in each case *we must be able to communicate to others what we have done and what we have learned, and that therefore the functioning of the measuring instruments must be described within the framework of classical physical ideas.* (emphasis added)

The properties of the quantum object, such as its position, charge, spin, and energy, can only be known by virtue of traces it leaves on an experimental system. It must be possible to interpret the results of the experiment in such a way that the visible and permanent traces on the apparatus can be explained as having been caused by an interaction with the object under investigation. If this were not the case—if no such causal inferences were possible—then such systems could not provide us with knowledge of the quantum object. To this end, we must implicitly presuppose a causal chain of events triggered by the object itself, through the apparatus, finally registering at the macroscopic scale, if the measuring apparatus is to serve its purpose. Or, to put it another way, it must be possible to trace this "sequence of cause and effect" from the observation of a spot on a photographic plate or an electric discharge in a counter device, back to the interaction with the object itself. In this sense, "the concept of causality underlies the very interpretation of each result of experiment" insofar as it forms the basis of any functional description of the apparatus (Bohr 1937, 293).

This point has generally not been well appreciated. Manuel Bächtold (2008, 627–8), for instance, argues on the basis of quantum mechanics that a classical physical description simply cannot be applied to the measurement apparatus "during the measurement process." Yet here the *functional-epistemological* account of the apparatus drops out of the picture entirely. If one could not say unequivocally that a visible macroscopic effect, such as a spot on a photographic plate, was caused by a particle striking the plate, then such an observation would in effect count for nothing. As Bohr put it in a letter to Schrödinger in 1935, "The description of any measuring arrangements must, in an essential manner, involve the arrangement of the instruments in space and their functioning in time, *if we shall be able to state anything at all about the phenomena*" (Bohr to Schrödinger, October 26, 1935, in Bohr 1996, 511; emphasis in the original).

Functional versus dynamical accounts of the experimental apparatus

Bohr's functional epistemology of the experimental apparatus, as outlined above, has often been misunderstood because of a failure to carefully distinguish it from the dynamical problem of the quantum-classical transition. As many commentators have pointed out, while Bohr insisted that we must employ the concepts of classical physics to describe whatever part of the system we have designated to function as a measuring instrument, it is always possible to give a quantum-mechanical description of the apparatus in its entirety. Bohr acknowledged that measuring instruments, like all systems, macroscopic or microscopic, are strictly speaking subject to the laws of quantum mechanics (though I would stress he did not assume that the apparatus could be described as an isolated dynamical system). As he explained at the Warsaw conference in 1938:

> In the system to which the quantum mechanical formalism is to be applied, it is of course possible to include any intermediate auxiliary agency employed in the measuring process. Since, however, all those properties of such agencies which, according to the *aim of measurements* have to be compared with the corresponding properties of the object, must be described on classical lines, their *quantum mechanical treatment will for this purpose be essentially equivalent with a classical description*. (Bohr 1939, 23–4; emphasis added)

Here again, we should note the use of normative language in referring to the aims and purposes of the experiment. While it is of course always possible to describe the apparatus as a quantum-mechanical system, Bohr insists that in doing so we would forfeit a functional account of the apparatus as a means of acquiring empirical knowledge. Measuring instruments, for Bohr, must admit of a classical description, otherwise they could not perform their epistemic function as measuring instruments. To this extent, any "quantum mechanical treatment" of a measuring instrument will, by virtue of its function as a measuring instrument, "be essentially equivalent with a classical description." However, this raises the further question of why, from a purely dynamical point of view, the quantum-mechanical treatment is, or can be treated as, essentially equivalent to a classical description. As Kroes (2003, 76) points out, from a purely structural (as opposed to a functional) point of view, the design and geometric configuration of the apparatus "is just some physical property" and "a completely contingent feature of the object."

Bohr's epistemological explanation for why we must use a classical description thus raises the question of what dynamical features of a macroscopic system entitle us to neglect the "quantum effects." Bohr (1935, 701) here appears to simply assume that there exists a macroscopic "region where the quantum-mechanical description of the process concerned is effectively equivalent with the classical description." Thus we are led to ask: How is it that classical physics can be employed, at least to a very good approximation, under certain dynamical conditions (typically those corresponding to measuring scenarios)? This is a salient question, given that, strictly speaking, the world, as Bohr recognized, is nonclassical.

Here I want to suggest that Bohr never provided a sustained or satisfactory dynamical explanation for the quantum-to-classical transition, leaving only fleeting remarks on the subject throughout his writings. He seems to have regarded the explanation as trivial, and on most occasions has been content to refer to the "massive" nature of macroscopic bodies serving as measuring instruments. A functional account of the experiment "is secured by the use, as measuring instruments, of rigid bodies sufficiently heavy to allow a completely classical account of their relative positions and velocities" (Bohr 1958b, 310). One finds a similar view expressed on several occasions in Bohr's later writings. As he put it in 1958, "all measurements thus concern bodies sufficiently heavy to permit the quantum [effects] to be neglected in their description" (Bohr 1958a, 170). One might well ask, how heavy is "sufficiently heavy"?

The question of why Bohr thought it was possible that a classical description can be "essentially equivalent to a quantum mechanical one" has been a source of much confusion. Yet these difficulties can be avoided, if we take care to distinguish between Bohr's functional-epistemological account of the experimental apparatus (on which he placed great emphasis) and his account of the quantum–classical transition (on which Bohr said very little). On the few occasions when Bohr did discuss the latter, he typically appealed to the "heaviness" or the "macroscopic dimensions" of the measuring apparatus.

By the 1950s, few of Bohr's contemporaries found this to be a satisfactory resolution. In a letter to Aage Petersen in May 1957, Hugh Everett said he could find no justification for the dogmatic assumption that "macrosystems are relatively immune to quantum effects" (Everett to Petersen, May 31, 1957, quoted in Osnaghi et al. 2009, 106). Here Everett stressed that such a view was nothing more than a "postulate," as it "most certainly does not follow from wave mechanics." Indeed some of Bohr's defenders were inclined to agree. Indeed two years earlier Heisenberg (1955, 23) had drawn attention to the fact that the "classicality" of macroscopic systems was "by no means trivial" and demanded further explanation. In his contribution to the volume commemorating Bohr's seventieth birthday, Heisenberg acknowledged, "there are many solutions of the quantum-mechanical equations" for macroscopic systems "to which no analogous solutions can be found in classical physics" (Heisenberg 1955, 23). In an intriguing anticipation of later developments, Heisenberg surmised that macroscopic systems behave "classically," which is to say, they do not exhibit quantum superposition, because they cannot be isolated from their environments.

For his part, Bohr appears to have simply assumed that any macroscopic system serving as a measuring instrument *must be* describable by means of a classical approximation—otherwise we could not rely on such system to perform an epistemic function. It is simply the case that without such a presupposition, it would not be possible to acquire empirical knowledge. Why such a presupposition holds dynamically, however, was not a question to which Bohr ever gave serious attention. The task then fell to Bohr's followers to resolve the problem of "what physical condition must be imposed on a quantum-theoretical system in order that it should show the features which we describe as 'classical'" (Weizsäcker 1971, 29). While many solutions were offered, based largely on thermodynamic considerations, it was not until the development of decoherence some decades later that physicists were able to provide anything like a satisfactory resolution to this problem (Camilleri and Schlosshauer 2015, 80–3). Bohr's functional-epistemological description of experiment, however, should be sharply distinguished from these later efforts, which gathered momentum in the 1960s, to provide a dynamical explanation of the quantum-classical transition.

Quantum objects and classical concepts

I want to now shift my attention from the description of the experimental apparatus to the description of the object of experimental inquiry. Here we confront the question of why classical concepts remain indispensable, not only for the description of the *functioning* of the experimental apparatus, but also for the interpretation of the *results* of measurement. Bohr's emphasis on the primacy of classical concepts in quantum theory has given rise to a host of different philosophical interpretations regarding his conception of the language (Petersen 1963 and 1968; Favrholdt 1993; MacKinnon 2011; Katsumori 2011). Undoubtedly Bohr's philosophical interpretation of quantum physics was shaped by his reflections on the limits of language and the problems of unambiguous communication. While Bohr's brief remarks on the subject do not amount to a fully developed philosophy of language, it is clear that he saw the importance of an epistemological focus on concepts in quantum physics.

Before proceeding, it is important to clarify a source of ambiguity that inevitably arises in Bohr's frequent use of the term "classical." This term, which is typically contrasted with the term "quantum," has been the source of misunderstanding, as it is frequently employed to refer variously to concepts, properties, phenomena, laws, or theories, without regard for these subtle but important distinctions. While Bohr often left it to his readers to decipher the precise meaning of ambiguous phrases such as "classical description," in his more careful moments he did distinguish between the use of classical concepts (such as position, charge, and momentum), employed in the description of quantum objects, and classical theories (such as Newtonian mechanics or Maxwell's field equations), used to describe the experimental apparatus. In his reply to the EPR paper, for example, Bohr (1935, 701) emphasized the necessity of using "classical *concepts* in the interpretation of all proper measurements, even though the classical *theories* do not suffice in accounting for the new types of regularities with which we are concerned in atomic physics" (emphasis added).

Bohr sometimes gave the impression he adopted a "verificationist conception of linguistic meaning" (Beller and Fine 1994, 18). However, such a misconception arises as a result of Bohr's somewhat ambiguous use of the term "definition" in the late 1920s and early 1930s.[17] Bohr's epistemological preoccupations with language lay less with the problems of meaning and reference, which would dominate analytic philosophy of language throughout the twentieth century, and more with the conditions of *applicability* of concepts. As Bohr (1937, 293) explained at the 1936 Unity of Science Congress in Copenhagen, it is simply the case that in dealing with quantum phenomena, we must "use two different experimental arrangements, of which only one permits the unambiguous use of the concept of position, while only the other permits the application of the concept of momentum." This situation, which demonstrates the complementary features of a quantum description, brings to light the previously "*unrecognized presuppositions for an unambiguous use of our most simple concepts*" (289–90; emphasis added). As Bohr (1949, 211) later explained:

> A sentence like "we cannot know both the momentum and the position of an atomic object" raises at once questions as to the physical reality of two such attributes of the object, which can be answered only by referring to the [experimental] conditions for the unambiguous use of space-time concepts, on the one hand, and dynamical conservations laws on the other hand.

As this passage suggests, Bohr understood perfectly well that beyond certain experimental conditions, the use of such concepts becomes ambiguous. While the indeterminacy relations are frequently taken to express a "nonclassical" feature of quantum objects, for Bohr, they were a constant reminder that we are simply forced to use the concepts of classical physics, albeit within certain limits of applicability. Whenever we speak of the indeterminacy of an electron's position or momentum, we use the words "position" and "momentum," and to this extent we invariably fall back on the *use* of classical concepts. Here we can see vividly that the *use*, or better still the *usefulness*, of concepts like position and momentum did not imply for Bohr that objects *have* such well-defined properties. It is merely that such concepts help us to "grasp" the phenomena. As Friedrich Steinle (2012) has pointed out, "Concepts do not have a truth-value." Unlike theories, concepts cannot be true or false; they "cannot be proved, or be confirmed or disconfirmed as such." One can only say they are useful or not, "appropriate or not" in different epistemic contexts (105). To this extent, concepts may "prove their worth" in certain limited domains of applicability.

But why did Bohr think that classical concepts are indispensable? Recall that for Bohr (1939, 23–4), it is inherent in the very "aim of measurements" that the relevant properties of the experimental apparatus must be "compared with the corresponding properties of the object." This seemingly innocuous statement holds an important clue to a deeper understanding Bohr's view. In order to perform a measurement, we must presuppose, even if only provisionally, that the object "interacts" with our measuring instrument somewhere in space and time. As Bohr explained, "A measurement can mean nothing else than the unambiguous comparison of some property of the object under investigation with a corresponding property of another system, serving as a measuring instrument" (19). If we did not operate under the assumption that the interaction between the object of investigation and the measuring apparatus involves an exchange of energy and momentum somewhere in space and time, it would not be possible to *design* or *perform* an experiment, nor could we *communicate what had occurred* during the experiment.

To this extent, Bohr (1939, 19) held that a measurable property is one that can be defined (in an operational sense) by means of "everyday language or in the terminology of classical physics." Here it might be objected that spin constitutes a "nonclassical" measurable quantity, but even in the case of the Stern-Gerlach experiment, spin measurements involve the deflection of particles by means of a magnetic field. While such experiments performed with a beam of particles display the phenomenon of spin quantization, to which there is no corresponding phenomenon in classical physics, the deflection of each individual particle is ultimately interpreted as a change in the particle's *momentum* caused by the magnetic field in *space* and *time*. Indeed it is difficult to associate any other meaning with the term "deflection." As Bohr ([1929] 1987b, 94) would put it in 1929: "It lies in the nature of physical observation that all experience must be ultimately expressed in terms of classical concepts." Note Bohr's careful use of phrases like "expressed in terms of" in this context. It is not the case that every measurable quantity *is* a classical property, but rather that any measurement must *make use of* classical concepts. In this sense, the very *aims of measurement* force upon us the need to use classical concepts, even though, as Bohr (1937, 293) explicitly recognized, "the whole situation in atomic physics deprives of all meaning such inherent attributes as the idealizations of classical physics would ascribe to the object."

Seen in this light, it becomes apparent that the fundamental question for Bohr, in formulating his doctrine of classical concepts, was not "what concepts are necessary to theoretically represent the quantum objects," as a Kantian might have asked, but rather "what concepts are necessary to investigate such objects by means of experiment." This is, strictly speaking, an epistemological question, though it is not one that attracted the attention of many of Bohr's contemporaries. The properties we ascribe to objects when we perform a measurement (such as position, momentum, charge, spin), or better still, the *concepts we use* in describing the results of such measurements, are inextricably tied to the *epistemic aims of experimental inquiry*. To this extent, I would suggest we can make better sense of Bohr's words once we realize that, for him, *classical concepts are necessary insofar as they form part of the normative structure of experimental practice*. This normative view of concepts comes close to the view recently articulated by the philosopher Ingo Brigandt (2012, 97):

> Concepts refer to the world and represent the world in a certain fashion. Consequently, concepts have usually been construed as consisting of some beliefs about the concept's referent; an intension, an inferential role, a definition, or an analytic statement. However, note that the *epistemic goal* pursued by a concept's use operates on a different dimension than reference and inferential role. For the epistemic goal does not consist in a belief *about states of the world*—not even in a desire as to how aspects of the world studied by science should be like. Instead its goal is *for scientific practice* . . . Such goals have to be taken into account to understand the dynamic operation of science, including the epistemology of scientific concepts. (Emphasis in the original)

This is an account of concepts I suspect Bohr would have found congenial. As we have already noted, in many of the passages in which he articulated the indispensability of classical concepts, Bohr (1949, 209) explicitly referred to the "aims of experiment," as well stressing that "by the word 'experiment' we refer to a situation where we can tell others what we have done and what we have learned." By the 1930s, Bohr was primarily interested in how concepts function in experimental practice, not how they function as linguistic elements in our theories. This is why he repeatedly insisted that the doctrine of classical concepts was inextricably intertwined with the pragmatics of communication and the conceptual presuppositions of experimental inquiry. It was in this sense that Bohr (1931, 692) remained convinced that "the language of Newton and Maxwell will remain the language of physics for all time."

Much of the confusion surrounding the doctrine of classical concepts, I suspect, stems from a tendency to misconstrue it as a doctrine pertaining primarily to theory, rather than experimental practice. Such misunderstandings have persisted in large part because Bohr frequently emphasized the relevance of the doctrine of classical concepts for any proper interpretation of the quantum formalism. Of course, formal considerations played an important role for Bohr, as is evident in his detailed analysis of the measurability of quantum fields in 1933 and his reply to the EPR paper in 1935. But such considerations were included primarily to explore the conditions of experimental inquiry. A deeper appreciation of this point may well provide an important clue to making sense of Bohr's reply to the EPR paper. As Beller and Fine (1994, 8) have noted, whereas Einstein, Podolsky, and Rosen characterized the state of a system by means of a wave functions, in his reconstruction of their argument, Bohr placed little emphasis on this way of formulating the problem of "completeness of quantum mechanics." For Bohr, the "object of knowledge" is not the object represented symbolically by a wave function, or a vector in Hilbert space, but rather, it is the object disclosed to us by means of the effects or traces registered on the experimental apparatus, which in turn must be interpreted by means of the concepts of classical physics. To put it in Kroes's (2003, 74) terms, "*the structural description of physical reality rests implicitly on a functional description of at least part of the world*" (emphasis added).

Here it remains to say a few words about Bohr's view of the relationship between classical concepts and the quantum formalism. As Manuel Bächtold (2008, 631) rightly points out, "the observables of quantum mechanics," defined by means of corresponding Hermitean operators, "are not strictly identical to the concepts of *classical physics*" (emphasis in the original). However, this poses no problem for Bohr's view of the indispensability of classical concepts. As we have seen, Bohr emphasized that the use of such concepts was indispensable for experimental praxis—that is, for conducting and interpreting experiments. He did not insist that they must appear explicitly as formal elements of the theory. But classical concepts serve another important function. As Bohr ([1929] 1987a, 16) explained, "It continues to be the application of these concepts alone which *makes it possible to relate the symbolism of the quantum theory to the data of experience*" (emphasis added). There is no suggestion here that the observables of quantum mechanics (or quantum electrodynamics for that matter) are *identical* to classical concepts. Rather, Bohr's point was that we can only connect these observables, as part of the formalism, to actual measurements by means of the concepts of classical physics.

4. Conclusions

"It would be a misconception," Bohr ([1929] 1987a, 16) wrote in 1929, "to believe that the difficulties of atomic theory may be evaded by eventually replacing the concepts of classical physics by new conceptual forms." In the wake of the development of quantum theory, it was no longer likely "that the fundamental concepts of classical theories will ever become superfluous for the description of physical experience" (16). Such remarks have often been interpreted as indicating some kind of Kantian element in Bohr's philosophy. Yet the real problem for Bohr was not so much that we lack the cognitive faculties or the capacity to develop a new conceptual framework or a new *theoretical language*, but that the very success of quantum mechanics, and for that matter, quantum field theory, lies precisely in accounting for a vast array of *experimental evidence*, which, by its very nature, depends on the use of classical concepts.

Here we might ask whether, or to what extent, the reading of Bohr's doctrine of classical concepts I have presented here reflects the influence of the Kantian or pragmatist philosophical traditions—both of which are visible in Høffding's work. The lack of clues within Bohr's published writings as to his philosophical influences has made this a difficult question to answer. While there is certainly a Kantian ring to Bohr's doctrine of classical concepts, and we do find evidence of the use of Kantian terminology at certain points in Bohr's writings,[18] Bohr rarely alluded to Kant in his writings, and when he did refer to philosophical systems, he often highlighted the extent to which the new developments in modern physics brought to light the need for new perspectives. As Bohr (1949, 239) himself made clear in his contribution to the 1949 Einstein volume, "Even in the great epoch of critical philosophy in the former century, there was only a question to what extent *a priori* arguments could be made for the adequacy of space-time coordination and causal connection of experience, but never a question of rational generalizations or inherent limitations of such categories of human thinking." Here Bohr noted the difficulties of reaching "mutual understanding not only between philosophers and physicists, but even between physicists of different schools" (240). My sense is that Bohr found these difficulties particularly acute, because his epistemological preoccupations were rather different from those of most of his contemporaries.

While I have not pursued a contextualist reading of Bohr here, I remain convinced that this is a fruitful avenue of historical inquiry in spite of the formidable difficulties it presents. Yet, such an approach needs to proceed in conjunction with careful reconstructive analysis of his texts, with due attention to the way in which Bohr may have spoken the language of early-twentieth-century philosophy, while investing it with new meaning. This remains a task for the future. My aim here has not been to present a systematic and coherent reading of Bohr's philosophy, but merely to bring out certain crucial features of his epistemological viewpoint, which have been neglected or obscured by many commentators. In doing so, I have taken my cues from Collingwood's "logic of question and answer."

Finally, it is worth emphasizing that it has not been my aim in this chapter to defend Bohr's doctrine of classical concepts. There are many places where I think Bohr's arguments are open to criticism. I have said nothing here about the arguments contained in his reply to the EPR paper, nor have I examined the formal analysis of complementarity that Bohr presented in several papers in the 1930s invoking the limitations forced on us by the quantum of action. Furthermore, even if one accepts that we *do* use classical concepts in making sense of measurements, one might doubt whether this entails the stronger conclusion that we *must* use classical concepts.[19] Ultimately this rests on whether one thinks Bohr was right in asserting that a functional-epistemological description of experiment in physics must, by its very nature, employ the conceptual framework of classical physics. Bohr's views are not immune to criticism. Yet, if we are to mount a serious critique of Bohr, I would maintain we must first understand him— as a philosopher of experiment.

Archival sources and abbreviations

AHQP Archive for the History of Quantum Physics. Microfilm. American Philosophical Society, Philadelphia.

LRP   Léon Rosenfeld Papers, Niels Bohr Archive, Copenhagen.

NBA   Niels Bohr Archive, Niels Bohr Institute, Copenhagen.

Notes

1 According to a classic story recounted by David Bohm: "Bohr's statements are designed to cancel out in the first approximation with regard to their meaning, and in the second approximation with regard to their connotation. It is only in the third approximation that one can hope to find what Bohr wants to say" (Bohm to Rosenfeld, December 13, 1966, LRP "Epistemology").

2 Even the physicists commonly associated with the "Copenhagen school" were often guilty of gross misrepresentations. A classic example can be found in Heisenberg's widely read *The Physical Principles of the Quantum Theory*, based on his 1929 Chicago lectures. In the preface Heisenberg (1930, x) announced, "The book contains nothing that is not to be found in previous publications, particularly in the investigations of Bohr." Yet, his account of wave-particle duality (which he later termed "wave-particle equivalence") and complementarity diverged in several critical respects from the view we find in Bohr's writings of the late 1920s (Camilleri 2009b, 77–84, 112–23). While Heisenberg presented himself as a faithful disciple of Bohr, by the mid-1930s crucial differences had emerged in their publications and private correspondence, which were not always acknowledged, and often went unnoticed.

3 As Catherine Chevalley (1999, 59) has argued, "what makes Bohr difficult to read is the fact that his views were identified with the so-called '*Copenhagen Interpretation of Quantum Mechanics*,' while such a thing emerged as a frame for philosophical discussion only in the mid-1950s." There are now good reasons to think that the "Copenhagen interpretation" was a retrospective postwar invention (Howard 2004; Camilleri 2009a).

4 Pauli to Heisenberg, May 13, 1954, in Pauli (1996, 620–1); Pauli to Rosenfeld, September 28, 1954, LRP "Correspondence particulière"; Born to Rosenfeld, January

21, 1953, Born sent Rosenfeld a ten-page typescript entitled "Dialectical Materialism and Modern Physics" on October 24, 1955, LRP "Correspondence particulière."

5 In an exchange of correspondence between Bohm and Rosenfeld in December 1966, we get a glimpse of the divisions among Bohr's spokesmen. Bohm had informed Rosenfeld that he had found Petersen's discussion of Bohr's views on language most illuminating. Yet Rosenfeld remained suspicious of the account Bohm had presented, and responded by saying, "I suspect that you have been badly misinformed by Petersen, or that you have perhaps misunderstood what Petersen told you" (Rosenfeld to Bohm, December 6, 1966; Bohm to Rosenfeld, December 13, 1966, LRP "Epistemology").

6 Abraham Pais (1991, 421) recalled that after attending a philosophical meeting, Bohr declared to Jens Lendahl: "I have made a great discovery, a very great discovery: all that philosophers have ever written is pure drivel."

7 There is now an extensive literature on the relationship between Bohr and Kantian philosophy (Honner 1982; Kaiser 1992; Folse 1985, 217–21; Murdoch 1987, 229–31; Folse 1978; Pringe 2009; Bitbol 2013).

8 The contributions by Favrholdt, Faye, Folse, Krips, McKinnon, and Murdoch in the 1994 volume edited by Faye and Folse, provide different perspectives on the extent to which Bohr's views are best construed as a "realist" or "antirealist" (Faye and Folse 1994).

9 It should be noted that the primary aim of many of the these works is not to provide a close reading of Bohr's own philosophical position, but rather to attempt to go "beyond Bohr," as it were, in using his work as a source of inspiration for developing new philosophical insights.

10 David Favrholdt has criticized Faye's thesis, in arguing that Bohr's notion of complementarity owes little to Høffding and to his Danish intellectual heritage, but was forged as an original response to the specific problems of quantum theory that Bohr confronted him as a physicist (Favrholdt 1992). See also Moreira (1994) and Faye (1994).

11 In addition to Chevalley both Steen Brock (2003) and Voetmann Christiansen (2006) have argued that Bohr's views reflect the influence of a particular strand of neo-Kantianism, which runs from Helmholtz through Herz to Høffding.

12 See, for example, Zinkernagel (2016).

13 As Plotnitsky (2012, 89) has argued, "Bohr saw quantum electrodynamics and other forms of quantum field theory as confirming his key ideas concerning the epistemology of quantum phenomena and quantum mechanics, and possibly giving these ideas more radical dimensions." But it is also evident that "quantum electrodynamics and quantum field theory had a shaping reciprocal impact on Bohr's work on quantum mechanics and complementarity."

14 Rosenfeld (1967, 115) later recalled that it was not until 1936 that Bohr felt he had brought to a conclusion the "task of deepening and consolidating the conceptual foundation of quantum theory." Both Murdoch and Faye have also argued that Bohr's thought undergoes a transformation in 1935. As Murdoch (1987, 145) puts it, "After 1935 Bohr expressed the indispensability thesis in what may be called 'semantic' as distinct from 'ontic' terms." In a similar vein Jan Faye (1991, 186) writes: "After 1935 his grounds for asserting complementarity were not so much epistemological as they were conceptual or semantical."

15 In his book on causality entitled *Determinism and Indeterminism in Modern Physics*, Cassirer ([1936] 1956, 194–5) declared: "There seems to be no return to the lost

paradise of classical concepts ... If it appears that certain concepts, such as those of position, velocity, or of the mass of an individual electron can no longer be filled with definite empirical content, we have to exclude them from the theoretical system of physics, important and fruitful though their function may have been."

16 See Zinkernagel (2016).

17 Mara Beller and Arthur Fine (1994, 18) have argued that Bohr adopted "an operational and verificationist conception of linguistic meaning" in his reply to the EPR paper. By contrast, Edward MacKinnon (1982, 271) has argued, "Bohr did not believe that the meaning of such terms as 'position' and 'trajectory' could be determined by such operational definitions ... Instead, the critical problem requiring analysis accordingly is how such already meaningful concepts function in quantum contexts." Much of the confusion arises as a result of Bohr's ambiguous use of the term "definition" in many crucial passages in the late 1920s. Yet, it is evident from several other passages that the term "definition" is used interchangeably with the term "application" or "use." As MacKinnon (1982, 274) has argued: "When Bohr speaks of 'definition,' he does not mean stipulating the meaning of some term of some context. He refers rather to rules governing the usage, in quantum contexts, of a term whose classical meaning is already determined." MacKinnon's reading is supported by a close reading of several of Bohr's writings in the 1930s, in which we find repeated reference to the conditions of applicability of classical concepts.

18 Bohr ([1929] 1987a, 5, 16) refers to "forms of perception" in his "introductory Survey" in 1929.

19 Weizsäcker (1952, 128, 130), for example, was inclined to adopt the weaker view: "We ought not to say, 'Every experiment that is even possible *must* be classically described,' but 'Every actual experiment known to us *is* classically described, and we do not know how to proceed otherwise.' This statement is not sufficient to prove that the proposition is *a priori* true for all, merely possible future knowledge; nor is this demanded by the concrete scientific situation. It is enough for us to know that it is *a priori* valid for quantum mechanics ... We have resolved not to say, 'Every experiment *must* be classically described' but simply, 'Every experiment *is* classically described.' Thus the factual, we might almost say historical situation of physics is made basic to our propositions" (emphasis in the original).

References

Bächtold, M. 2008. "Are all measurement outcomes classical?" *Studies in History and Philosophy of Modern Physics* 39: 620–33.

Barad, K. 2007. *Meeting the Universe Halfway: Quantum Physics and the Entanglement of Matter and Meaning*. Durham, NC: Duke University Press.

Beller, M. 1999. *Quantum Dialogue: The Making of a Revolution*. Chicago: University of Chicago Press.

Beller, M., and A. Fine. 1994. "Bohr's response to EPR." In *Niels Bohr and Contemporary Philosophy*, edited by J. Faye and H. J. Folse, 1–31. Dordrecht: Kluwer Academic Publishers.

Bitbol, M. 2013. "Bohr's complementarity and Kant's epistemology." *Séminaire Poincaré* 17: 145–66.

Bohr, N. 1931. "Maxwell and modern theoretical physics." *Nature* 128: 691–2.

Bohr, N. 1935. "Can quantum-mechanical description of physical reality be considered complete?" *Physical Review* 48: 696–702.

Bohr, N. 1937. "Causality and complementarity." *Philosophy of Science* 4: 289–98.

Bohr, N. 1939. "The causality problem in atomic physics." In *New Theories in Physics*, 11–30. Paris: International Institute of Intellectual Cooperation.

Bohr, N. 1948. "On the notions of causality and complementarity." *Dialectica* 2: 312–19.

Bohr, N. 1949. "Discussions with Einstein on epistemological problems in atomic physics." In *Albert Einstein, Philosopher-Scientist: The Library of Living Philosophers, Vol. 7*, edited by P. A. Schilpp, 201–41. Evanston, IL: Open Court.

Bohr, N. 1958a. "On atoms and human knowledge." *Dædalus* 87: 164–75.

Bohr, N. 1958b. "Quantum physics and philosophy: Causality and complementarity." In *Philosophy at Mid-century*, edited by R. Klibanksy, 308–14. Florence: La Nuova Italia Editrice.

Bohr, N. [1929] 1987a. "Introductory survey." In *Atomic Theory and the Description of Nature: The Philosophical Writings of Niels Bohr, Vol. 1*, 1–24. Woodbridge, CT: Ox Bow Press.

Bohr, N. [1929] 1987b. "The quantum of action and the description of nature." In *Atomic Theory and the Description of Nature: The Philosophical Writings of Niels Bohr, Vol. 1*, 92–101. Woodbridge, CT: Ox Bow Press.

Bohr, N. [1938] 1987c. "Natural philosophy and human cultures." In *Atomic Physics and Human Knowledge: The Philosophical Writings of Niels Bohr, Vol. II, Essays 1932–1957*, 23–31. Woodbridge, CT: Ox Bow Press.

Bohr, N. 1996. *Collected Works, Vol. 7, Foundations of Quantum Mechanics II (1933–1958)*, edited by J. Kalckar. Amsterdam: Elsevier.

Bohr, N. 2005. *Collected Works, Vol. 11, The Political Arena (1934–1961)*, edited by F. Aaserud. Amsterdam: Elsevier.

Bohr, N., and L. Rosenfeld. [1933] 1979. "On the question of measurability of electromagnetic field quantities." In *Selected Papers of Léon Rosenfeld*, edited by R. S. Cohen and J Stachel, 357–400. Dordrecht: Boston. D. Reidel.

Brigandt, I. 2012. "The dynamics of scientific concepts: The relevance of epistemic aims and values." In *Scientific Concepts and Investigative Practice*, 75–103. Berlin: de Gruyter.

Brock, S. 2003. *Niels Bohr's Philosophy of Quantum Physics in the Light of the Helmholtzian Tradition of Theoretical Physics*. Berlin: Logos Verlag.

Camilleri, K. 2009a. "Constructing the myth of the Copenhagen interpretation." *Perspectives on Science* 17: 26–57.

Camilleri, K. 2009b. *Heisenberg and the Interpretation of Quantum Mechanics: The Physicist as Philosopher*. Cambridge: Cambridge University Press.

Camilleri, K., and M. Schlosshauer. 2015. "Niels Bohr as philosopher of experiment: Does decoherence theory challenge Bohr's doctrine of classical concepts?" *Studies in History and Philosophy of Modern Physics* 49: 73–83.

Cassirer, E. [1936] 1956. *Determinism and Indeterminism in Modern Physics: Historical and Systematic Studies of the Problem of Causality*, translated by O. T. Benrfey. New Haven: Yale University Press.

Cassirer, E. [1929] 1957. *The Philosophy of Symbolic Forms, Vol. 3 The Phenomenology of Knowledge*, translated by R. Manheim. New Haven: Yale University Press.

Chevalley, C. 1991. "Glossaire." In *Physique atomique et conaissance humaine*, N. Bohr, 345–567. Paris: Gallimard.

Chevalley, C. 1994. "Niels Bohr and the Atlantis of Kantianism." In *Niels Bohr and Contemporary Philosophy*, edited by J. Faye and H. J. Folse, 33–55. Dordrecht: Kluwer Academic Publishers.

Chevalley, C. 1999. "Why do we find Bohr obscure?" In *Epistemological and Experimental Perspectives on Quantum Physics*, edited by D. Greenberger, 59–73. Dordrecht: Kluwer Academic Publishers.

Christiansen, F. V. 2006. "Heinrich Herz' neo-Kantian philosophy of science, and its development by Harald Høffding." *Journal for General Philosophy of Science/Zeitschrift für allgemeine Wissenschaftstheorie* 37: 1–20.

Collingwood, R. G. 1939. *An Autobiography*. Oxford: Oxford University Press.

Cuffaro, M. 2010. "The Kantian framework of complementarity." *Studies in History and Philosophy of Science* 41: 309–17.

Favrholdt, D. 1992. *Niels Bohr's Philosophical Background*. Copenhagen: Munksgaard.

Favrholdt, D. 1993. "Niels Bohr's views concerning language." *Semiotica* 94: 5–34.

Faye, J. 1991. *Niels Bohr: His Heritage and Legacy: An Anti-realist View of Quantum Mechanics*. Dordrecht: Kluwer Academic Publishers.

Faye, J. 1994. "Once more: Bohr-Høffding." *Danish Yearbook of Philosophy* 29: 106–13.

Faye, J. 2007. "Niels Bohr and the Vienna circle." In *The Vienna Circle and the Nordic Countries: Networks and Transformation of Logical Empiricism, Vienna Circle Institute Yearbook*, edited by J. Manninen and F. Stadler, 30–45. Dordrecht: Springer.

Faye, J., and H. J. Folse. 1994. *Niels Bohr and Contemporary Philosophy*. Dordrecht: Kluwer.

Feyerabend, P. 1968. "On a recent critique of complementarity I." *Philosophy of Science* 35: 309–31.

Feyerabend, P. 1969. "On a recent critique of complementarity II." *Philosophy of Science* 36: 82–105.

Fjelland, R. 2002. "The Copenhagen interpretation of quantum mechanics and phenomenology." In *Hermeneutic Philosophy of Science*, edited by B. E. Babich, 53–65. Dordrecht: Kluwer.

Fock, V. 1957. "On the interpretation of quantum mechanics." *Czechoslovak Journal of Physics* 7: 643–56.

Fock, V. 1958. "Remarks on Bohr's article on his discussions with Einstein." *Soviet Physics Upsehki* 66: 208–10.

Folse, H. J. 1978. "Kantian aspects of complementarity." *Kant-Studien* 69: 58–66.

Folse, H. J. 1985. *The Philosophy of Niels Bohr: The Framework of Complementarity*. Amsterdam: North-Holland.

Folse, H. J. 1994. "Bohr's framework of complementarity and the realism debate." In *Niels Bohr and Contemporary Philosophy*, edited by J. Faye and H. J. Folse, 119–39. Dordrecht: Springer.

Frank, P. 1975. *Modern Science and Its Philosophy*. New York: Arno Press.

Graham, L. 1988. "The Soviet reaction to Bohr's quantum mechanics." In *Niels Bohr: Physics and the World: Proceedings of the Niels Bohr Centennial Symposium*, edited by H. Feschbach, T. Matsui, and O. Oleson, 305–17. London: Harwood Academic.

Heisenberg, W. 1930. *The Physical Principles of the Quantum Theory*. New York: Dover.

Heisenberg, W. 1934. "Wundlagen der Grundlagen der exakten Naturwissenschaften in jungster Zeit." *Angewandte Chemie* 47: 697–702.

Heisenberg, W. 1955. "The development of the interpretation of quantum theory." In *Niels Bohr and the Development of Physics: Essays Dedicated to Niels Bohr on the Occasion of His Seventieth Birthday*, edited by W. Pauli and V. Weisskopf, 12–29. New York: McGraw-Hill.

Hermann, G. 1935. *Die Naturphilosophischen Grunlagen Der Quantenmechanik.* Berlin: Öffentliches Leben.

Hermann, G. 1937a. "Die naturphilosophische Bedeutung des Übergangs von der klassischen zur modernen Physik." In *Travaux Du Ixe Congrès Internationale De Philosophie 'Congrès Descrates*, edited by R. Bsyer, 99–101. Paris.

Hermann, G. 1937b. "Über die Grundlagen physikalischer Aussagen in den älteren und der modernen Theorien." *Abhandlungen der Fries'schen Schule, Neue Folge* 6: 309–98.

Honner, J. 1982. "The transcendental philosophy of Niels Bohr." *Studies in History and Philosophy of Modern Physics* 13: 1–29.

Honner, J. 1994. "Description and deconstruction." In *Niels Bohr and Contemporary Philosophy*, edited by J. Faye and H. J. Folse, 141–53. Dordrecht: Kluwer Academic Publishers.

Hooker, C. A. 1972. "The nature of quantum mechanical reality: Einstein vs Bohr." In *Paradigms and Paradoxes: The Philosophical Challenge of the Quantum Domain*, edited by R. G. Colodny and A. Fine, 67–302. Pittsburgh: University of Pittsburgh Press.

Hooker, C. A. 1994. "Bohr and the crisis of empirical intelligibility. An essay on the depth's of Bohr's thought and our philosophical ignorance." In *Niels Bohr and Contemporary Philosophy*, edited by J. Faye and H. J. Folse, 155–99. Dordrecht: Kluwer Academic Publishers.

Howard, D. 1994. "What makes a classical concept classical?" In *Niels Bohr and Contemporary Philosophy*, edited by J. Faye and H. J. Folse, 201–29. Dordrecht: Kluwer Academic Publishers.

Howard, D. 2004. "Who invented the 'Copenhagen interpretation?' A study in mythology." *Philosophy of Science* 71: 669–82.

Jacobsen, A. 2007. "Léon Rosenfeld's Marxist defense of complementarity." *Historical Studies in the Physical and Biological Sciences* 37: 3–34.

Jacobsen, A. 2012. *Léon Rosenfeld: Physics, Philosophy, and Politics in the Twentieth Century.* Singapore: World Scientific.

Jordan, P. 1934. "Über den positivistischen Begriff der Wirklichkeit." *Naturwissenschaften* 22: 485–90.

Jordan, P. 1936. *Anschauliche Quantentheorie: eine Einführung in die moderne Auffassung der Quantenerscheinungen.* Berlin: Springer.

Jordan, P. 1944. *Physics of the Twentieth Century.* New York: Philosophical Library.

Kaiser, D. 1992. "More roots of complementarity: Kantian aspects and influences." *Studies in the History and Philosophy of Science* 23: 213–39.

Katsumori, M. 2011. *Niels Bohr's Complementarity: Its Structure, History, and Intersections with Hermeneutics and Deconstruction.* Dordrecht: Springer.

Kroes, P. 2003. "Physics, experiments and the concept of nature." In *The Philosophy of Scientific Experimentation*, edited by H. Radder, 68–86. Pittsburgh: Pittsburgh University Press.

Lakatos, I., and P. K. Feyerabend. 1986. *For and against Method: Imre Lakatos and Paul Feyerabend*, edited by M. Motterlini. Chicago: University of Chicago Press.

MacKinnon, E. 1982. *Scientific Explanation and Atomic Physics.* Chicago: University of Chicago Press.

MacKinnon, E. 1985. "Bohr on the foundations of quantum theory." In *Niels Bohr: A Centenary Volume*, edited by A. P. French and P. J. Kennedy, 101–20. Cambridge, MA: Harvard University Press.

MacKinnon, E. 2011. *Interpreting Physics: Language and the Classical/Quantum Divide.* Dordrecht: Springer.

Myer-Abich, K. M. 1965. *Korrespondenz, Individualität und Komplementarität: eine Studie zur Geistesgeschichte der Quantentheorie in den Beiträgen Niels Bohrs.* Wiesbaden: F. Steiner.

Moreira, R. 1994. "Høffding and Bohr: Waves or particles." In *Waves and Particle in Light and Matter*, 395–410. Heidelberg: Springer.

Murdoch, D. 1987. *Niels Bohr's Philosophy of Physics.* Cambridge: Cambridge University Press.

Osnaghi, S., F. Freitas, and O. Freire Jr. 2009. "The origin of the Everettian heresy." *Studies in History and Philosophy of Modern Physics* 40: 97–123.

Pais A. 1991. *Niels Bohr Times, in Physics, Philosophy, and Polity.* Oxford: Oxford University Press.

Pauli, W. 1996. *Wissenschaftlicher Briefwechsel mit Bohr, Einstein, Heisenberg: 1950–1952 Vol. 4, Part I.* New York: Springer.

Petersen, Aa. 1963. "The philosophy of Niels Bohr." *Bulletin of the Atomic Scientists* 198: 14.

Petersen, Aa. 1968. *Quantum Physics and the Philosophical Tradition.* Cambridge: MIT Press.

Plotnitsky, A. 1994. *Complementarity: Anti-Epistemology after Bohr and Derrida.* Durham: Duke University Press.

Plotnitsky, A. 2006. *Reading Bohr: Physics and Philosophy.* Dordrecht: Springer.

Plotnitsky, A. 2012. *Niels Bohr and Complementarity.* Dordrecht: Springer.

Pringe, H. 2009. "A transcendental account of correspondence and complementarity." In *Constituting Objectivity*, edited by M. Bitbol, P. Kerszberg, and J. Petitot, 317–23. Dordrecht: Springer.

Röseberg, U. 1994. "Hidden historicity. The challenge of Bohr's philosophical thought." In *Niels Bohr and Contemporary Philosophy*, edited by J. Faye and H. J. Folse, 325–43. Dordrecht: Kluwer Academic Publishers.

Röseberg, U. 1995. "Did they just misunderstood each other? Logical empiricists and Bohr's complementary argument." In *Physics, Philosophy and the Scientific Community: Essays in the Philosophy and History of the Natural Sciences and Mathematics in Honor of Robert S. Cohen*, edited by K. Gavroglu, J. Stachel, and M. Wartofsky, 105–23. Dordrecht: Kluwer Academic Publishers.

Rosenfeld, L. 1967. "Niels Bohr in the thirties. Consolidation and extension of the conception of complementarity." In *Niels Bohr: His Life and Work as Seen by His Friends and Colleagues*, edited by S. Rozental. 114–36. Amsterdam: North-Holland.

Rosenfeld, L. [1953] 1979. "Strife about complementarity." In *Selected Papers of Léon Rosenfeld*, edited by R. S. Cohen and J. Stachel, 465–83. Dordrecht: Boston. D. Reidel.

Rosenfeld, L. [1957] 1979. "Misunderstandings about the foundations of quantum mechanics." In *Selected Papers of Léon Rosenfeld*, edited by R. S. Cohen and J. Stachel, 465–83. Dordrecht: Boston. D. Reidel.

Scheibe, E. 1973. *The Logical Analysis of Quantum Theory.* Oxford: Pergamon.

Shimony, A. 1985. "Review of *The Philosophy of Niels Bohr: The Framework of Complementarity*, by Henry J. Folse." *Physics Today* 38: 108–109.

Steinle, F. 2012. "Goals and fates of concepts: The case of magnetic poles." In *Scientific Concepts and Investigative Practice*, 105–25. Berlin: de Gruyter.

von Weizsäcker, C. F. 1936. "Die naturphilosophischen Grundlagen der Quantenmechanik." *Physikalische Zeitschrift* 37: 527–8.

von Weizsäcker, C. F. 1941a. "Die Verhältnis Der Quantentenmechanik zur Philosophie Kants." *Die Tatwelt* 17: 66–98.

von Weizsäcker, C. F. 1941b. "Zur Deutung der Quantenmechanik." _Zeitschrift fur Physik_ 118: 489–509.

von Weizsäcker, C. F. 1952. _The World View of Physics_, translated by M. Grene. London: Routledge and Kegan Paul.

von Weizsäcker, C. F. [1966] 1994. "Kant's theory of natural science according to P. Plaass." In _Kant's Theory of Natural Science_, P. Plaass, translated by A. E. Miller and M. G. Miller, 167–87. Dordrecht: Kluwer.

von Weizsäcker, C. F. 1971. "The Copenhagen interpretation." In _Quantum Theory and Beyond_, edited by T. Bastin, 25–31. Cambridge: Cambridge University Press.

Zinkernagel, H. 2016. "Niels Bohr on the wave function and the classical/quantum divide." _Studies in the History and Philosophy of Modern Physics_ 53: 9–19.

On Bohr's Transcendental Research Program

Michel Bitbol

1. Introduction

Even during the heyday of the so-called Copenhagen interpretation (spanning approximately between 1927 and 1952) Bohr's views on quantum mechanics did not gain universal acceptance among physicists. The "orthodox" reading of quantum mechanics derived from the latter interpretation was a mixture of elements borrowed from Heisenberg, Dirac, and Von Neumann, with a few words quoted from Bohr and due reverence for his pioneering work, but with no unconditional allegiance to his ideas. One reason for this lukewarm reception of Bohr's views was that physicists (such as Enrico Fermi) usually found them "too philosophical" (Holton 1978, 162). Physicists needed an expedient theoretical scheme and some guiding representations, whereas Bohr rather developed a general reflection about the epistemological status of such schemes, and stressed the limits of representations in science. In addition, it proved somehow uneasy to reach definite conclusions as to the true nature of Bohr's philosophy. Some commentators detected a positivistic (Faye 2009) or pragmatist (Murdoch 1987, 225) flavor in Bohr's thought, especially in view of his overt interest for William James and his apparently operationalist definition of microphysical phenomena. A few historians also aptly insisted on his personal and academic connections with the Danish philosopher Harald Høffding who transmitted some of the abovementioned philosophical influences to Bohr.

But amid this mixture of philosophical styles, a strong Kantian pattern can easily be recognized in Bohr's thought. Even though Bohr himself did not bother to pinpoint his debt toward Kant, and despite the critical stance of the "Copenhagen group" toward some aspects of the philosophy of Kant, a very important component of his theory of knowledge had clearly, albeit indirectly, made its way in Bohr's interpretation of quantum mechanics. It had filtered through Bohr's scattered philosophical background, to reemerge clandestinely but very recognizably in his analysis of quantum knowledge. Early neo-Kantian readings of quantum mechanics, such as Carl-Friedrich von Weizsäcker's (Heisenberg 1971), Chapter X. Grete Hermann's (1996), and Ernst Cassirer's (1956 [1936]), thus took advantage of Bohr's position instead of distancing themselves from it.

Later on, the Kantian element in Bohr's thought was recognized by many historians and philosophers of physics (Hooker 1972, 135–72; Honner 1982; Chevalley 1991; Kaiser 1992; Brock 2003) almost at the same time as it was minimized or denied by many others (Folse 1978 and 1985; Murdoch 1987; Pais 1991). The latter disagreement is easy to understand: on the one hand, the Kantian component of Bohr's epistemology is buried in his idiosyncratic remarks about the role of measuring devices and the boundaries of theoretical domains; and on the other hand, after more than two and a half centuries of erudite commentaries, the gist of Kant's philosophy remains a matter of debate. A twofold clarification, initiated by several authors (Held 1995; Kauark-Leite 2012; Pringe 2007; Cuffaro 2010) but still perfectible, is thus needed to give a definite answer about what *is*, after all, the epistemology of this major physicist.

The present chapter is conceived as a step toward such clarification, and it will develop in four parts. The first section concentrates on a central enigma which was present from the very beginning of Bohr's reflections about quantum theory, namely, the status of classical physics and classical concepts in it; indeed, this is likely to have been a central motivation for the Kant-like features of Bohr's philosophy. The second section attempts to sort out the various dimensions of Kant's theory of knowledge, and to identify what has been retained and what has been left out by Bohr. The third section develops on Bohr's concept of "complementarity," and points out that several strata of Kant's theory of knowledge are needed to make sense of it. The fourth and final section expands the field of the discussion by documenting a debate between transcendental and naturalized conceptions of the measurement problem (together with a precise definition of these terms).

2. Should classical physics be a permanent component of the foundations of quantum physics?

Bohr's first model of the atom includes a nonconventional compromise between classical mechanics and electrodynamics on the one hand, and Planck's "quantum rules" on the other hand. It has often been regarded as "incoherent" due to its baroque combination of electronic stationary orbits ruled by classical mechanics, quantized transitions from one orbit to another, and overt violation of certain theorems of classical electrodynamics (since electrons on stationary orbits were supposed to emit no radiation despite their angular acceleration). Yet, the accusation is excessive, insofar as the rules for using each one of the elements of the model in succession are mostly well stated. The patchwork-structure of Bohr's model of the atom does not necessarily entail inconsistency, provided the pieces of this patchwork are ascribed a precise but limited use in certain restricted theoretical contexts (Vickers 2007). In other terms, Bohr's model lacks unity rather than logical and practical consistency. Of course, unity is also a major regulative ideal of physics. The natural assumption of those who realized with amazement the efficiency of Bohr's clumsy model for predicting a large range of atomic spectral lines was that this could only be a *provisional* compromise which would sooner or later be replaced by a fully unified theory. But, to the surprise of many physicists, Bohr's path toward the new quantum theory constantly postponed the banishment of classical physics. During the intermediate phase that separated the old quantum theory from the advent of modern quantum mechanics, say between 1913 and 1925, Bohr was looking for new ways of articulating classical physics with quantum postulates, rather than trying to eliminate classical features altogether. The "correspondence principle" between classical and quantum physics acted as a pivotal element of this strategy. It went far beyond the usual retrospective requirement that the old physical theory be a limiting case of the new theory, and rather irrupted in the very fabric of the new theory. It was taken as a prospective guide toward new theoretical structures, by way of "generalization" and extrapolation. And it was also used as a sort of spare wheel for assessing the value of certain variables that were absent from the hybrid model, such as the line amplitudes. The correspondence principle here worked as a meta-theoretical structure that enabled one theoretical structure (the classical one) to serve as analogic scaffolding for the elaboration and completion of another theoretical structure (the quantum one) (Darrigol 1992). Thus, according to Bohr (1922, 24),

> Although the process of radiation cannot be described on the basis of the ordinary theory of electrodynamics ... there is found, nevertheless, to exist a far-reaching correspondence between the various types of possible transitions between the stationary states on the one hand, and the various harmonic components.... . This correspondence is of such a nature that the present theory of spectra is in a certain sense to be regarded as a rational generalization of the ordinary theory of radiation.

This nonconventional approach to theorizing has cogently been compared with Kant's central prescription in his *Critique of Judgment* (1987): to bring seemingly heterogeneous theoretical structures together into a unique *system* irrespective of their differences, by way of a hypothetical use of reason (Pringe 2009). Nevertheless, Bohr was less and less satisfied by the original compromise, and throughout this period he was systematically varying the boundaries between the classical and quantum elements of his theory in order to find the most appropriate locus of the sought articulation. His strategy was a permanent negotiation between what should be retained and what should be excluded of the old patterns of physics. One crucial (though unsuccessful) step was taken in 1924 when Bohr tried to find a new way to reconcile the continuous spatiotemporal orbits and the discontinuity of quantum jumps in his model. There, he decided to retain the "picture" of orbits, but to jettison somehow the principle of causality and the principles of conservation of energy and momentum, when the connection between each electronic motion and the emission/absorption of radiation is at stake. While being no longer valid for individual processes of interaction between radiation and matter, the principles of conservation were still enforced at the statistical level (Bohr, Kramers, and Slater 1924).

However, this new compromise was soon perceived as unsuccessful, and pressure was exerted toward eliminating the very classical "picture" of orbital trajectories in favor of a thoroughly "quantum" kinematics and dynamics. Such revolutionary advance was advocated by Pauli in a letter to Bohr of December 1924 (Darrigol 1992, 208), and it was sketched by Born under the newly invented label "*Quantenmechanik*" (quantum mechanics) (Born 1924). The final accomplishment in this direction was, of course, Heisenberg's (1925) "matrix mechanics," which completely dispensed with the concept of a continuous trajectory, replacing it with a law-like structure applying to spectral observables. Here, a systematic procedure of symbolic translation of classical laws into quantum laws was undertaken, by way of replacing ordinary, continuous variables by non-commuting matrices of measurable discontinuous quantities. Yet, Bohr (1925) hailed Heisenberg's quantum mechanics as a "precise formulation of the tendencies embodied in the correspondence principle," and Heisenberg himself called his theory "a quantitative formulation of the correspondence principle" (Darrigol 1992, 276). The vestige of classical physics was still utterly active in the new theory, and the only thing that remained obscure was the *reason* for this stubborn presence.

In a move whose *style* is unmistakably Kantian, Bohr later attempted to clarify this reason by turning away from ontological considerations about nature, and relying on epistemology instead. This turnabout was likely to be the true birth certificate of the "Copenhagen interpretation" of quantum mechanics. From then on, the footprint of classical physics was construed not as a manifestation of the (possibly semiclassical) essence of micro-objects, but as a constraint that is inherent to us qua knowing and speaking subjects able to ascribe meaning to what we observe. One of Bohr's (1934, 16) earliest statements of this cognitive status ascribed to classical physics reads thus:

> No more is it likely that the fundamental concepts of the classical theories will *ever* become superfluous for the description of physical experience.... It continues to be the application of these concepts alone that makes it possible to relate the symbolism of the quantum theory to the data of experience.

In other words, no acceptable account of what is observed in a laboratory, and no acceptable description of the instruments that are used for that purpose, can be given in nonclassical terms according to Bohr. Classical physics here remains the crypto-epistemological substrate of quantum mechanics, its basic and too often unrecognized background. Not, to repeat, because this old stratum of the history of physics has captured some sort of definitive truth about the world, but because it embodies a set of even more primitive preconditions of knowledge. These preconditions were later expressed by Bohr ([1958] 1987a, 39) in a celebrated statement:

> *However far the phenomena transcend the scope of classical physical explanation, the account of all evidence must be expressed in classical terms.* The argument is simply that by the word "experiment" we refer to a situation where we can tell to others what we have done and what we have learned and that, therefore, the account of the experimental arrangement and of the results of the observations must be expressed in unambiguous language with suitable application of the terminology of classical physics. (Bohr's emphasis)

It becomes clear from the former sentences that what is needed in full-blown quantum physics is not the integral theoretical contents of classical physics, but its *terminology*, and above all its logical two-valued unambiguous pattern of *communication*. Classical physics is here taken as the proxy of the most elementary conditions of possibility of intersubjective agreement about experimental results, as well as about the way to build and to calibrate measuring instruments. This Kant-like status of classical descriptions immediately disposes of so many criticisms of Bohr according to which the latter was a "macro-realist" (Leggett 1988) who ascribed an upper level of reality to macroscopic objects while he downgraded the reality of microscopic objects, or according to which he believed that nature is made of two realms of beings: the classical macro-entities and the quantum micro-entities. Indeed, there is no such ontological divide in Bohr's conception of quantum mechanics, but only a practical and epistemological divide between (i) the referred to quantum objects and (ii) the logico-linguistic tools of reference that happen to be implemented in classical physics. Even in the most advanced attempts at avoiding any mixture of classical elements with quantum physics, the logico-linguistic *function* of the former elements must be fulfilled. More about that in Section 4.

3. Kant's and Bohr's Copernican revolutions

It is our task in this section to document some essential features of Kant's theory of knowledge, and to find equivalent moves or clear differences in Bohr's philosophy of quantum mechanics.

The first major feature of Kant's epistemology is the adoption of a radically new stance that he called the "Copernican revolution." Its name is inspired from Copernicus's decision to base the explanation of the apparent motion of planets on the *relations* between the motion of our Earth and the orbits of the planets, instead of sticking to a theory of their *intrinsic* kinematics. Kant was especially fascinated by Copernicus's reflective stance that pushed him to pay attention to the situation of the astronomer on planet Earth and to the way this contributes to shaping cosmological knowledge, rather than to remain exclusively fascinated by his object of study (namely, the other planets). The generalization of this stance in scientific knowledge is described by Kant (1996, 21) in striking terms: "Thus far it has been assumed that all our cognition must conform to objects ... Let us try to find out by experiment whether we shall not make better progress ... if we assume that objects must conform to our cognition." The latter statement does *not* mean that objects are so to speak *created* by our cognition, but (i) that one cannot dispense with a proper analysis of our own faculties of knowing, if knowledge is to be understood at all; and (ii) that the form of objects is predetermined by a set of cognitive conditions enabling us to overcome the variegation of fleeting subjective appearances, and to circumscribe some invariant phenomena which can be intersubjectively recognized and designated. In this case, "object" is the name of such experiential invariants, not of something beyond experience. Kant then tried to identify what, in the structure of our cognition, makes the identification of unified invariant phenomenal patterns possible. And he found two classes of such structures. The first one is the given continuity of the forms of our sensory intuitions, namely, space and time. The second one is the table of general concepts of our understanding (or categories) that are used to bring a variety of sensory appearances under a common organization. Among the latter, we find: (i) the category of *substance*, which permanently unifies a set of attributes; and (ii) the category of *causality*, which gives us the possibility of making a difference between unruly subjective successions and law-like sequences of phenomena that any subject can pick out.

That Bohr also performed a sort of "Copernican revolution" is already clear from our former reflection about the epistemological rather than ontological status of classical concepts in his mature interpretation of quantum mechanics. Bohr (1934) was thoroughly concerned with the resources of our own cognition, and he thought that we cannot dispense with studying them, if we want to make sense of the theoretical by-product of our scientific inquiry. After all, "the task of science is both to extend the range of our experience and to reduce it to order" (1).

Realizing this restrictive mission of science is no little achievement, and might well be indispensable for going through true scientific revolutions. Indeed, as long as a current paradigm is taken to be valid, the project of science can be equated with characterizing better and better the *objects* of this paradigm. But when an old paradigm is crumbling, even its objects come under suspicion; even the standard procedures for extracting invariant phenomena must be questioned. Once this point of extreme doubt has been reached, one is bound to fall back on the only firm ground left, which is experience itself together with the experimental means of getting definite measurement values on which the manifold of subjective experiences can concur. So, just as Kant did, Bohr undertook a reflective analysis of the generic structure of our capability to know. However, unlike Kant, Bohr distanced himself from a study of mental faculties such as sensibility and understanding. He rather focused on a technological counterpart of sensibility, namely, the measuring apparatus, and on an intersubjective counterpart of understanding, which is common language.

Another important feature of Kant's epistemology is the organic articulation of the two sources of knowledge, which he took to be sensibility and understanding. He considered that the unity brought by concepts could only concern the data of sensory intuition; and he then looked for a proper locus of connection between the forms of understanding and the contents of intuition. This locus was found in what he called the "schematism of pure imagination" (Kant 1996, 210). Schematism can conveniently be figured out as the ability to elaborate pictures guiding possible actions that purport to anticipate in a systematic way certain contents of sensory intuition. The most celebrated example of this kind of connection between intellectual structures and sensory data is offered by *causality*. The latter concept of pure understanding applies, according to Kant, to sensory contents preordered by a spatiotemporal structure. And its application is mediated by the scheme of succession according to a rule (a generic term for the trajectories of a dynamics) that allows one to anticipate later phenomena from the knowledge of an appropriate set of earlier phenomena.

It is precisely at this point that Bohr parts company with Kant. As we have seen in the previous section of this chapter, Bohr was increasingly diffident about many components of the Kant-like classical pattern. He questioned the universal applicability of the category of causality in 1924, deeming that this could well prove incompatible with "picturing" of trajectories in space-time; and in view of the failure of his challenge to the conservation principles at the individual level, he tended to believe that, conversely, no single "picture" could be meaningfully used to represent the detail of atomic processes. Pictures were accordingly restricted to the role of purely "symbolic," not to say "poetical," approaches to the atomic domain (Heisenberg 1971). Later on, Bohr held an even more fully articulated position according to which, in the microscopic domain, the category of causality and the spatiotemporal coordination of phenomena are mutually exclusive (as we will see more precisely in Section 3). Anticipation of phenomena was no longer performed on the basis of a continuous spatiotemporal representation of phenomena, but rather by using "a purely symbolic scheme permitting only predictions … as to results obtainable under conditions specified by means of classical concepts" (Bohr 1987a, 40). Therefore, a major keystone of Kant's theory of knowledge was removed, and to many physicists of that time the whole building looked like it was doomed to crumbling.

This divergence between Bohr and Kant can be characterized in a few words. Kant claimed that the forms of sensibility and understanding he had identified were a priori conditions of possibility of objective knowledge in general. This suggested that they were fixed forever by way of necessity, and that no vicissitude of scientific research could alter them. As for Bohr (1934), he accepted the typically Kantian idea of prior forms of knowledge. He stressed that "in spite of their limitations, we can by no means dispense with those forms of perception which colour our whole language and in terms of which all experience must ultimately be expressed" (5). But Bohr (1934, 1) also considered that, in view of these limitations, "we must always be prepared ... to expect alterations in the points of views best suited for the ordering of our experiments." The idea of modifying the so-called a priori forms of human knowledge according to the advances of scientific research was so averse to Kant that this alone justified the rejection by many philosophers of science of any similarity between Bohr's and Kant's theory of knowledge (Kaiser 1992). However, this point of disagreement can easily be shown to be less serious than it appears at first sight. Kant's epistemology and Bohr's philosophy of physics host in themselves enough resources to be able to come together in the long term.

To begin with, Kant's epistemology should not be restricted to the precise personal position of Kant. In practice, it worked as a *research program*, and was later developed by an entire school of neo-Kantian philosophy whose major move was to historicize and relativize the so-called *a priori* forms of knowledge. The function of these forms, which is to unify and extract invariants in the midst of appearances, could easily be ascribed to plastic "symbolic forms" (Cassirer 1965), or to historically relative "principles of coordination" (Reichenbach 1965), as it has been repeatedly documented in recent scholarship (Friedman 1992; Bitbol 1996a, 1998; Pradelle 2013).

Furthermore, on Bohr's side, there was no denial of the enduring pragmatic value of Kant's forms of knowledge at the macroscopic scale. Bohr's insistence on the perennial need for classical concepts can easily be read this way, since Kant's original a priori forms were explicitly adapted to the case of classical mechanics. As we have seen in Section 1, Bohr considered classical concepts as a condition for the possibility of intersubjectively shared knowledge in Kant's style. The only proviso added by Bohr is that the sphere of validity of this latter condition of possibility is severely hampered (Bitbol 2009a). It does not extend to the whole domain of microphysics, but only to the direct environment of mankind, at the macroscopic scale. It is an anthropological condition of possibility for the technological conditions of possibility of microphysical research. Therefore, as Heisenberg (1990, 78) pointed out, it can be taken as a *second-order* condition for the possibility of microphysical knowledge: "What Kant had not foreseen was that these *a priori* concepts can be the conditions for science and at the same time have a limited range of applicability." The forms that directly precondition knowledge of our direct environment and of the domain of classical physics have limited range of applicability, but at the same time they indirectly precondition knowledge of any research *beyond* this domain. This was Bohr's simultaneous vindication and confinement of Kant's a priori.

One more relevant couple of features of Kant's epistemology must be pinpointed at this stage: (i) his insistence on the purely relational nature of phenomena, and (ii) his claim that the ideally conceived "thing in itself," as it *is* independently of any relation with our faculties of knowledge, is doomed to remain undisclosed forever. These two features are deeply related to one another, as we will soon realize. According to Kant (1996, 340),

Things ... are given in intuition with determinations that express mere relations without being based on anything intrinsic; for such things are not things in themselves, but are merely appearances. Whatever [characteristics] we are acquainted with in *matter* are nothing but relations (what we call its intrinsic determinations is intrinsic only comparatively); but among these relations there are independent and permanent ones, through which a determinate object is given to us. (Emphasis in the original)

What we call "proper-ties of material objects" are only the expression of the cognitive relations that we establish with our environment; they are not "proper" to some object, but rather arise as an unanalyzable by-product of our interaction with "it" (the quotation marks surrounding the word "it" is justified by the fact that in Kant's approach, there are no such things as preexisting objects waiting for sensorial or experimental study, but that "object" is the name we give to the common focus of a consistent set of "independent and permanent" relations with *what there is*). What exists beyond these cognitive relations and independently of them is *in principle* unreachable, since reaching it would precisely mean establishing a cognitive relation with it. So much so that some commentators concluded that "thing in itself" is only Kant's catchword for the impossibility of disentangling ourselves completely from the content of our knowledge (Hintikka 1974); it is only a word for the impossibility to subtract the contribution of the knower from the mass of what is to be known.

This probably represents the point on which Bohr was closest to Kant, not because he drew inspiration from the texts of the philosopher of Königsberg, but because he rediscovered Kant's position about properties of objects under the pressure of quantum physics. The new element that generated this pressure is the *indivisibility of the quantum of action*, together with the circumstance "that the properties of atoms are always obtained by observing their reactions under collisions or under the influence of radiation" (Bohr 1934, 95). Indeed, the fact that there is no other way to disclose atomic properties than by provoking a quantized and uncontrollable interaction with some colliding object implies a "limitation on the possibilities of measurement" (95). But at this point, Bohr's originality is manifested. He did not conclude, as many other physicists of the Copenhagen group (including Heisenberg), that atomic properties are unfortunately *disturbed* by the measuring agent, and that therefore these properties are partially unknown to us, or blurred, or unsharp. He rather insisted on the "impossibility of a strict separation of phenomena and means of observation" (96), and accordingly on the unworkability of defining attributes that are *proper* to the atomic object (which conditions the use of the name "property"). Since "interaction forms an inseparable part of the phenomena" (Bohr 1987b, 4), any discourse about phenomena going on in nature independently of any measuring interaction is meaningless according to Bohr; and the very concept of independent properties of objects accordingly loses any ground. This later gave rise to the "interactionality conception of microphysical attributes" (Jammer 1974, 160), which was developed at length by Grete Hermann (1996), and endorsed by Vladimir Fock (Jammer 1974, 202) as a good substitute for the "idealism" of those views involving the conscious observer, such as Pauli's, London and Bauer's, or Von Neumann's. Now, the consequences of the inseparability thesis are exactly the same as those of Kant's thorough relationism. The two available options are: (i) accepting that there is something like a "micro-object in itself" that we can know only obliquely, by means of successive interactive approaches; (ii) declaring that any reference to some such obliquely knowable "micro-object in itself" is the fake name one gives to the impossibility of disentangling the wholeness of phenomena. A discussion of this point is presented in Section 3, around the typically Bohrian concept of "complementarity."

To conclude this section, I will mention another more general similarity between Kant and Bohr. Kant documented two distinct standpoints and attitudes that human beings must successively adopt, in order to deal with two different kinds of challenges in their lives. In his *Critique of Pure Reason*, which addresses our interest for *nature*, he explained how, without truly disentangling ourselves with respect to what there is, we can nevertheless elaborate a form of knowledge that works *as if* we were separated from nature. This approach, which ascribes us a role that mimics that of an external *spectator* of nature, is called "objective knowledge." By contrast, in his *Critique of Practical Reason*, which deals with the issues of freedom, action, and morals, any such separation is precluded and we are ascribed the role of true *actors* of our own deeds (Beck 1963, 31). This dialectic of actor and spectator was later taken over by Schopenhauer, in his famous book *The World as Will and Representation*. The will is experienced from the point of view of a living actor, whereas the representation of the world is obtained from a point of view that superficially resembles that of a spectator. But the latent lesson of both Kantianism and post-Kantianism is that there is no *true* spectator's standpoint (except in a minimalist "as if" sense); that in view of our deep insuperable entanglement with what there is, the standpoint of a spectator of nature is extrapolated out of the only available standpoint, which is the actor's.

This is exactly what Bohr concluded from his reflections on quantum mechanics, with some additional radicality. He soon discovered that our apparent disentanglement from nature in classical knowledge is only a limiting case, and that we are actually immersed in it as any actor is; he pointed out that in the quantum domain we usually cannot even find circumstances in which we can behave *as if* we were spectators; but he also insisted that we must still play the role of a classical pseudo-spectator at the macroscopic scale of the laboratory, to enable "unambiguous communication." This is why "the new situation in physics has so forcibly reminded us of the old truth that we are both onlookers and actors in the great drama of existence" (Bohr 1934, 119).

4. Complementarity about what?

According to Bohr, a crucial consequence of the "interactionality" conception of microphysical attributes, of the fact that experimental modes of access cannot be separated from phenomena, is *complementarity*. Let us follow Bohr's reasoning. In classical physics, where the influence of the measuring procedure can easily be subtracted from its outcome, data obtained by using various instruments "supplement each other and can be combined into a consistent picture of the behaviour of the object under consideration" (Bohr 1987b, 4). But in quantum physics, changing the experimental arrangement is tantamount to changing the holistic phenomenon itself, which therefore turns out to be incompatible with other holistic phenomena. "Combination into a single picture" (4) of various experimental data then automatically yields contradictions. Such contradictions are overcome only at the cost of loosely articulating the information derived from various experimental arrangements by a new nonpictorial method called "complementarity." Indeed, in complementarity, these different pieces of information are taken to be mutually exclusive, yet jointly indispensable to characterize a certain micro-object: "evidence obtained under different experimental conditions cannot be comprehended within a single picture, but must be regarded as complementary in the sense that only the totality of the phenomena exhausts the possible information about the objects" (Bohr 1987a, 40).

But Bohr's complementarity is no simple or unambiguous concept. One may list *three* (disputable) applications of the concept of complementarity in quantum mechanics (Faye 1991; Held 1994; Bitbol 1996b):

C1—*The complementarity between incompatible variables.* "In quantum physics … evidence about atomic objects obtained by different experimental arrangements exhibits a novel kind of complementary relationship" (Bohr 1987a, 4). A standard example is the complementarity of the archetypal couple of conjugate variables, namely, position and momentum.

C2—*The complementarity between causation and spatiotemporal location of phenomena.* This was the first explicit statement of complementarity in Bohr's (1934, 54) Como lecture of 1927: "The very nature of quantum theory … forces us to regard the space-time coordination and the claim of causality, the union of which characterizes the classical theories, as complementary but exclusive features of the description, symbolizing the idealization of observation and definition respectively."

C3—*The complementarity between the continuous and discontinuous pictures of atomic phenomena, that is, between the wave model and the particle model:* The individuality of the elementary electrical corpuscles is forced upon us by general evidence. Nevertheless, recent experience, above all the discovery of the selective reflection of electrons from metal crystals, requires the use of the wave theory superposition principles in accordance with the original ideas of L. de Broglie.… . In fact, here again we are not dealing with contradictory but with complementary pictures of the phenomena which only together offer a natural generalization of the classical mode of description. (Bohr 1934, 56)

Let us discuss these three versions of complementarity in turn, starting with the last one, C3. This is the most popular but also the most controversial version of complementarity, since it involves remnants of classical *representations* (rather than classical *variables*, or classical terminology for the description of measuring apparatuses) (Murdoch 1987). Moreover, after 1935, Bohr "tacitly abandons the idea of wave-particle complementarity" (Held 1994). Indeed, from this moment on, the concept of complementarity only allows indirect "clarification" of the dilemma of wave-particle dualism (Bohr 1987b); it does not operate directly on the corresponding couple of representations.

The C1 version of complementarity is likely to be closer to the roots of the concept, yet it is not without difficulties of its own. The most benign difficulty is that, from the incompatibility of experimental arrangements, one cannot derive *immediately* the incompatibility of the corresponding value ascriptions to an object. For nothing precludes that it is only our experimental *knowledge* of position which is incompatible with our experimental *knowledge* of momentum, whereas the object intrinsically "possesses" both properties. But, as we have seen previously, this narrowly epistemic interpretation of incompatibility was ruled out at once by Bohr, when he declared that the experimental arrangement is part of the *definition* of the corresponding variable, rather than an instrument for reaching its intrinsic value.

A much more serious difficulty bears on the claim that the values of incompatible variables are *jointly* indispensable to exhaust knowledge about the object. How can it be made compatible with Bohr's rejection of the classical thesis according to which objects are *simultaneously* endowed with definite values of both incompatible variables? A way to bypass *simultaneous* possession consists in saying that joint indispensability concerns *successive* measurements actually performed in order to exhaust knowledge about the object. But this solution is not satisfactory either, because the values can vary according to the *order* of measurements. Another alternative, more satisfactory, interpretation is that exhaustiveness concerns the full list of *possibilities* of observation one has, before any experimental arrangement has been chosen (Held 1994). Along this line of thought, one may reconcile mutual exclusion and completion of conjugate variables without logical inconsistency: mutual exclusion pertains to *actual* experimental arrangements, whereas completion (or exhaustiveness) refers to *possible* measurements.

Nevertheless, even in the framework of the latter (partly satisfactory) interpretation, a problem remains unaddressed. What *reason* do we have to impose (and try to justify) the idea that conjugate variables are *jointly indispensable*, potentially or actually? For what exactly are they indispensable? Bohr's answer is classical-like, or rather common-sense-like: only the complete list of conjugate variables, he declares, can exhaust information *about* the objects under investigation. This assumption that experimental information is about something, its "*aboutness*" so to speak, is repeatedly mentioned by him as a crucial feature of microphysical knowledge: "evidence obtained under different experimental conditions ... must be regarded as complementary in the sense that only the totality of the phenomena exhausts the possible information *about* the objects" (Bohr 1987a, 40; my emphasis); and he adds, "together (these phenomena) exhaust all definable knowledge *about* the objects concerned" (90); and "such phenomena together exhaust all definable information *about* the atomic objects" (99; my emphasis). But this insistence on aboutness is highly problematic. It presupposes that despite the unanalysability (or interactional nature) of micro-phenomena, it is still possible to *refer to* micro-objects as if they were somehow independent of any experimental procedure. It also suggests that these putative atomic objects have the same general predicative structure as the objects of classical mechanics, even though they are not ascribed predicates as such. Just like classical particles, "atomic objects" are indeed construed by Bohr as points of convergence of two families of conjugate features such as position *and* momentum. Conversely, if one did *not* presuppose that experimental outcomes are *about* objects endowed with the same predicative structure as the moving bodies of classical mechanics, one could perfectly well accept that position and momentum are mutually exclusive without being *jointly* indispensable for some *thing*. Once again, we are facing the dilemma of the "object in itself": should we refer to some such "micro-object in itself" characterized by successive but mutually incompatible interactive probings; or should we look for a new mode of objectification that retains *nothing* (not even the adumbration of a predicative structure) of the classical corpuscularian concept?

If we follow Bohr in adopting the first (conservative) strategy, we must ascribe a highly nonconventional status to micro-objects. Expressing this status in a Kantian idiom, Bohr's cloudy "micro-object" should be considered only as a unifying *symbol* used as a regulative-heuristic device (Kant 1987, 227; Pringe 2007), not as a tangible *something*. But if we adopt the "revolutionary" (neo-Kantian) strategy of looking for novel forms of objectification instead of sticking to old object-like patterns, we become free to consider appropriate elements of the quantum formalism as denotations of previously inconceivable objects of knowledge. Schrödinger's option, when he considered the wave function as a description of "reality," can be interpreted this way (Bitbol 1996b) as a quest for a new focus for the constitution of a stable objective domain, from a corpus of fleeting phenomena. Similarly, Heisenberg's initial proposal of a matrix mechanics can be understood as an invitation to change the space in which the constitution of objectivity (Bitbol, Kerszberg, and Petitot 2009) is performed: from ordinary space to its Fourier transform (i.e., to the spectral domain) (Petitot 1997).

Let us finally examine complementarity C2 between causality and the use of space-time concepts. According to Bohr (1987b, 11), "The use of any arrangement suited to study momentum and energy balance—decisive for the account of essential properties of atomic systems—implies a renunciation of detailed space-time coordination of their constituent particles." This suggests that, in the same way as the complementarity of conjugate variables, the complementarity between causality and space-time coordination is related to the incompatibility of experimental arrangements, combined with the interactionality thesis. However, other texts of the earlier period,[1] as well as Heisenberg's lucid interpretation of Bohr's position, favor another conception of the complementarity between causality and space-time coordination, which does *not* reduce it to the complementarity of pairs of conjugate variables. In his *Physical Principles of the Quantum Theory*, Heisenberg (1930, 63) argues that the measurement of *any* variable whatsoever involves spatiotemporal aspects. Thus, it is not only space-time coordination, but any description of phenomena in space-time, which is incompatible with causality. Indeed, one cannot observe *any* kind of spatiotemporally circumscribed phenomena without influencing them; and such an influence makes causal laws inapplicable to the said phenomena. Such a Heisenbergian interpretation of the complementarity between causality and space-time location has rapidly spread. C. F. Von Weizsäcker (1985) and P. Mittelstaedt (1976) thus interpreted this version of complementarity as a relation of mutual exclusiveness and joint completion between the abstract deterministic law of evolution of ψ-functions (i.e., the Schrödinger equation) and the measurement of *any* variable by an apparatus located in space-time.

As suggested in Section 2, this dismantling of Kant's articulation between the category of causality and spatiotemporally shaped sensory experience is likely to have momentous consequences. Not only does it challenge the special architecture of the *Critique of Pure Reason* and its alleged a priori conditions of any possible knowledge, but also, at first sight, it threatens the global Kantian strategy of finding how the chaos of subjective impressions is progressively ordered into an objective pattern. Kant's first step requires embedding these impressions into the spatiotemporal web, and his second step requires connecting the spatiotemporally located phenomena with one another according to a *law* of succession. How can one objectify phenomena, if the former procedure is no longer available? Bohr's answer is daring, but in line with the spirit (if not the letter) of Kant's epistemology. It consists in turning complementarity into a connecting device that *supplants* causality, beyond ascribing causality a limited role within its own scheme of mutual exclusion and joint exhaustivity. According to Bohr (1937), quantum physics "forces us to *replace* the ideal of causality by a more general viewpoint called 'complementarity'" (emphasis mine). Insofar as the standard (causal) connection of phenomena is prevented by the incompatibility of the various modes of experimental access, a new mode of (complementary) connection is offered as a substitute. "Complementarity is called for to provide a frame wide enough to embrace the account of fundamental *regularities* of nature which cannot be comprehended within a single picture" (Bohr 1987b, 12; emphasis mine).

But once again, I wish to point out that this replacement of a causal mode of connection with a complementary mode of connection is aimed at extrapolating the mode of existence of standard classic-like objects endowed with mutually exclusive attributes *well beyond their standard domain of relevance*. The revolutionary mode of connection here compensates for a conservative ontology. A diametrically opposed strategy has been proposed, and it consists in sticking to the standard (causal) mode of connection at the cost of redefining the object of microphysics (Mittelstaedt 1976).

5. Transcendental versus naturalized approaches to measurement

As we have seen in Section 1, Bohr ascribed a kind of extraterritorial status to measuring instruments, measurement outcomes, and experimental activities by imposing a classical mode of description onto them and excluding them from the quantum mode of description. This was not meant to assert the uncommon *nature* of such macroscopic objects and laboratory procedures, but only to ascribe to them an indispensable *function* in the system of knowledge: that of a condition of possibility of intersubjective agreement about experimental results and activities. Such functional rather than substantial status of the classical mode of description is confirmed by Bohr's insistence that the boundary between the classical and quantum domains is by no means fixed, and that we even have a "free choice" (Bohr 1935) of the exact location of this boundary. An advantage of ascribing to classical concepts the function of a condition of possibility of quantum knowledge is that this automatically dissolves the measurement problem from the outset. Indeed (i) insofar as a classic-like realistic description of a relevant part of the measuring instruments is taken as a fundamental presupposition of any quantum account of phenomena, there is no need to figure out a mechanism for the transition between potentialities and actuality; and (ii) since the quantum description is prevented from extending to the totality of the measurement chain, there is no such thing as a superposition of pointer macrostates.

Saying that Bohr's classic-like extraterritorial status of measuring procedures is not very popular among physicists would be an understatement. In fact, this approach to the measurement problem is frequently *abhorrent* to them, because it suggests that their discipline has fundamental limits that cannot be overcome, and because it acknowledges that physics cannot account (even retrospectively) for the experimental conditions of its own assessment. The archetypal bearer of this rejection of Bohr's view is Bell's (1990) celebrated paper "Against 'Measurement.'" What Bell criticizes eagerly in this article is the decision to stick to propositions that are valid only "for all practical purposes (FAPP)." Being a standard example of such propositions, the classic-like status of measuring devices, together with the split between classical apparatuses and the quantum world, is the central target of Bell's attack. Bell thus called for a mature quantum mechanics in which the "shifty split" would be completely omitted, and its symbols would "refer to the world as a whole." Bohm's theory and Everett's theory are good candidates for such a status: they both claim to be alternative formulations of quantum physics, yet to provide us with descriptions of the world as a whole, including the measuring apparatus. Bell (1987, 136) was leaning toward Bohm's theory, while he criticized Everett's for its defense of what he called *solipsism of the present*. More recently, decoherence theories have become the most widely accepted realizations of the project of describing the world as a whole, since they claim to derive classical behaviors from the quantum formalism, instead of presupposing them (Joos et al. 2003).

So, can we now bury safely Bohr's alleged "cut" between the domain of study and the experimental means of investigation; can we just avoid the presupposition of classical concepts in the foundations of quantum physics? My answer is negative: the function of the "cut" *has* to be fulfilled somehow. I must repeat that the function of Bohr's cut was to avoid any confusion between the predictive formalism of quantum mechanics and the elementary presuppositions needed for putting it to the test. These two items are logically distinct; they appertain respectively to a theoretical and pre- (or meta-) theoretical level of thought (Mittelstaedt 1998, 8). True, no intrinsic feature of the measuring apparatus forces one to treat it on the pre-theoretical rather than the theoretical level (104), but if the theoretical level is chosen for describing the apparatus, the pre-theoretical level *has* to irrupt at some earlier or later stage, in the statement of the project of measurement and the measuring outcome. As a consequence, the shadow of this distinction remains visible, and uneliminable, in *all* the abovementioned theories. Thus, the privilege of space coordinate variables in Bohm's theory is the theoretical shadow cast by the spatiotemporally located pre-theoretical pointer variables. Similarly, Everett's frequent changes of standpoint between (i) a purely external description of the "universal state vector" and (ii) an internal description of what is recorded in the "memory bracket" of some given counterpart of the observer on one "branch" of this state vector is the shadow of the two levels (theoretical and pre-theoretical, predictive and descriptive, potential and actual) of any physical description. As for decoherence theories, their ability is only to derive quasi-classical probabilistic structures (namely, approximately diagonal reduced density matrices of systems), but not a full degree of classicality (Schlosshauer and Camilleri 2011; Bitbol 2009b); moreover, in order to do so, they rely on a deliberate selection of the relevant degrees of freedom of the system within a larger domain of environmental degrees of freedom, which can be seen as the shadow of the system-context cut.

We must ponder some more time on this issue of the shadowy presence of the system-context or theoretical-pre-theoretical cut, because it is especially tricky. In classical physics, the indispensability of the pre-theoretical level was usually hidden, because in this very special case, the pre-theoretical treatment of the measurement process could easily be made *isomorphic* to its theoretical description. So much so that isomorphism could be taken as a case of identity, and "the measurement process could be understood as a special case of the general laws applying to the entire universe" (Bohm and Hiley 1993, 13). Nothing seemed to hinder the grand project of an entirely *naturalized* theory of knowledge, whereas Kant's intimation of a *transcendental* conception of knowledge was neglected or taken as optional. By contrast, all we have seen until now inclines us to think that no attempt at elaborating a fully *naturalized* theory of the knowledge process has been successful in the quantum paradigm, and that a *transcendental* approach to knowledge (whose archetype has been formulated by Bohr) becomes more clearly unavoidable in this theory than anywhere else. Let us make one more little pause at this point: what is the difference between a naturalized and a transcendental theory of knowledge? What is the meaning of these two words, "naturalized" and "transcendental"?

A naturalized theory purports to treat the process of acquisition of knowledge as if it were an object of the sciences of nature (of the *knowledge* of nature!) among many others. The measuring apparatus, for instance, is supposed to obey the same laws of nature as the measured object, so that the measurement interaction can be treated as a standard natural process. More generally, the whole cognitive operation, including the *decision* to perform an experiment and the *realization* of the outcome of that experiment, is considered as a topic for natural sciences such as neurobiology, biochemistry, optics, and so on. At first sight, this option of naturalization looks quite reasonable and innocuous. Nothing seems to prevent the continuous expansion of the domain of validity of the sciences of nature toward an increasing disclosure of the knowing process: after all, a measuring apparatus, a sense organ, and a brain are quite standard material objects. Nothing therefore seems to threaten the ideal of sciences, which is to understand in their own terms each element of *what there* is. Unfortunately, a little (but momentous) logical imprecision decisively undermines the project. This logical problem was expressed as a slogan in a well-known article about the measurement problem of quantum mechanics: quantum theory can describe *anything*, but not *everything* (Peres and Zurek 1982; Fuchs and Peres 2000). The ideal of science has then been abusively extrapolated, from the ability to account for any particular phenomenon, to the ability to account for the whole display (let alone the whole *Being*). As one of the most accomplished realizations of this ideal, quantum theory can describe anything known thus far, but it cannot avoid leaving *the preconditions for this very description* outside its scope. Such preconditions can be given the form of a classical subset of objects, or of a privilege ascribed to spatial coordinate variables, or of a memory bracket, or of a conventional cut between the system and the environment, but in every case it is *functionally excluded* from the quantum domain.

These remarks challenge the universalist dream of a naturalized theory of knowledge, but they are at the same time the central tenet of a transcendental theory of knowledge. The word "transcendental" is a variation of the more common word "transcendent." Both terms share a common component of meaning, which is "exceeding experience." But "exceeding" can be achieved in two antithetical ways. A transcendent *object* exceeds experience insofar as it is said to exist *beyond* experience, as a remote (and intellectually reconstructed) external cause of experienced phenomena. Conversely, a transcendental structure exceeds experience because it is a *precondition* of experience: it shapes experience without being part of experience. Moreover, as long as the act of knowing develops, such a transcendental structure is bound to remain in the silent background of this act. Here again, we bump into the extraterritorial status of the preconditions of knowledge. The said extraterritoriality is considered by Kant, who created the very concept of a transcendental epistemology, as a strong logical requirement. In his own terms, how the condition of possibility of knowledge "is (itself) possible, will not admit of any further solution or answer, because we invariably require it for all answers and for all thought of objects" (Kant 1994, §36). To sum up, what plays the role of the knower cannot be known in the very process of this knowledge; what preconditions the possibility of a quantum description cannot be described quantum-mechanically in the very process of describing.

Interestingly, this debate between naturalized and transcendental epistemologies overtly took place during a remarkable discussion between Bohr, Wheeler, and Everett at the end of the 1950s, about the latter's interpretation of quantum mechanics. Wheeler presented Everett's interpretation to Bohr as a new development of some ideas of the "Copenhagen school." But this interpretation was explicitly rejected by Bohr and his Copenhagen group (that included Leon Rosenfeld and Aage Petersen) as a deep misunderstanding of their ideas (Osnaghi, Freitas, and Freire 2009). The main reason for this rejection was precisely that Bohr and his team could not accept Everett's rejection of the classical-like extraterritorial status of (part of) the measurement process. Against Everett's claim that basing quantum physics upon classical physics was only a provisional step, Rosenfeld replied: "To try (as Everett does) to include the experimental arrangement into theoretical formalism is perfectly hopeless, since this can only shift, but never remove, this essential use of unanalyzed concepts which alone makes the theory intelligible and communicable." Here, the irreducibly *transcendental* status of the measurement process, manifested by its being described with classical concepts, was clearly reasserted against Everett's dream of naturalizing everything (including the act of measuring) within the range of validity of quantum physics. This transcendental status can be ascribed to any part of the process, its field of application can be extended or shrunk, but it can by no means be eliminated altogether (in Rosenfeld's terms, it can be "shifted" but not "removed").

6. Conclusion

If one compares a few statements deemed to represent Bohr's philosophy of science with a frozen description of Kant's original epistemology, there are more reasons to pinpoint the differences than to discern the resonances. But this way of doing comparative philosophy does not do justice to the status of the two positions at stake. None of them should be taken as a definitive statement about what knowledge is, but rather as an ongoing epistemological *research program*. This is especially true of transcendental epistemology, whose basic principles have been formulated by Kant, but which has been transformed sometimes beyond recognition by several generations of post-Kantian and neo-Kantian thinkers. The most appropriate question to be raised is then the following: does Bohr's construal of quantum physics respect the minimal rules and orientations that define the *transcendental research program*?

As it is well known, the concept of a research program has been elaborated by Imre Lakatos (1978) to insist on the dynamic, rather than static, nature of scientific theories and paradigms. Applying this concept reflectively, to characterize the dynamic aspects of a philosophy of science instead of a scientific theory, represents a nonconventional use. As a preliminary step, let's then list the main features of scientific research programs in order to assess their applicability to epistemological research programs.

Lakatos's concept of a scientific research program was initially designed to go beyond naive falsificationism by accommodating the possibility for a given scientific paradigm to adapt itself (to a certain extent) in the face of apparently incompatible empirical evidence. This implies a redefinition of what a scientific theory or a scientific paradigm consists of: a scientific theory is not made of a single set of rigidly formulated axioms or rules, but rather involves a two-layered structure with a gradient of flexibility. The first layer is called the "hard core" of the scientific theory construed as a research program; it *defines* the latter and should be kept intact throughout its history, lest one trivialize the said program by denying it any precise identity. The second layer is called the "protective belt" of the program; it is progressively elaborated in response to new (potentially threatening) empirical evidence. But this further elaboration is submitted to strict criteria meant to avoid "ad hoc" adaptations and incoherent aggregation of arbitrary assumptions or models. Thus, "protective belts" are constrained by the norm of a "progressive problem shift," according to which any new assumption or model must be both (i) compatible with the "hard core" of the program, and (ii) capable of making new predictions that can be submitted to empirical test, rather than just accommodating old empirical data. However, the distribution of roles between the "hard core" and the "protective belt" is not fixed from the outset; it is part of what has to be discovered (or shaped) during the development of the scientific research program.

One example analyzed at length by Lakatos is precisely the expansion of Bohr's atom model construed as a research program. In 1913, Bohr's model for the hydrogen atom had to be taken as a block. But in the course of its advancement, until its partial collapse in 1924, it became clear that initial simplifying assumptions such as circular orbits, nonrelativistic mechanics, and location of the barycenter of the atomic system at the center of the nucleus could easily be modified in order to elaborate a "protective belt" of new hypotheses. The true nucleus of Bohr's research program was progressively circumscribed by contrast: it consisted in a systematically controlled compromise between the "quantum postulate" and the underlying classical postulates, thus giving rise to a sequence of increasingly precise formulations of the correspondence principle.

The same can be shown to hold for the research program of a transcendental epistemology. When it was first formulated, in the *Critique of Pure Reason*, transcendental epistemology looked like a rigid doctrine with its clear-cut distinction between phenomena and things in themselves, fixed "faculties," a priori forms of sensibility and understanding, and the subsequent table of categories. But in the course of the development of this research program, the successors of Kant learned to extract its true "hard core." It then appeared that what defines the program of a transcendental epistemology includes neither the concept of a thing in itself (see its criticism by Fichte, Schopenhauer, etc.) nor an immutable set of a priori forms, which is challenged by the progress of scientific research. What truly circumscribes the identity of such program is: (i) the fundamental attitude of the "Copernican revolution," namely, trying to disclose the mark left by the activity of knowing upon the structures of knowledge, (ii) the attempt to pinpoint a set of background assumptions that preconditions the delimitation of a domain of objects in the course of the activity of knowing (Hans Reichenbach), (iii) the correlative definition of each object as an invariant complex of phenomena according to the rules established by a "symbolic form" (Ernst Cassirer), and (iv) more generally, the quest for "unity" beyond the variegation of phenomena (Hermann Cohen). Now, we have seen that Bohr's philosophy of quantum mechanics offers precise equivalents of each one of these fundamental stances or moves: (i) the "Copernican revolution" of ascribing an epistemological rather than ontological necessity to classical concepts; (ii) (iii) the shaping of quantum objects by the symbolic forms of complementarity; (iv) the quest for symbolic unity of physical knowledge beyond the dismantling of properties by their insuperable contextuality. This is enough to classify unambiguously Bohr's philosophy of physics as a variety of transcendental philosophy of science.

Notes

1 See, for example, Bohr (1934, 54): "If in order to make observation possible we permit certain interactions with suitable agencies of measurement, not belonging to the system, an unambiguous definition of the state of the system is naturally no longer possible, and there can be no question of causality in the ordinary sense of the word. The very nature of the quantum theory thus forces us to regard the space-time co-ordination and the claim of causality, the union of which characterizes the classical theories, as complementary but exclusive features of the description, symbolizing the idealization of observation and definition respectively."

2 L. Rosenfeld, letter to S. Bergmann, December 21, 1959, quoted and commented in Osnaghi, Freitas, and Freire (2009).

References

Beck, L. W. 1963. *A Commentary on Kant's Critique of Practical Reason*. Chicago: The University of Chicago Press.

Bell, J. 1987. *Speakable and Unspeakable in Quantum Mechanics*. Cambridge: Cambridge University Press.

Bell, J. 1990. "Against 'measurement'." *Physics World* 8: 33–40.

Bitbol, M. 1996a. *Mécanique quantique, une introduction philosophique*. Paris: Flammarion.

Bitbol, M. 1996b. *Schrödinger's Philosophy of Quantum Mechanics*. Dordrecht: Kluwer Academic Publishers.

Bitbol, M. 1998. "Some steps towards a transcendental deduction of quantum mechanics." *Philosophia naturalis* 35: 253–80.

Bitbol, M. 2009a. "Reflective metaphysics: Understanding quantum mechanics from a Kantian standpoint." *Philosophica* 83: 53–83.

Bitbol, M. 2009b. "Decoherence and the constitution of objectivity." In *Constituting Objectivity: Transcendental Perspectives on Modern Physics*, edited by M. Bitbol, P. Kerszberg, and J. Petitot. Berlin: Springer.

Bitbol, M., P. Kerszberg, and J. Petitot (eds). 2009. *Constituting Objectivity: Transcendental Perspectives on Modern Physics*. Berlin: Springer.

Bohm, D., and B. Hiley. 1993. *The Undivided Universe*. London: Routledge.

Bohr, N. 1913. "On the constitution of atoms and molecules." *Philosophical Magazine* 26: 1–25.

Bohr, N. 1922. *The Theory of Spectra and Atomic Constitution*. Cambridge: Cambridge University Press.

Bohr, N., H. Kramers, and J. C. Slater. 1924. "The quantum theory of radiation." *Philosophical Magazine* 47: 785–822.

Bohr, N. 1925. "Atomic theory and mechanics." Supplement to *Nature* 116: 845–52.

Bohr, N. 1934. *Atomic Theory and the Description of Nature*. Cambridge: Cambridge University Press.

Bohr, N. 1935. "Can quantum-mechanical description of physical reality be considered complete." *Physical Review* 48: 696–702.

Bohr, N. 1937. "Causality and complementarity." *Philosophy of Science* 4: 289–98.

Bohr, N. [1958] 1987a. *Essays 1933–1957 on Atomic Physics and Human Knowledge*. New York: Wiley. Reprinted as *The Philosophical Writings of Niels Bohr*, Vol. 2. Woodbridge, CT: Ox Bow Press.

Bohr, N. 1987b. *Essays 1958–1962 on Atomic Physics and Human Knowledge*. New York: Wiley. Reprinted as *The Philosophical Writings of Niels Bohr*, Vol. 3. Woodbridge, CT: Ox Bow Press.

Born, M. 1924. "Über Quantenmechanik." *Zeitschrift für Physik* 26: 379–95.

Brock, S. 2003. *Niels Bohr's Philosophy of Quantum Physics*. Berlin: Logos Verlag.

Cassirer, E. 1956. *Determinism and Indeterminism in Modern Physics*. New Haven: Yale University Press. (The original German version was published 1936.)

Cassirer, E. 1965. *Philosophy of Symbolic Forms*. New Haven: Yale University Press.

Chevalley, C. 1991. *Le dessin et la couleur*. In *N. Bohr, Physique atomique et connaissance humaine*. Paris: Gallimard.

Cuffaro, M. E. 2010. "The Kantian framework of complementarity." *Studies in History and Philosophy of Modern Physics* 41: 309–17.

Darrigol, O. 1992. *From c-Numbers to q-Numbers*. Oakland: University of California Press.

Hooker, C. A. 1972. "The nature of quantum mechanical reality." In *Paradigms and Paradoxes*, edited by R. G. Colodny, 135–72. Pittsburgh: University of Pittsburgh Press.

Faye, J. 1991. *Niels Bohr, His Heritage and Legacy*. Dordrecht: Kluwer Academic Publishers.

Faye, J. 2009. "Niels Bohr and the Vienna circle." In *The Vienna Circle in the Nordic Countries: Networks and transformation of logical empiricism*, edited by J. Manninen and F. Stadler, 33–45. *Vienna Circle Institute Yearbook*. Dordrecht: Springer.

Folse, H. J. 1978. "Kantian aspects of complementarity." *Kant-Studien* 69: 58–66.

Folse, H. J. 1985. *The Philosophy of Niels Bohr: The Framework of Complementarity*. Amsterdam: North-Holland.

Friedman, M. 1992. *Kant and the Exact Sciences*. Cambridge, MA: Harvard University Press.

Fuchs, C., and A. Peres. 2000. "Quantum theory needs no 'interpretation.'" *Physics Today* 53: 70–1.

Heisenberg, W. 1925. "Über die Quantentheoretische Umdeutung kinematischer und mechanischer Beziehungen." *Zeischrift für Physik* 33: 879–93.

Heisenberg, W. 1930. *The Physical Principles of the Quantum Theory*. Chicago: University of Chicago Press.

Heisenberg, W. 1971. *Physics and Beyond*. New York: Harper and Row.

Heisenberg, W. 1990. *Physics and Philosophy*. London: Penguin.

Held, C. 1994. "The meaning of complementarity. *Studies in History and Philosophy of Science* 25: 871–93.

Held, C. 1995. "Bohr and Kantian idealism." In *Proceedings of the Eighth International Kant Congress*, edited by H. Robinson and G. Brittan. Milwaukee: Marquette University Press.

Hermann, G. 1996. *Les fondements philosophiques de la mécanique quantique*. Paris: Vrin. (The original German version was published 1935.)

Hintikka, J. 1974. "Dinge an sich revisited." In *Knowledge and the Known*, J. Hintikka. Dordrecht: Reidel.

Holton, G. J. 1978. *The Scientific Imagination: Case Studies*. Cambridge: Cambridge University Press.

Honner, J. 1982. "The transcendental philosophy of Niels Bohr." *Studies in the History and Philosophy of Sciences* 13: 1–30.

Jammer, M. 1974. *The Philosophy of Quantum Mechanics*. New York: John Wiley.

Joos, E., H. D. Zeh, C. Kiefer, D. Giulini, J. Kupsch, and I.-O. Stamatescu. 2003. *Decoherence and the Appearance of a Classical World in Quantum Theory*. New York: Springer.

Kaiser, D. 1992. "More roots of complementarity: Kantian aspects and influences." *Studies in the History and Philosophy of Science* 23: 213–39.

Kant, I. 1987. *Critique of Judgment*. Cambridge, MA: Hackett Publishing Co.

Kant, I. 1994. *Prolegomena to Any Future Metaphysics*. New York: Pearson.

Kant, I. 1996. *Critique of Pure Reason*. Cambridge, MA: Hackett Publishing Co.

Kauark-Leite, P. 2012. *Théorie quantique et philosophie transcendantale: dialogues possibles*. Paris: Hermann.

Lakatos, I. 1978. *The Methodology of Scientific Research Programmes, Philosophical Papers*, Vol 1. Cambridge: Cambridge University Press.

Leggett, A. J. 1988. "Quantum mechanics and macroscopic realism." AIP Conference Proceedings 180: 229.

Mittelstaedt, P. 1976. *Philosophical Problems of Modern Physics*. Dordrecht: Reidel.

Mittelstaedt P. 1998. *The Interpretation of Quantum Mechanics and the Measurement Process*. Cambridge: Cambridge University Press.

Murdoch, D. 1987. *Niels Bohr's Philosophy of Physics*. Cambridge: Cambridge University Press.

Osnaghi, S., F. Freitas, and O. Freire. 2009. "The origin of the Everettian heresy." *Studies in History and Philosophy of Modern Physics* 40: 97–123.

Pais, A. 1991. *Niels Bohr's Time in Physics, Philosophy, and Polity*. Oxford: Oxford University Press.

Peres, A., and W. Zurek. 1982. "Is quantum theory universally valid?" *American Journal of Physics* 50: 807–10.

Petitot, J. 1997. "Objectivité faible et philosophie transcendantale." In *Physique et réalité: un débat avec Bernard d'Espagnat*, edited by M. Bitbol and S. Laugier. Paris: Frontières-Diderot.

Pradelle, D. 2013. *Généalogie de la raison*. Paris: Presses Universitaires de France.

Pringe, H. 2007. *Critique of the Quantum Power of Judgment. A Transcendental Foundation of Quantum Objectivity*. Berlin: de Gruyter.

Pringe, H. 2009. "A transcendental account of correspondence and complementarity." In *Constituting Objectivity: Transcendental Perspectives in Modern Physics*, edited by M. Bitbol, P. Kerszberg, and J. Petitot. Berlin: Springer.

Reichenbach, H. 1965. *The Theory of Relativity and a priori Knowledge*. Oakland: University of California Press.

Schlosshauer, M., and K. Camilleri. 2011. "What classicality? Decoherence and Bohr's classical concepts." *Advances in Quantum Theory: AIP Conference Proceedings* 1327: 26–35.

Vickers, P. J. 2007. "Bohr's Theory of the Atom: Content, Closure and Consistency." http://philsci-archive.pitt.edu/4005/.

von Weizsäcker, C. F. 1985. *Aufbau der physik*. Cincinnati: Hanser.

3

Transcendental versus Quantitative Meanings of Bohr's Complementarity Principle

Patricia Kauark-Leite

This chapter has two aims: first, to analyze the different formulations of Niels Bohr's principle of complementarity, from a broad philosophical point of view that cannot be reduced to wave-particle duality, as related to mutually exclusive experimental arrangements; and second, to discuss the arguments for and against the alleged violation of the principle of complementarity in new quantum experimental contexts, especially the Afshar experiment. In the first section, I present some general considerations concerning Bohr's principle of complementarity. In the second section, I critically examine attempts to restrict the complementarity principle and its application to so-called intermediate situations. In the third section, I critically reconstruct arguments in the context of the Afshar experiment for its alleged violation of the principle of complementarity. And in the fourth and final section, I critically evaluate the philosophical implications of "which-way" experiments for testing complementarity.

1. Some general considerations concerning Bohr's principle of complementarity

The immediate historical background of the complementarity principle can be traced back to two important scientific events in 1927: The *International Conference on Physics* in September at Como, Italy, and the *5th Solvay Conference on Electrons and Photons* in October at Brussels, Belgium. It was at the Italian meeting that Bohr presented his complementarity principle for the first time. A revised and expanded version of his Como lecture was published the following year in the journal *Nature*, under the title "The Quantum Postulate and the Recent Development of Atomic Theory." In the same year, 1928, a German version of the Como lecture was published in *Naturwissenschaften* and also a French version in the *Proceedings* of the 5th Solvay Conference. There are no differences between these three versions as regards the scientific and philosophical content of the lecture. Indeed, the importance of the 1927 Solvay Conference for the intellectual roots of Bohr's conception of complementarity was not due to a new clarification of the principle in relation to his Como lecture, but instead due to a very important dispute between Bohr and Einstein regarding the application of the complementarity principle in interference experiments. However, even despite this importance, as Bacciagaluppi and Valentini (2009) have shown, "no record of any such debate appears in the published proceedings, where both Bohr and Einstein are in fact relatively silent." As they further point out, "The famous exchange between Bohr and Einstein [at the 5th Solvay Conference] actually consisted of informal discussions, which took place semi-privately (mainly over breakfast and dinner), and which were overheard by just a few of the participants, in particular Heisenberg and Ehrenfest" (268). The content of this debate, moreover, became public only in 1949 after Bohr's paper "Discussion with Einstein on Epistemological Problems in Atomic Physics," and it marks a turning point in the way people conceive the complementarity principle, restricting it to wave-particle duality. In the rest of this section, I will first sketch a general overview of the three meanings of Bohr's complementarity in his Como lecture and other essays, second, justify the general philosophical meaning in a transcendental framework, and then, third, evaluate Einstein's objection to this principle when it is applied to quantum interference experiments, and also Bohr's reply to him.

The three meanings of Bohr's complementarity

According to Bohr and his principal supporters, the principle of complementarity is required for a consistent interpretation of quantum theory. As Bohr ([1929] 1934, 11) pointed out: "[The] feature of complementarity is essential for a consistent interpretation of the quantum-theoretical methods." In this sense, in order to avoid confusion, we should separate the theory and its formal structure from its interpretation. So the principle of complementarity does not belong to the formal level of the theory but to the interpretative level. This second level, although independent of the first one, cannot, however, be neglected since we do not have a full understanding of the theory without both levels "working" together. Rosenfeld (1969, 696) puts this point as follows: "Bohr's conception of complementarity in quantum mechanics is not the expression of a 'specific philosophical position,' but an inherent part of the theory which has the same validity as its formal aspect and is inseparable from it." And Pauli ([1958] 1980, 7) emphasizes the same essential and inherent complementarity feature of quantum theory saying: "We might call modern quantum theory 'The Theory of Complementarity' (by analogy with the term 'Theory of Relativity')."

For a more comprehensive account of complementarity, we need now turn to the three different meanings of it that are presented by Bohr in his Como lecture:

C1—the complementarity between space-time and causality;

C2—the complementarity between wave and particle behaviors; and

C3—the complementarity between incompatible observables, like position and momentum, or energy and time.

I will refer to these three different notions of complementarity by the following names: C1 = *space-time/causality* complementarity, C2 = *wave/particle* complementarity, and C3 = complementarity *between incompatible observables*. What is the meaning of each of these three notions, and what is the relationship between them? There is great confusion about the meaning and the proper use of the first notion. The other two notions are easier to understand in the light of quantum theory, since they are directly related to the quantum mathematical formalism.

C2 or wave/particle complementarity is essentially related to the wave/particle duality, and can be grounded, according to Bohr, in both the Einstein relation ($E=h\nu$) that establishes the corpuscular nature of a photon of light, and in the de Broglie relation ($p=h/\lambda$) that establishes the wave-like nature of a particle of matter.

C3, or the complementarity between incompatible observables, is essentially related associated in the theoretical formalism to the uncertainty relations as expressed by the mathematical expressions of $qp-pq=ih/2\pi$ (where q and p represent the canonical position and momentum variables of a particle and h denotes Planck constant) or $tE-Et=ih/2\pi$ (where t and E represent time and energy variables).

In contrast to the second and third notions of complementarity, the first one cannot be directly related to any specific part of the mathematical structure of the quantum theory. This has led many scholars of Bohr's thought to interpret the space-time/causality complementarity in highly contradictory and confusing ways. As Mara Beller (1999, 118) has pointed out, some interpreters related causal description to particle behavior and space-time to wave propagation while others, on the contrary, related space-time description to particles and causality to waves. According to Beller, the source of these contradictory readings lies in Bohr's Como lecture itself. As she puts it, "Complementarity between space-time and causality is an imprecise umbrella concept." This, in turn, expresses a widely held view of this Bohrian notion. Even Einstein (1949, 649) admitted that he didn't understand it, referring to "Bohr's principle of complementarity, the sharp formulation of which, moreover, I have been unable to achieve despite much effort which I have expended on it." As Beller (1999, 119) further emphasizes: "In the history of scientific thought it is hard to find another contribution about which opinions continued to differ sharply more than half a century after its appearance."

In the face of this difficulty, Dugald Murdoch (1987, 58–108) considers only two meanings of complementarity: first, wave-particle complementarity (i.e., C2), and second, a hybrid of C1 and C3 that he calls *kinematic–dynamic* complementarity and describes as a complementarity between *spatiotemporal descriptions* and *momentum-energy descriptions*. So, according to Murdoch, the complementarity between space-time and causality (i.e., C1) has the same meaning as the complementarity between incompatible observables, such as position (space) and momentum, or time and energy (i.e., C3). In Murdoch's view, the "claim of causality" in Bohr's sense means "knowledge of energy and momentum of the object" (58).

MacKinnon (1982), however, in his essay "The Consolidation of Bohr's Position" argues that there are two distinct phases in Bohr's position about complementarity. In Phase 1, from 1927 to 1929, Bohr defended space-time-causality complementarity (i.e., C1). But after this very short period, according to MacKinnon, Bohr changed his original formulation because it "was inadequate and not altogether self-consistent" (276), and began defending, after 1929, in Phase 2, wave-particle complementarity (i.e., C2). I completely disagree with this thesis. There is no textual support for MacKinnon's two-phase interpretation of Bohr's thinking about complementarity. Bohr continued to defend space-time/causality complementarity (i.e., C1) even after 1929, as is clear in his 1948 paper "On the Notions of Causality and Complementarity," or in his 1950 paper "Some General Comments on the Present Situation in Atomic Physics": "In proper quantum mechanics all paradoxes find their straightforward explanation in the complementary relationship between phenomena described in terms of space-time coordination and phenomena accounted for by means of dynamical conservation laws" (Bohr 1950, 377).

Von Weizsäcker (1955), in his turn, as Max Jammer (1974, 102–104) highlighted, parceled out the three notions of complementarity into two general types: *parallel* complementarities, and *circular* complementarity. Parallel complementarities are the wave/particle complementarity (i.e., C2) and the complementarity between incompatible observables (i.e., C3). They are "parallel" because the concepts involved are mutually exclusive. In the case of wave-particle complementarity (i.e., C2), both concepts belong to different intuitive pictures. By contrast, in the case of complementarity between incompatible observables (i.e., C3), both concepts, as position and momentum, or as time and energy, belong to the same intuitive picture. C1, or the space/time-causality complementarity, is classified by Von Weizsäcker as being of an essentially different kind from the other two, which is why he labels it "circular" complementarity. It is circular because the two alternatives involved belong to different contexts of description—spatiotemporal description and causal description—that are related one to another in a circular way. The spatiotemporal description corresponds to the experimental situation, and the causal description corresponds to the theoretical domain of dynamical laws. Heisenberg ([1930] 1949, 65), in turn, took circular complementarity to be Bohr's original conception of complementarity. Heisenberg justified this view by calling attention to the fact that whereas in classical physics, we have "causal relationships of phenomena described in terms of space and time," in quantum theory we have "either phenomena described in terms of space and time (but uncertainty relations) or causal relationships expressed by mathematical laws (but physical description of phenomena in space-time [is] impossible)" and "the two alternatives [are] related statistically".

I follow von Weizsäcker in his classification of types of complementarity, and also Heisenberg in his (I believe) accurate understanding of the fundamental basis of this notion. In my view, nevertheless, the three notions of complementarity in Bohr's Como lecture and also in many of his essays do not have the same epistemic status. Wave/particle complementarity (i.e., C2) and the complementarity between incompatible observables (i.e., C3) can be taken as physical-theoretical principles that can be derived from the formalism of quantum theory. In this sense, we can say that they belong to the object-language. By contrast, space-time/causality complementarity (i.e., C1) is a meta-theoretical principle that cannot be derived from the formalism itself. In this sense space-time/causality complementarity belongs to the meta-language and is best understood as a kind of transcendental principle in a Kantian sense. More specifically,

I think that there is only one general principle of complementarity, space-time/causality complementarity (i.e., C1), and that it functions as a meta-theoretical principle. In turn, the wave/particle complementarity (i.e., C2) and the complementarity between incompatible observables (i.e., C3) are just two different ways of instantiating the same general meta-theoretical principle.

Indeed, there are also, as Bohr pointed out, other ways of instantiating the same general meta-theoretical principle beyond the realm of physics: in biology, anthropology, psychology, and so on. As a meta-theoretical principle, then, this first notion, the space-time/causality complementarity (i.e., C1), is an "imprecise umbrella concept," as Beller said, *only* in the light of physical theory, at the object-language level: but it becomes fully intelligible in a philosophical sense at the meta-language level. More precisely, as a transcendental principle, in Kantian terms, C1-complementarity is not a principle about the world we experience, but instead about our *cognition* of the world we experience. Again, it's not a principle about the world, at the object-language level, but instead about our knowledge of the world, at the meta-language level.

The transcendental framework of complementarity

A great deal has been written about the relation between Bohr's and Kant's epistemologies (e.g., Hooker 1972 and 1994; Folse 1985, Honner 1982; Kaiser 1991 Chevalley 1994; Pringe 2007 and 2009; Kauark-Leite 2009, 2010a, b, 2012, and 2015; Cuffaro 2010; Bitbol 2010; Bitbol and Osnaghi 2013) and it is not my intention here to add to what has been already said about it. On the contrary, my point is simply to emphasize that the Kantian epistemic and metaphysical framework can provide a very clear way of making the principle of C1-complementarity intelligible. So in order to avoid taking this notion as an "imprecise umbrella concept," we need to postulate the truth of Kant's thesis about the ideality of the space and time, and, correspondingly, Kant's distinction, on the one hand, between empirical intuition (i.e., sense perception) and empirical concepts, and on the other, between spatiotemporal forms of empirical intuition and conceptual definition. In this Kantian framework, the representations of space and time are not concepts but instead a priori forms of our sensibility, namely, pure or a priori intuitions; space and time themselves are "nothing other than" these formal representations (this is Kant's thesis of the transcendental ideality of space and time); and causality is an a priori form of our understanding, namely, a pure or a priori concept (this is Kant's thesis of the transcendental ideality of causality). Although Kant ([1786] 2004) acknowledges the physical concepts of space and time, that is, those that are measurable and defined according to physical theories, he has to presuppose the more basic thesis that space and time are nothing other than a priori forms of our sensible intuition, with the aim of explaining how pure concepts like causality can be intelligibly and meaningfully applied to our experience of the world. So I propose that it is only in this Kantian framework that we can understand straightforwardly and consistently Bohr's ideas about causality and complementarity. For Bohr, in the light of this Kantian distinction, the unity of spatiotemporal and causal knowledge has a sharply different status in classical physics and quantum physics. In classical physics, it is thought of as a *conjunction of* the essentially mutually independent a priori notions of space-time and causality, but in quantum physics, it is thought of as a *disjunction between* space-time and causality, whereby the disjoined notions of space-time and causality are nevertheless essentially related by the general principle of complementarity. In classical physics, space-time facts in experience like the trajectory of a moving object, and causal relations like the dynamical laws, expressed in terms of differential equations, can be separately and precisely determined, and, at the same time, correlated in a such a way that the empirical measurement of some variables of the moving object coincides fairly well with theoretical predictions based on the causal laws. For this reason, space-time intuition and the causality principle can be unified in the same comprehensive intuitive picture, and we can say, from a classical physics point of view, that every space-time fact is also a causal fact. But in quantum physics, precisely determining a space-time fact in some experimental situation, such as the recording of one precise position of some moving object on a photosensitive surface, cannot be precisely determined on the basis of a causal law like the Schrödinger equation; and conversely, precisely determining a causal law like the Schrödinger equation cannot be made to correspond with a determinate space-time perception of, for example, the path that a moving object follows through space. For this reason, from a quantum point of view, a space-time fact is not a causal fact, and thus space-time facts and causal relations *can be precisely determined only in complementary and mutually exclusive association to one another, relative to human experimental cognizers.*

As I have already pointed out in previous works (see Kauark-Leite 2009; 2012), from a Kantian standpoint, Bohr's claim about causality, expressed by means of dynamical conservation laws, is better understood in terms of Kant's principle of the *Anticipations of Perception* rather than in terms of Kant's principle of causality, the *Second Analogy of Experience.* In the Second Analogy, in formulating his principle of the temporal sequencing of events according to the law of causality, Kant ([1781/1787] 1998, B232) asserts that "all alterations occur in accordance with the law of the connection of cause and effect." By contrast, he presents the principle of the Anticipations of Perception, that is, the mathematical-transcendental principle related to the category of quality, as follows:

> [*In the first edition:*] The **principle**, which anticipates all perceptions, as such, run thus: In all appearances the sensation, and the **real**, which corresponds to it in the object (*realitas phaenomenon*) has an **intensive magnitude**, i.e., a degree. (A166; emphasis mine)

> [*In the second edition:*] Its principle is: **In all appearances the real, which is an object of the sensation, has intensive magnitude**, i.e., a degree. (B207; emphasis mine)

By "anticipation" Kant means "all cognition through which I can cognize and determine *a priori* what belongs to empirical cognition" (B207) and by "continuity" he means the "property of intensive magnitudes on account of which no part of them is the smallest (no part is simple)" (B211). As Hermann Cohen (1885, 1883) pointed out, the principle of intensive magnitude is essentially grounded in the concept of the "differential," hence the metaphysical importance of the principle of Anticipations actually derives

from the fact that it contains the transcendental foundation of the differential calculus. So the possibility of anticipating the instantaneous values of physical quantities—
which change continuously in time—lies in the fact that they are expressed by the
differential of a continuous series or function. But, despite the difference between these
two Kantian principles (the Anticipations of Perception, and the Second Analogy of
Experience—i.e., the Principle of Causality), both principles belong to the same over-
arching Kantian conceptual framework of understanding. Then, the disjunction, presupposed by Bohr, between an intuitional representation (of space and time) and a
conceptual representation (of, e.g., the Hamiltonian function of energy) remains.

From this perspective, in Bohr's (1928a, 580) Como lecture, there is one and only
one general notion of complementarity—the space-time/causality complementarity (i.e., C1)—which asserts that "[the quantum] postulate implies a renunciation as
regards the causal space-time co-ordination of atomic processes." This general transcendental principle states the condition of the possibility for conceiving the wave/
particle duality, as Bohr explicitly argues:

> [here is] the simple formula which forms the common foundation of the theory of
> light quanta and of the wave theory of material particles:
>
> $$E\tau = I\lambda = h, \, (1)$$
>
> In this formula the two notions of light and also of matter enter in sharp contrast.
> While energy (E) and momentum (I) are associated with the concept of particles,
> and hence may be characterized according to the classical point of view by definite
> space-time co-ordinates, the period of vibration (τ) and wave-length (λ) refer to a
> plane harmonic wave train of unlimited extent in space and time. (581)

And because it is possible to associate the two mutually exclusive intuitive pictures
(i.e., particle and wave) to the same differential equation (i.e., Schrodinger's equation),
its precise spatiotemporal representation is impossible. As Bohr puts it:

> In the characteristic vibrations of Schrodinger's wave equation we have, as men-
> tioned, an adequate representation of the stationary states of an atom allowing an
> unambiguous definition of the energy of the system by means of the general quan-
> tum relation (1) [$E\tau = I\lambda = h$, (1)]. This entails, however, that in the interpretation
> of observations, a fundamental renunciation regarding the space-time description
> is unavoidable. (587)

Murdoch (1987, 61) correctly observes that the two crucial notions of Bohr's conception of complementarity are *joint completion* and *mutual exclusiveness*, as Bohr himself
says: "The very nature of the quantum theory thus forces us to regard the space-time
co-ordination and the claim of causality, the union of which characterizes the classical
theories, as *complementary* but *exclusive features* of the description, symbolizing the
idealization of observation and definition, respectively" (Bohr 1928a, 580; emphasis
mine). Here the word "idealization" is best understood in the framework of Kant's

transcendental idealism. The notion of "space-time description" is essentially related to the ideality of observation (intuition and sense perception) and the notion of "causal description" is essentially related to the ideality of conceptualization (definition). By "ideality," in a transcendental sense, Kant means a specific cognitive attitude regarding the external world, which does not assume that things exist independently of our experience (which would imply a form of noumenal realism), but one which assumes that things are as they are presented to us in our experience. Thus, this ideality refers to what enables to us to produce some objective sense of a world that can only be presented to us subjectively, that is, mind-dependently, according our subjective or transcendentally ideal forms of sensibility and understanding. In this sense, Bohr can say that transcendentally ideal space-time intuition enables us to constitute forms of visualization as particle-like or as wave-like, and that causality as a form of conceptual or definitional ideality is what enables us to constitute the laws of conservation of energy and momentum. So for Bohr (Bohr [1929] 1934, 11), transcendentally ideal space and time are essentially related to pure perceptual forms of visualization, and transcendentally ideal causality is essentially related to the conceptually formulated laws of conservation of energy and momentum, as he explicitly asserts in his 1929 paper: "This indeterminacy exhibits, indeed, a peculiar complementary character which prevents the simultaneous use of space-time [representations] and the laws of conservation of energy and momentum, which is characteristic of the mechanical mode of description." Or in his 1931 paper:

> We have thus either space-time description or description where we can use the laws of conservation of energy and momentum. They are complementary to one another. We cannot use them both at the same time. If we want to use the idea of space-time we must have watches and rods which are outside and independent of the object under consideration, in the sense that we have to disregard the interaction between the object and the measuring rod used. (Bohr [1931] 1985, 369)

Thus, the fundamental meaning of Bohr's conception of quantum theory implies the complementarity between the phenomena described in terms of space-time coordination (pure or empirical intuition) and the phenomena accounted for by means of dynamical conservation laws (conceptually represented causality).

Furthermore, as Henry Folse (1978) pointed out, Bohr's formulation and defense of complementarity depends on two postulates:

i. The "quantum postulate" that supposes that physical action is quantized
ii. The "postulate of objectivity" that assumes that the non-metaphysical claim of scientific knowledge requires that the description of nature be cast in unambiguous terminology communicable from one experimenter to another.

Folse, however, has argued that there are no substantive arguments in support of Kantian aspects of Bohr's complementarity. He claimed that only the second postulate "is of interest to the question of Bohr's alleged Kantianism." And because "it has a certain Kant-like appearance" it gives rise to a misleading interpretation that does not resist a deeper analysis. My view, on the contrary, is to argue that the two postulates can be interpreted in a broader transcendental sense that does not necessarily coincide with strict Kantianism. From this perspective, I propose to rename the second, and also to add a third postulate, and then to classify them as follows:

i. The quantum postulate (construed as a metaphysical postulate in Kant's sense)
ii. Instead of Folse's "postulate of objectivity," *the postulate of intersubjectivity* (construed as a pragmatic postulate)
iii. *The postulate of contextuality*, which says that the observable phenomena depend on the circumstances in which they are observed (construed as an epistemic postulate)

In my view, the first postulate (i.e., the quantum postulate) should be interpreted as a metaphysical idea of reason in the Critical sense. As an idea of reason, it can be only hypothetically used like a regulative principle, not as a constitutive one, *as if* (*"als ob"*) the reality were heuristically constituted in such-and-such way, but never asserting that reality is actually, objectively, and truly in accordance with this idea. This kind of assertion is neither false nor true; hence it is not descriptive but prescriptive and therefore cannot be realistically interpreted. It is projected by human reason into the nature in order to provide an organic and systematic unity for scientific theories (see Kant [1790] 2000, KU, Ak III, 429).

In its turn, the second postulate presupposed by the complementarity principle is better called "the postulate of intersubjectivity," instead of "the postulate of objectivity," since "intersubjectivity" does not mean "existing independently of the subject," but rather refers to the conditions of possibility of an unambiguous communication of the experimental results between subjects who use the same common natural language, and the professional vocabulary of classical physics. This postulate can be classified as "pragmatic" in the sense that the meaning of the symbols of a scientific theory depends on their use by the interpreters in specific contexts. As Bohr points out, the experience can be said to be objective if and only if it can be communicated in an unambiguous way: "This is a simple logical demand, since by the word 'experiment' we can only mean a procedure regarding which we are able to *communicate* to others what we have done and what we have learnt" ([1958] 1963, 3; emphasis mine). "Our task must be to account for [human] experience in a manner independent of individual subjective judgment and therefore objective in the sense that it can be unambiguously communicated in the common human language" (Bohr [1960] 1963, 10).

The pragmatic perspective adopted by Bohr (1949) in characterizing the objectivity of experience in terms of the intersubjectivity of an unambiguous communication presupposes that this communication is expressed in classical-physical terms. In his words, "It is decisive to recognize that, however far the phenomena transcend the scope of classical physical explanation, the account of all evidence must be expressed in classical terms" (249). Thus, Bohr's thesis of the necessary use of classical concepts in the experimental situation should be anchored in his pragmatic postulate.

I added a third basic postulate about complementarity and called it "the postulate of contextuality," in order to capture Bohr's antirealist position about the impossibility of a sharp distinction between the independent behavior of the object measured, and measuring devices. It actually implies a more radical epistemic "revolution" than Kant's "Copernican revolution" in metaphysics, namely, (i) Kant's radical agnosticism concerning the objective existence of things in themselves independently of their relation to our subjective forms of intuition and understanding, (ii) the necessary conformity of the forms of cognized objects to the forms of our cognitive capacities, and (iii) the transcendental ideality of space, time, and causality. In Bohr's view, another limitation is introduced in quantum theory, not regarding the existence of the things in themselves, but regarding the existence of the phenomenon in itself, that prevents us from even considering *the experimental phenomenon* independently of the experimental context in which we measure and observe it. This, in turn, is the reason why we need to introduce a third postulate, namely, an epistemic-contextualist postulate asserting that the observable phenomenon depends on the circumstances it is observed. As Bohr ([1929] 1934, 11–12) argues: "*the finite magnitude of the quantum of action prevents altogether a sharp distinction being made between a phenomenon and the agency by which it is observed*, a distinction which underlies the customary concept of observation and, therefore, forms the basis of the classical ideas of motion" (emphasis in the original). And also: "The discovery of the quantum of action shows us, in fact, not only the natural limitation of classical physics, but, by throwing a new light upon the old philosophical problem of the objective existence of phenomena independently of our observations, confronts us with a situation hitherto unknown in natural science" (115).

Now that we have situated the general meaning of the principle of complementary in a broader philosophical and transcendental framework, let us consider its instantiation in the specific case of a double-slit interference experiment, in which wave/particle complementarity (i.e., C2) plays a special role. But, before we go on, let me say a few words regarding the instantiation of the general principle in the case of the complementarity between incompatible observables (i.e., C3). Some scholars have argued that this formulation of the principle of complementarity coincides with Heisenberg's uncertainty principle and for this reason it does not constitute a third meaning, but instead it is only a special way of expressing the same content expressed by the several uncertainty relations. And I think that, in a sense, they are right. As I am trying to show, there is only one fully general meta-principle of complementarity, namely, that expressed by space-time/causality complementarity (i.e., C1). In this perspective, neither the wave/particle complementarity (i.e., C2) nor the complementarity between incompatible observables (i.e., C3) is, strictly speaking, a principle. C2 corresponds to the wave/particle duality and C3 to the uncertainty principle, which can be expressed by each one of the uncertainty relations. The important idea introduced by my Kantian approach to Bohr's complementarity interpretation lies in the fact that the general meta-principle of complementarity (i.e., C1) can explain both the wave/particle duality and the uncertainty relations. This is why I am taking the two meanings of complementarity, C2 and C3, as different special instantiations, in the object-language, of the general meta-principle C1. As Bohr (1948, 315) explicitly asserts for one of the uncertainty relations:

Thus, a sentence like "we cannot know both the momentum and the position of an electron" raises at once questions as to the physical reality of such two attributes, which can be answered only by referring to the mutually exclusive conditions for the unambiguous use of space-time coordination, on the one hand, and dynamical conservation laws, on the other.

Let's now see how the general meta-principle C1 can be instantiated in some familiar interference experiments, as Einstein suggested, in order to provide a kind of thought experiment (*Gedankenexperiment*) for the purpose of refuting C2, that is, wave/particle complementarity in the same experimental arrangement, but *not* refuting C3, that is, the complementarity regarding two incompatible observables.

Einstein's objection

As I mentioned at outset of this section, even though there is no explicit reference in the *Proceedings* of 5th Solvay Conference to the semiprivate Bohr-Einstein discussions concerning Bohr's viewpoint on complementarity, Bohr later described the main points of this debate in his 1949 paper and also made explicit Einstein's discomfort with the renunciation of the causal space-time coordination by quantum theory. In Bohr's (1949, 212) words: "During the discussions, ... Einstein expressed however, a deep concern over the extent to which causal account in space and time was abandoned in quantum mechanics." The bulk of the discussion between them, reported by Bohr in this paper, was focused on the analysis of the double-slit interference experiment, as illustrated in Figure 3.1 reproduced from Bohr (212).

In this kind of experimental arrangement, we have a first barrier with one narrow slit, then a second barrier parallel to the first one with two narrow slits, and then finally a photographic plate that can record each photon or each electron that arrives on it coming from a source on the left side. We can observe the interference pattern on the right side both for the electron and for the photon alike, and using our classical concepts we should say that both quantum entities behave like waves. Nevertheless, if we have correct information about which path the electron or the photon has taken, then the interference

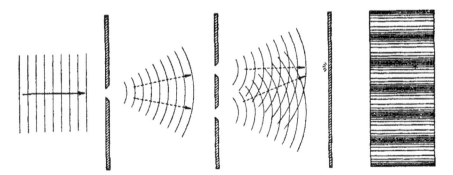

Figure 3.1 Illustration of double-slit experiment, taken from Bohr's 1949 paper.

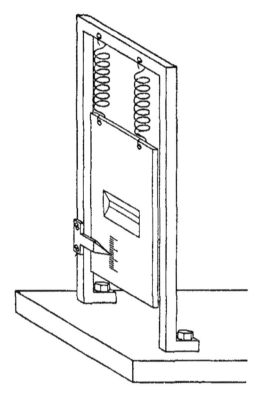

Figure 3.2 Schematic diagram of the device used as the first barrier in the double-slit experiment, taken from Bohr's 1949 paper.

pattern disappears, and we should say that the electron or the photon behaves like a particle. So a "same object" may present itself as a wave or as a particle depending on the specific experimental set-up. But the "same object" cannot manifest mutually exclusive features (wave-like and particle-like) in the same experimental arrangement. This is a familiar example of how the complementary phenomena appear in mutually exclusive experimental contexts, and thus a case of the instantiation of the general meta-principle of complementarity. In this particular situation, the complementarity principle is formulated by saying that photons or any quantum entities can behave as either waves or particles, but cannot be observed as both during the same experimental arrangement. This formulation is often called "the wave/particle complementarity" (i.e., C2).

Einstein, who was not satisfied with the approach taken by Bohr, conceived an idea of a double-slit experiment in which the first barrier could move in a certain way, such that, in view of the laws of conservation of energy and momentum, the path of the photon may be traced after passing through the first slit, as Bohr (1949, 220) illustrated in Figure 3.2. In this case, it would be possible to determine, indirectly, the path that each photon has followed, without destroying the interference pattern. If it is the case, then wave/particle complementarity principle would be violated: we would have in the same experimental arrangement both wave-like and also particle-like behaviors.

But against Einstein, based on the uncertainty relations, Bohr argued that the interaction between the recoiling slit and the photon would introduce a random phase shift to the light beam, which in turn would bring about the destruction of the interference pattern. So wave/particle complementarity is still valid.

This discussion reported by Bohr allowed us to formulate the wave/particle complementarity (i.e., C2) in interferometer experiments as follows:

i. If we have an interference pattern (IP) that is perfectly visible, then a *sharp wave behavior* (*w*) can be ascribed to the quantum, *as if* the quantum had passed through the two slits (i_w).
ii. If we have complete information about which path the quantum followed, that is, complete which-way information (WWI), then *a sharp particle behavior* (*p*) can be ascribed to the quantum, *as if* the quantum had passed only through one slit (i_p).
iii. C2: Therefore, it is logically impossible to observe *w* and *p* behaviors in the same experimental arrangement, given that i_w and i_p are *mutually exclusive* logical inferences.

We will see in the next section how this formulation of wave/particle complementarity in interferometer experiments has recently received a quantitative treatment and even a precise mathematical expression, thereby rekindling the debate about this possible violation of Bohr's principle.

2. Quantitative complementarity and intermediate situations

A significant step toward quantifying wave/particle complementarity was taken by W. K. Wootters and W. H. Zurek in 1979, which resulted in the mathematical formulation of an inequality proposed by B.-G. Englert in 1996 for two-path interferometers.

In their 1979 paper, Wootters and Zurek proposed analyzing complementarity in a quantitative way that applies to intermediate situations where some partial interference patterns can be observed. As they pointed out in the abstract of their paper:

A detailed analysis of Einstein's version of the double-slit experiment, in which one tries to observe both wave and particle properties of light, is performed. Quantum nonseparability appears in the derivation of the interference pattern, which proves to be surprisingly sharp even when the trajectories of the photons have been determined with fairly high accuracy. An information-theoretic approach to this problem leads to a quantitative formulation of Bohr's complementarity principle for the case of the double-slit experiment. A practically realizable version of this experiment, to which the above analysis applies, is proposed. (473)

More precisely, they argued that even when some accurate information can be obtained about the photon's path, a clear interference pattern can still be observed, and in this case both complementary aspects of light would simultaneously manifest themselves in the same experimental arrangement.

In his 1996 paper, Englert proposed replacing wave/particle duality by an interferometric duality, using the visibility of interference pattern as a measure of the wave nature of quantum particle, and interferometric information about which path the particle followed as a measure of the corpuscular behavior. He thereby derived a mathematical expression of the interferometric complementarity, as he summarized in his abstract:

> An inequality is derived according to which the fringe visibility in a two-way interferometer sets an absolute upper bound on the amount of which-way information that is potentially stored in a which-way detector. In some sense, this inequality can be regarded as quantifying the notion of wave-particle duality. The derivation of the inequality does not make use of Heisenberg's uncertainty relation in any form. (2154)

The inequality is expressed by the simple formula $D^2 + V^2 \leq 1$, where D denotes the distinguishability of the paths and V the fringe visibility. "$V = 1$" means that the interference fringes are perfectly visible, and "$V = 0$" means that there is no discernible interference pattern. In turn, "$D = 0$" means that there is no information about the quantum path and "$D = 1$" means that the information about the path is perfectly known. From the inequality, we can easily conclude that:

i. If $V = 1$, then $D = 0$. So we have a typical case of *sharp wave-like behavior.*
ii. If $V = 0$, then $D = 1$. So we have a typical case of *sharp particle-like behavior.*

As Wootters and Zurek (1979) had already pointed out, quantum mechanics does not rule out the presence of *unsharp wave-particle behaviors* in the same experimental arrangement for intermediate situations, where partial visibility and partial information can be obtained. So according to the mathematical expression proposed by Englert, in the cases of intermediate situations, we have $0 < V < 1$ and $0 < D < 1$.

What is ruled out by the new formalism, however, in complete agreement with Bohr's complementarity principle, is the presence of *sharp wave/particle behaviors* in the same experimental arrangement: that is, the state of affairs such that $V = 1$ and $D = 1$ is impossible. So the interferometric complementarity expressed by Englert's mathematical inequality may be interpreted as follows: the sharp interference pattern observation ($V = 1$) and the sharp information acquisition path on which passed the particle ($D = 1$) are mutually exclusive in the same experimental arrangement.

3. The Afshar experiment and the alleged violation of the principle of complementarity

On March 23, 2004, a seminar at Harvard was announced by the experimental physicist Shahriar Afshar under the highly evocative title "Waving Copenhagen

Good-bye: Were the Founders of Quantum Mechanics Wrong?" In the abstract of this seminar he asserted:

> In a "welcher weg" or "which-way" experiment we can obtain perfect knowledge about the origin and path of quantum particles (particle-like behavior), but this action must lead to a complete destruction of the interference pattern (wave-like behavior). I will report a novel quantum optical "which-way" experiment, currently being duplicated at HEPL, which violates the predictions of PC. *It seems that we can now finally settle the Bohr-Einstein debate in favor of Einstein!* (Emphasis mine)

Then in two different papers, "Violation of the Principle of Complementarity, and Its Implications" (2005) and "Paradox in Wave-Particle Duality" (Afshar et al. 2007), Afshar provides details of the famous experiment in which he purports to refute both Bohr's complementarity principle and also the quantitative complementarity expressed by Englert's inequality. Through a special double-slit interferometer experiment, Afshar proposes to determine both sharp wave and particle behaviors of a photon beam in the same experimental arrangement, which would constitute a clear violation of Bohr's complementarity principle and Englert's inequality ($V = 1$ and $D = 1$) alike. In contrast to Wootters and Zurek's experiments, where unsharp wave and particle patterns were simultaneously observed in the same experimental arrangement, the Afshar experiment intends not to be a case of intermediate situations for the double behaviors. Instead, in the experimental arrangements Afshar proposed, a new non-perturbative measurement technique is introduced that prevents the violation of the uncertainty relations, but not the violation of the complementarity principle. The Afshar experiment generated an intense controversy about the possible violation, or not, of Bohr's principle of complementarity, that I will briefly summarize in the rest of this section.

The Afshar experiment was performed in three different experimental arrangements. In Experimental Arrangement 1, we are presented with a classical two-slit interferometer by means of which a sharp wave-like behavior of a photon beam is observed, as shown in Figure 3.3 (Afshar 2005, 231).

The sharp wave-like behavior is identified with the interference pattern I_{12} that is observed at plane σ_1. In this case we have maximum visibility of the interference pattern and no distinguishability as to which slit (1 or 2) each single photon that contributes to the formation of the interference pattern passed through, that is, $V = 1$ and $D = 0$. So it is a case of wave-like behavior of the photon beams.

In Experimental Arrangement 2, the photosensitive surface is removed at σ_1 and a suitable converging lens (L) is placed just before it, as shown in Figure 3.4 (Afshar 2005, 232).

In this case, two well-resolved images ($1'$ and $2'$) of the corresponding pinholes (1 and 2) are formed at the plane σ_2. No interference pattern is observed at σ_2, and therefore it can be said that the path of each photon is known, that is, $V = 0$ and $D = 1$. So here we have a typical case of particle-like behavior.

In Experimental Arrangement 3, a wire grid is introduced that is supposed to have dark interference fringes. Because there is no evidence of diffraction through the grid

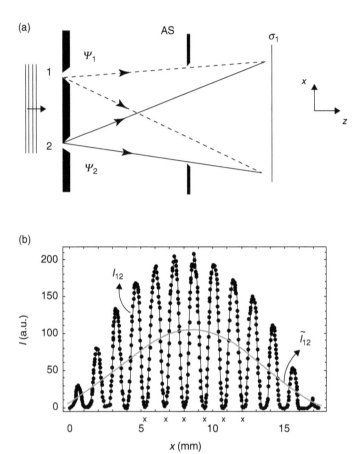

Figure 3.3 Afshar's first experimental arrangement.

wires, it can be said that an interference pattern is detected at σ_1, and hence a wave-like behavior can be ascribed to the photon beam. The particle-like behavior is also simultaneously preserved, as is shown by the two well-resolved images at the plane σ_2, according to Figure 3.5 (Afshar 2005, 238).

In this case, we have both sharp wave behavior (maximum interference pattern) and sharp particle behavior (maximum distinguishability detected) in the same experimental arrangement, that is, $V = 1$ (in σ_1) and $D = 1$ (in σ_2). This conclusion is clearly in opposition to Bohr's complementarity principle, and seems also to be a violation of Englert's complementarity inequality, apparently demonstrating that the complementarity principle is not an essential and fundamental ground of quantum phenomena.

Many criticisms were made of Afshar's interpretation, some of which I will briefly summarize below. All of them, on the basis of physical reasons, assert that the third experimental arrangement does *not* in fact constitute a violation either of Bohr's wave/particle complementarity or of Englert's inequality.

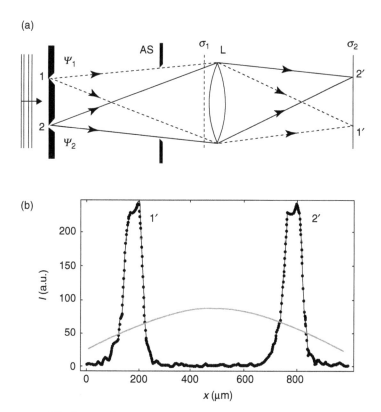

Figure 3.4 Afshar's second experimental arrangement.

One group of physicists, especially including Unruh (2004), Kastner (2009), Reitzner (2007), and Qureshi (2012), claim that the interference pattern is preserved ($V = 1$) even with the introduction of the wire grid, but deny that we can have any "which-way" information about which slit each single photon passed through, even if a two-peaked distribution image is obtained at σ_2. Kastner (2009), for instance, argues that the Englert's inequality does not apply to Afshar's experiment and "that it is wrong to make any use of it in support of claims for or against the bearing of this experiment on Complementarity." The measurement of the photosensitive surface (σ_2) "doesn't give any physically meaningful 'which-slit' information since the particle already went through both slits" (1144). And as Reitzner (2007 says: "The two-peaked distribution is an interference pattern and the photon behaves as a wave and exhibits no particle properties until it hits the plate." Thus it appears that Afshar's claim that "which-way" information can be obtained in his experiment when the two slits are open is completely false. To the same effect, Qureshi (2012) asserts that "although the modified two-slit experiments do have genuine interference, as shown by introducing thin wires, the detectors detecting the photons behind the converging lens, do not yield any which-way information."

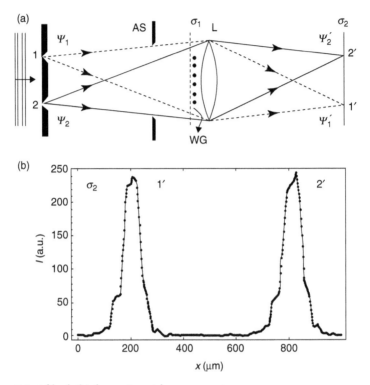

Figure 3.5 Afshar's third experimental arrangement.

Another group of physicists, especially Drezet (2005; 2011) and Steuernagel (2007), claim that we can obtain "which-way" information in Afshar's experiment, but deny that we have a sharp interference pattern on the plane of the wire grid. As Steuernagel asserts:

> The presence of the interference pattern in Afshar's experiment is only inferred, but according to the orthodox interpretation of quantum mechanics it has to be measured in order to be described by the quantum mechanical operator formalism. This requirement is well captured by Wheeler's famous dictum to the effect "that only a registered event is a real event." When studying the complementary aspects of a quantum particle's behavior both aspects have to be measured simultaneously for every particle.

Another group of experimental physicians (Jacques et al. 2008) at Quantum Optics Laboratories, at Cachan's *Ecole Normale Supérieure*, and at Palaiseau's *Institut d'Optique* created a small-scale modified version of the Afshar experiment in order to show that the presence of the wire grid destroys the exact correspondence between the emission pinhole and the measured spot on the screen. The introduction of the grid has then partially erased the interference pattern. All this is very similar to the familiar "moving screen" argument originally used by Einstein against Bohr's wave/particle complementarity. In fact, this is a purely classical effect (Fourier optics). As they argued: "However, the grid has an unavoidable effect due to diffraction, which redirects some light from path 1 to detector P2 and, reciprocally, from path 2 to detector P1. The introduction of the grid has then partially erased the WPI since it becomes impossible to univocally associate each output detector to a given path of the interferometer" (Jacques et al. 2008). On this basis, they claim that there are two wave patterns (wave and particle) in the experiment, but in fact it is a case of unsharp measurements that are realized in the context of intermediate situations. Then, by varying the thickness of grid, they found the usual complementarity between the "which-way" information and the interference's fringe contrast. In this way, they show that Afshar's argument is simply wrong. His mistake is to think that a simple introduction of the wire grid into the black fringes, without any measuring act, allows obtaining sharp information about the interference's fringe. But a correct interpretation must consider the perturbation added by the grid.

4. By way of conclusion

By way of concluding this study, I would like to restate my main points in order to evaluate the implications of the recent experimental contexts and arguments for Bohr's theory. The general meta-philosophical principle of complementarity, formulated in terms of space-time and causality, is not reducible to wave/particle complementarity. Indeed, as a meta-philosophical principle, Bohr's principle of complementarity is best understood in Kantian terms as a transcendental principle that expresses in quantum contexts the complementarity connection between space-time intuition and conceptual definition. In turn, Bohr's claim about conceptual definition is best understood in terms of Kant's principle of the Anticipations of Perception rather than in terms of Kant's Second Analogy of Experience, the principle of empirical causality, as I have also argued in previous work (Kauark-Leite 2009; 2012).

As regards the differences between the three meanings of complementarity that we can find in Bohr's Como lecture (C1, C2, and C3), the general meta-philosophical principle of space-time/causality complementarity (i.e., C1) is itself rooted in three philosophical postulates: the quantum postulate (which is metaphysical), the postulate of intersubjectivity (which is pragmatic), and the postulate of contextuality (which is epistemic). Correspondingly, the wave/particle complementarity (i.e., C2), including Englert's inequalities, and the complementarity between incompatible observables (i.e., C3), including Heisenberg's inequalities, are ultimately rooted in the formal structure of quantum mechanics. Nevertheless, I want to reemphasize that Englert's inequality is not a quantitative expression of Bohr's general principle. Rather, it is restricted to one particular form of instantiation of the complementarity principle, namely, C2, the wave/particle instantiation.

Despite all the recent experiments concerning quantitative complementarity, the conceptual framework of quantum theory, expressed in terms of the Schrödinger equation or other formulations (i.e., the conceptual level), is still not given in a space-time intuition. But the experimental arrangements, including the Afshar experiment and all the other complementarity experiences, are still given in a space-time framework (i.e., intuitional level). In all these new experiments about complementarity, the three fundamental postulates of complementarity are maintained: the quantum postulate, the postulate of intersubjectivity, and the postulate of contextuality. We still use the fundamental hypothesis of the quantum of action (the quantum postulate). We still use the classical vocabulary, in terms of *wave-like* or *particle-like* behaviors, in order to communicate the empirical results (the postulate of intersubjectivity). And we still continue to use the contextual definition of phenomena (the postulate of contextuality). As Bohr (1947, 1) rightly said: "We have to do with phenomena where no sharp separation is possible between an independent behavior of the objects and their interaction with the measuring agencies necessary for the definition of the observable phenomena."

Acknowledgment

I am grateful to Maria Carolina Nemes (in memorian), Romeu Rossi, Pablo Saldanha, Ana Paula Gomes Pereira, Raul Corrêa, Saulo Moreira, Júlia Parreira, Diego Henrique Melo, Osvaldo Pessoa Jr., Michel Bitbol, Jean Petitot, Jan Faye, Stefano Osnaghi, Philippe Grangier, Frank Laloë, and Hervé Zwirn for very fruitful discussions concerning the topics covered in this chapter.

References

Afshar, S. S. 2005. "Violation of the principle of complementarity, and its implications." In *The Nature of Light: What Is a Photon?* Vol. 5866 of *Proceedings of SPIE,* edited by C. Roychoudhuri and K. Creath, 229–44. San Diego: SPIE. doi: 10.1117/12.638774.

Afshar, S. S., E. Flores, K. F. McDonald, and E. Knoesel. 2007. "Paradox in wave-particle duality." *Foundations of Physics* 37 (2): 295–305.

Bacciagaluppi, G., and A. Valentini. 2009. *Quantum Theory at the Crossroads: Reconsidering the 1927 Solvay Conference.* New York: Cambridge University Press.

Beller, M. 1999. *Quantum Dialogue: The Making of a Revolution.* Chicago: The University of Chicago Press.

Bitbol, M. 2010. "Reflective metaphysics: Understanding quantum mechanics from a Kantian standpoint." *Philosophica* 83: 53–83.

Bitbol, M., P. Kerszberg, and J. Petitot (eds). 2009. *Constituting Objectivity: Transcendental Perspectives on Modern Physics.* Berlin: Springer.

Bitbol, M., and S. Osnaghi. 2013. "Bohr's complementarity and Kant's epistemology." In *Séminaire Poincaré 2013: Bohr, 1913–2013*, edited by Institut Henri Poincaré. Vol. XVII, 145–66. Palaiseau: Éd. de l'école Polytechnique.

Bohr, N. 1928a. "The quantum postulate and the recent developments of atomic theory." *Nature (Suppl.)* 121: 580–90.

Bohr, N. [1929] 1934. "Introductory survey." In *Atomic Theory and the Description of Nature*, 1–24. Cambridge: Cambridge University Press.

Bohr, N. [1931] 1985. "Space-time-continuity and atomic physics." H.H. Wills Memorial Lecture, given at the University of Bristol on October 5, 1931, UNPUBLISHED MANUSCRIPT, 1–14. In *Niels Bohr Collected Work: Foundations of Quantum Physics I (1926–1932)*, vol. 6, edited J. Kalckar, 363–70. Elsevier: Amsterdam.

Bohr, N. 1934. *Atomic Theory and the Description of Nature*. Cambridge: Cambridge University Press.

Bohr, N. 1947. "Problems of elementary-particle physics." In *Report of an International Conference on Fundamental Particles and Low Temperatures*, edited by Physical Society and Cavendish Laboratory, Vol. 1, *Fundamental Particles*, 1–4. London: The Physical Society.

Bohr, N. 1948. "On the notions of causality and complementarity." *Dialectica* 2 (3–4): 312–19.

Bohr, N. 1949. "Discussion with Einstein on epistemological problems in atomic physics." In *Albert Einstein: Philosopher-Scientist. The Library of Living Philosophers*, Vol. VII, edited by P. A. Schilpp, 199–241. Evanston, IL: Northwestern University Press.

Bohr, N. 1950. "Some general comments on the present situation in atomic physics." In *Les particules elementaires: rapports et discussions; huitième conseil de physique tenu à Bruxelles du 27 septembre au 2 octobre 1948*, edited by Conseil de Physique, Institut International de Physique Solvay, 376–80. Bruxelles: R. Stoops.

Bohr, N. [1958] 1963. "Quantum physics and philosophy: Causality and complementarity." In *Essays 1958–1962 on Atomic Physics and Human Knowledge*, 1–7. New York: Wiley & Sons.

Bohr, N. [1960] 1963. "The unity of human knowledge." In *Essays 1958–1962 on Atomic Physics and Human Knowledge*, 8–16. New York: Wiley & Sons.

Bohr, N. 1963. *Essays 1958–1962 on Atomic Physics and Human Knowledge*. New York: Wiley & Sons.

Chevalley, C. 1994. "Niels Bohr's words and the Atlantis of Kantianism." In *Niels Bohr and Contemporary Philosophy*, edited by J. Faye and H. J. Folse, 33–55. Boston Studies in the Philosophy of Science, Vol 153. Dordrecht: Kluwer Academic Publishers.

Cohen, H. 1883. *Das Prinzip der Infinitesimal-Methode and seine Geschichte: Ein Kapitel zur Grundlegung der Erkenntniskritik*. Berlin: Dümmler.

Cohen, H. 1885. *Kants Theorie der Erfahrung*, Zweite neubearbeitete Auflage. Berlin: Dümmler.

Cuffaro, M. E. 2010. "The Kantian framework of complementarity." *Studies in History and Philosophy of Modern Physics* 41 (4): 309–17.

Drezet, A. 2005. "Complementarity and Afshar's experiment." arXiv: quant-ph/0508091.

Drezet, A. 2011. "Wave particle duality and the Afshar experiment." *Progress in Physics* 1: 57–67.

Einstein, A. 1949. "Reply to criticisms." In *Albert Einstein: Philosopher-Scientist. The Library of Living Philosophers*, Vol. VII, edited by P. A. Schilpp, 663–88. Evanston, IL: Northwestern University Press.

Englert, B.-G. 1996. "Fringe visibility and which-way information: An inequality." *Physical Review Letters* 77 (11): 2154–57.

Faye, J., and H. J. Folse (eds). 1994. *Niels Bohr and Contemporary Philosophy*. Boston Studies in the Philosophy of Science, Vol 153. Dordrecht: Kluwer Academic Publishers.

Folse, H. J. 1978. "Kantian aspects of complementarity." *Kant-Studien* 69: 58–66.

Folse, H. J. 1985. *The Philosophy of Niels Bohr: The Framework of Complementarity*. Amsterdam: North-Holland.

Heisenberg, W. [1930] 1949. *The Physical Principles of the Quantum Theory,* translated by
    Carl Eckart and Frank C. Hoyt. New York: Dover Publications.

Honner, J. 1982. "The transcendental philosophy of Niels Bohr." *Studies in the History and
    Philosophy of Science* 13: 1–29.

Hooker, C. A. 1972. "The nature of quantum mechanical reality: Einstein versus Bohr."
    In *Paradigms & Paradoxes*, edited by R. G. Colodny, 67–302. Pittsburgh: University of
    Pittsburgh Press.

Hooker, C. A. 1994. "Bohr and the crisis of empirical intelligibility: An essay on the depth
    of Bohr's thought and our philosophical ignorance." In *Niels Bohr and Contemporary
    Philosophy*, edited by J. Faye and H. J. Folse, 155–99. Boston Studies in the Philosophy
    of Science, Vol 153. Dordrecht: Kluwer Academic Publishers.

Jacques, V., N. D. Lai, A. Dréau, D. Zheng, D. Chauvat, F. Treussart, P. Grangier, and J.-F.
    Roch. 2008. "Illustration of quantum complementarity using single photons interfering
    on a grating." *New Journal of Physics* 10 (12): 123009. doi:10.1088/1367–2630/10/12/
    123009.

Jammer, M. 1974. *The Philosophy of Quantum Mechanics: The Interpretations of Quantum
    Mechanics in Historical Perspective*. New York: Wiley & Sons.

Kaiser, D. 1991. "More roots of complementarity: Kantian aspects and influences." *Studies
    in History and Philosophy of Science* 23 (2): 213–39.

Kant, I. [1781/1787] 1998. *Critique of Pure Reason*, translated by P. Guyer and A. W.
    Wood. New York: Cambridge University Press.

Kant, I. [1786] 2004. *Metaphysical Foundations of Natural Science*, translated by M.
    Friedman. Cambridge: Cambridge University Press.

Kant, I. [1790] 2000. *Critique of the Power of Judgment*, translated by P. Guyer and E.
    Matthews. New York: Cambridge University Press.

Kastner, R. E. 2009. "On visibility in the Afshar two-slit experiment." *Foundation of Physics*
    39: 1139–44.

Kauark-Leite, P. 2009. "The transcendental role of the principle of anticipations of
    perception in quantum mechanics." In *Constituting Objectivity: Transcendental
    Perspectives on Modern Physics*, edited by M. Bitbol, P. Kerszberg, and J. Petitot, 203–
    13. Berlin: Springer.

Kauark-Leite, P. 2010a. "Transcendental philosophy and quantum physics."
    Manuscrito: Rev. Int. Phil. 33 (1): 243–67.

Kauark-Leite, P. 2010b. "Classical reason and quantum rationality: Transcendental
    philosophy face contemporary physics." In *Klassische Vernunft und Herausforderungen
    der modernen Zivilisation*, bd. 2, edited by W. N. Bryuschinkin, 237–46.
    Kaliningrad: Baltische Föderale Immanuel-Kant-Universität.

Kauark-Leite, P. 2012. *Théorie quantique et philosophie transcendantale: dialogues possibles*.
    Paris: Éditions Hermann.

Kauark-Leite, P. 2015. "Redefining the curvature of the arc: Transcendental aspects of
    quantum rationality." In *Kant and the Metaphors of Reason*, 561–77. Hildesheim: Olms.

MacKinnon, E. M. 1982. *Scientific Explanation and Atomic Physics*. Chicago: The
    University of Chicago Press.

Murdoch, D. 1987. *Niels Bohr's Philosophy of Physics*. Cambridge: Cambridge
    University Press.

Pauli, W. [1958] 1980. *General Principles of Quantum Mechanics*, translated by P. Achuthan
    and K. Venkatesan. Berlin: Springer.

Pringe, H. 2007. *Critique of the Quantum Power of Judgment. A Transcendental Foundation
    of Quantum Objectivity*. Berlin: De Gruyter.

Pringe, H. 2009. "A transcendental account of correspondence and complementarity." In *Constituting Objectivity: Transcendental Perspectives on Modern Physics*, edited by M. Bitbol, P. Kerszberg, and J. Petitot, 317–27. Berlin: Springer.

Qureshi, T. 2012. "Modified two-slit experiments and complementarity." quant-ph/0701109v3.

Reitzner, D. 2007. "Comment on Afshar's experiments." quant-ph/0701152.

Rosenfeld, L. 1969. "Review of Max Jammer: The conceptual development of quantum mechanics." *Nuclear Physics* A126: 696.

Steuernagel, O. 2007. "Afshar's experiment does not show a violation of complementarity." *Foundation of Physics* 37: 1370–85.

Unruh, W. G. 2004. "Shahriar Afshar—quantum rebel?" http://www.theory.physics.ubc.ca/rebel.html.

Von Weizäcker, C. F. 1955. "Komplementarität und logik." *Die Naturwissenschaften* 42: 521–9, 545–55.

Wootters, W. K., and W. H. Zurek. 1979. "Complementarity in the double slit experiment: Quantum nonseparability and a quantitative statement of Bohr's principle." *Physical Review D* 19 (2): 473–84.

4

Complementarity and Pragmatic Epistemology: A Comparison of Bohr and C. I. Lewis

Henry J. Folse

1. Pragmatism—influences and affinities

Many scholars have commented on the pragmatist views that seem to characterize Bohr's philosophical outlook, leading inevitably to a search for evidence connecting Bohr to the American pragmatists. Aside from an occasional nod to Peirce's pragmatic criterion of meaning, the spotlight has focused almost exclusively on William James. This is natural enough: not only did Bohr express a great enthusiasm for James and single out his chapter "The Stream of Thoughts" from his *Principles of Psychology* for special praise (Bohr 1962), but also his mentor Harald Høffding was familiar with both the man William James and his thought, and indeed had been a guest in the James household (Faye 1991). Høffding's own eclectic philosophy included many themes properly regarded as pragmatic. It is also quite plausible that the topics in the philosophical discussions carried on by the scientific men in Bohr's father's circle in Copenhagen, which included Høffding, and which we know to have influenced the young Bohr, would not have been so terribly dissimilar from the topics in contemporaneous philosophical discussions going on among scientific men in Cambridge, Massachusetts. For these reasons it is reasonable to conjecture that some pragmatic seeds cast forth by James may have found their way across the Atlantic and taken root in young Niels Bohr's imagination.

Given this historical context when Bohr speaks of the complementarity between kinematic and dynamic descriptions it is not absurd to imagine that we hear a distant echo of James's (1890, 224–90) well-known metaphor of the perchings and flightings of consciousness.[1] Indeed, throughout the chapter "The Stream of Thoughts" James is grappling with the very issue of continuity versus discontinuity in the psychologists' attempts to describe consciousness, much as Bohr was grappling with continuity and discontinuity in the physicists' attempts to describe the coupling between atomic systems and the electromagnetic field. Just as James saw in pragmatism a way of mediating between the counterclaims of the "tender" and "tough-minded syndromes," Bohr believed complementarity mediated between the inevitable subjectivism of a positivistic phenomenalism/instrumentalism, which abandons the objectivity of physics, and a classical realism, which will inevitably find the quantum description incomplete.

However, no matter how tempting James may be as the ultimate historical *source* of pragmatism's influences on Bohr, it is worth noting that James is hardly the most obvious pragmatist to consider in exploring the *affinities* between Bohr's thinking and pragmatism. James's leading interests were in psychology, religious belief, and a place for values in the world of science; none of these were motivating interests for Bohr's philosophy.

Thus if we set aside the issue of an actual historical connection, it is important to realize that there are many aspects of Bohr's thinking which exhibit themes properly considered essential to "pragmatism," and which reiterate analogous themes in the writings of the American pragmatists of Bohr's own generation. Reflecting on the fact that Bohr refers to his main philosophical point as an *"epistemological* lesson," then the obvious pragmatist to examine would not be James, who had nothing approaching a systematic epistemology, but C. I. Lewis, whose dates, 1883–1964, make him almost an exact contemporary of Bohr, 1885–1962.

Although Lewis is not conspicuous in contemporary philosophy of science, he "was perhaps the most important American academic philosopher active in the 1930s and 1940s … [who] made major contributions in epistemology and logic" (Hunter 2014, 1). The mainstream of American pragmatism runs through his life; he was the student of both James and Royce, and the teacher of such subsequent epistemological luminaries as Quine, Chisholm, Goodman, and Wilfred Sellars, all of whom inherited the cartography of epistemology as drawn by Lewis. His epistemology includes positions and arguments that today's philosophers of science are more likely to associate with the later Carnap, Quine, and Popper, all of whom wrote after Lewis. He was among the first in America to realize the significance of *Principia Mathematica* and bring the methods and implications of symbolic logic to American philosophers and students. Like Bohr, his early intellectual life was much influenced by the conceptual revolution wrought by Einstein's theories of relativity. His first systematic exposition of the epistemology he called "conceptual pragmatism" in *Mind and the World Order*, published in 1929, was written at approximately the same time that Bohr was working out his argument for complementarity (Lewis 1929). If we ask what aspects of Bohr's thinking deserve to be called "pragmatist," Lewis can serve as the historically most appropriate paradigm of a pragmatist to serve as our standard. All of these facts make a reasonable circumstantial case for taking a look at how Bohr's views relate to Lewis's epistemology.

2. Kantianism and pragmatism in Bohr and Lewis

But there is another *philosophical* reason for hoping that Lewis might enhance our understanding of Bohr's epistemological lesson: Lewis explicitly presents his philosophy as the product of a synthesis of James's pragmatism and Kantianism. In fact he freely acknowledged his indebtedness to Kant and was (by all accounts justly) famed for his course on Kant, which he taught nearly every year of his career at Harvard

University.[2] Of course it is indisputable that Kantian influences were widespread on the thinking of the period in which Bohr matured, so quite naturally scholars have looked for, and found, Kantian themes in Bohr's thinking. Indeed, this is just what we would naturally expect, because the philosophical influences acting through Høffding on the young Bohr included strong elements of not only pragmatic ways of thinking but also Kantian influences.

Pragmatists accept Kant's basic point of departure—his Copernican Revolution—that scientific knowledge is the product of a synthesizing activity in which an empirically given element is combined with a conceptual form provided by the active mind in its understanding of the world. The mind does not merely passively experience an objective world existing apart from us, but actively participates in forming the world as we experience it. Thus both Kantians and pragmatists are agreed in rejecting a Cartesian spectator theory of knowledge with its ideal of a passive mind that "mirrors" an objective world.

However these two approaches are diametrically opposed (at least that's how pragmatists originally saw themselves) when it comes to what it is that conditions the mind's contribution to knowledge. For the Kantian these must be a priori demands, because only such considerations can yield the universality and necessity that are the essential goals of the quest for certainty. For the pragmatist these are contingent, historical demands that reflect both the state of human knowledge and our evolutionarily determined capacities and needs. In contrast, Kant was a true defender of the Enlightenment ideal of universal and necessary knowledge, leading him to try to prove that the categories he distilled from Reason were absolutely universal and necessary. This set him on the path of his "transcendental deduction."

Pragmatists emphatically reject this Enlightenment conception of knowledge and its ensuing quest for certainty. Instead of considering the categories as dictated by the demands of "pure reason," pragmatists regard them as originating in human needs and interests. The actual categories employed in the scientific understanding of the empirical world depend not only on those goals that arise in human nature and our capacities for reaching them, but also on our history and the dialectical journey that science has taken up to the present day. Our actual way of conceptualizing the world could in principle be replaced by another quite different framework, just as the outlook of today has replaced that of earlier times.

This was precisely the view of the conceptual element that the mind imposes on the sensory "given" that Lewis defended under the name of "conceptual pragmatism." As a pioneer in the discussion and dissemination of symbolic logic he was acutely sensitive to the distinction between a formal deductive system and its possible interpretations. Moreover, the development of general relativity had given Bohr's and Lewis's generation an exciting example of a formal system originally being developed with no application as Riemannian geometry, and then given an interpretation by Einstein that transformed it into a revolutionary new theory of gravitation. Thus Lewis was quick to emphasize the *arbitrary* conventional nature of any rules in an uninterpreted formal system, and in fact he mounted a considerable campaign against the *Principia* definition of "material implication" as adequate for expressing the logical structure of hypothetical propositions.

In 1926, the year before Bohr's Como paper, Lewis (1926, 242) gave the following précis of his conception of scientific knowledge:

> If human knowledge at its best, in the applications of mathematics and in the well-developed sciences, is typical of knowledge in general, then the picture we must frame of it is this: that there is in it an element of conceptual interpretation, theoretically always separable from any application to experience and capable of being studied in abstraction. When so isolated, concepts are like the Platonic ideas, purely logical entities constituted by the pattern of their systematic relations. There is another element, the sensuous or given, likewise always separable by abstraction … This given element, or stream of sensation, is what sets the problem of interpretation, when we approach it with our interests of action. The function of thought is to mediate between such interests and the given. Knowledge arises when we can frame the data of sense in a set of concepts which serve as guides for action, just as knowledge of space arises when we can fit a geometrical interpretation upon our direct perception of the spatial.

According to Lewis the heart of pragmatism is its theory of the conceptual interpretation of the given, but close to the heart of Lewis's epistemology is his theory of the pragmatic a priori. Lewis shares with Kant a passion for the conceptual scheme which the mind imposes on experience, as well as Kant's starting point of distinguishing a priori knowledge from empirical knowledge and assigning the former to the conceptual form that the mind imposes on the sensory "given."

However pragmatists reject anything resembling the Enlightenment appeal to Reason as the guarantor of this conceptual element. Lewis (1923, 287) tells us that "traditional conceptions of the a priori have proved untenable. That the mind approaches the flux of immediacy with some godlike foreknowledge of principles which are legislative for experience, that there is any natural light or any innate ideas, it is no longer possible to believe." Where Kant spoke of the understanding as "synthesizing" the presentations (*vorstellungen*) of experience, Lewis speaks of the mind as "interpreting the given"; both appeal to an active mind. However he parts company with his great Enlightenment predecessor in regarding the conceptual element as purely a consequence of the meanings of the concepts we employ, so he gives no defense parallel to Kant's case for synthetic a priori knowledge:

> The empirical generalization is forever at the mercy of future experience, and hence probable only, while the a priori proposition is forever certain. But … this does not represent any greater assurance about the content of future experience, or of nature, in the one case than in the other; it represents only an intention of interpretation or classification which maintains a connection between two concepts regardless of experience in the one case but not in the other. Since the a priori in general is definitive and analytic, not synthetic, the case is the same for all a priori propositions. (Lewis 1929, 303)

In liberating a priori knowledge from the fixed certainties of Enlightenment epistemology, Lewis can lay the foundation for a priori knowledge on "only an intention of interpretation or classification," because he wants to stress that insofar as definitions are arbitrary, alternative conceptual schemes are possible. We are "in some sense" free to choose which one we want to use:

> What can be known a priori must meet the apparently contradictory requirements that it may be known in advance to hold good for all experience and that it have alternatives. The principle of classification and interpretation meets these require-ments, because the alternative to a definition is not its falsity but merely its aban-donment in favor of some other. Thus the determination of the a priori is in some sense like free choice and deliberate action. (232)

Because this choice of our classificatory and interpretive principles is, like the choice of a definition, "free," it follows that we have no "absolute assurance" that any a priori propositions are necessarily true; alas Kant was laboring under an Enlightenment "delusion": "No substantive conception, determined a priori, is able to confine particu-lar experiences within its conceptual embrace with absolute assurance; that all identifi-cations of objects and all material truth about future experience remains probable only. The supposition that any theory may secure for the a priori a different significance is a delusion" (305). The pragmatists thus jettison the Enlightenment goals and accept a thoroughgoing fallibilism, which in turn allows anchoring the conceptual element in the interpretation of the given in human "purpose and interest":

> The concepts which we thus impose upon given experience are almost always such as we have formulated only as the need for them arose. Experience itself has instigated our attitudes of interpretation. The secret of them lies in purpose or interest... .The mathematician has a whole cupboardful of such conceptual sys-tems for which nobody as yet has found any useful application. *All* concepts have intrinsically the possibility of such separate status; and all truth or knowledge rep-resents an order which is capable of being considered, like mathematical systems, *in abstrcto.* (Lewis 1926, 240–1; emphasis in the original)

The logical structure of the conceptual element by which we interpret experience is fixed by their definitions and their logical interrelations, giving us the sort of certainty characteristic of pure mathematics, but the "brute-fact of given experience" deter-mines the "truths of nature." Thus once they are used to interpret experience, at least some concepts must have more than intensional meaning defined in terms of other concepts; they must also have a denotation to that which is interpreted expressed in terms of sense and imagery:

> The pragmatic element in knowledge concerns the choice in application of con-ceptual modes of interpretation. On the one side, we have the abstract concepts themselves, with their purely logical implications. The truth about these is abso-lute, in the fashion in which pure mathematics offers the typical illustration.... On other side is the absolute brute-fact of given experience. Though in one sense inef-fable, yet the given is [in] its own fashion determinate; once the categorical system, in terms of which it is to be interpreted, is fixed, and concepts have been assigned

a denotation in terms of sensation and imagery, it is this given experience which determines the truths of nature. (Lewis 1929, 272)

While it is true that pragmatists assert that in one sense truth is "made" by human beings, once a conceptual system is chosen the denotation or sense meaning of the concepts makes the truth of nonanalytic statements contingent on particular experience. In this respect we are not "free" to believe what we like; the brute-fact of the given highly constrains justifiable belief:

> Pragmatists have sometimes neglected to draw the distinction between the concept and immediacy, between interpretation and the given, with the result that they may seem to put all truth at once at the mercy of brute-fact experience and within the power of human choice or in a relation of dependence upon human need. But this would be an attempt to have it both ways. The sense in which facts are brute and given cannot be the sense in which the truth about them is alterable to human decision. The separation of the factors is essential. (Lewis 1926, 244)

Not every choice of concepts for interpreting experience will be successful, but we do have considerable latitude in the choice of concepts by which we interpret experience. That choice of concepts is a matter of trial and error with no certainty that the interpretation of the given they make possible will succeed in leading to expectations satisfied in future experience:

> It is between these two [i.e., the conceptual and the given], in the choice of conceptual system for application, and in the assigning of sensuous denotation to the abstract concept, that there is a pragmatic element in truth and knowledge. In this middle ground of trial and error, of expanding experience and the continual shift and modification of conception in our effort to cope with it, the drama of human interpretation and the control of nature is forever being played. That the issues here are pragmatic; that they do not touch that truth which still is absolute and eternal-this is the only thing that would save those who appreciate the continually changing character of this spectacle from skepticism. (Lewis 1929, 272–3)

The "drama of human interpretation" implies that what is true in our interpretation of experience at one point may come to be false later on; this is the essence of pragmatism's "dynamic" theory of truth. In freeing a priori knowledge from the criteria of universality and necessity, pragmatism opens up the possibility of viewing the growth of human knowledge in terms of altering and replacing the conceptual element in our understanding of the world: "With this complex system of interrelated concepts, we approach particular experiences and attempt to fit them, somewhere and somehow, into its preformed patterns. Persistent failure leads to readjustment; the applicability of certain concepts to experiences of some particular sort is abandoned, and some other conceptual pattern is brought forward for application" (306). This mention of a "readjustment" of our conceptual scheme is our cue for Bohr's entrance into the discussion.

3. Bohr and Lewis on the growth of knowledge

Bohr (1934, 1) begins *Atomic Theory and the Description of Nature* with the following observation:

> The task of science is both to extend the range of our experience and to reduce it to order, and this task presents various aspects, inseparably connected with each other. Only by experience itself do we come to recognize those laws which grant us a comprehensive view of the diversity of phenomena. As our knowledge becomes wider, we must always be prepared, therefore, to expect alterations in the points of view best suited for the ordering of our experience. In this connection we must remember, above all, that, as a matter of course, all new experience makes its appearance within the frame of our customary points of view and forms of perception.

These four sentences strike themes that are essential to any pragmatic account of scientific knowledge: (i) all scientific knowledge of nature is empirical, there is no synthetic a priori knowledge; (ii) science aims to extend our experience and reduce it to order; (iii) this knowledge is presented in a conceptual framework "best suited for the ordering of our experience" (or in Lewis's vocabulary, by which we "interpret the given"); (iv) as experience extends into new domains that framework may prove too narrow to "reduce to order" new phenomena; and (v) that framework may be altered to suit the purpose of incorporating these new phenomena into our ordered and comprehensive account of the empirical world.

These same themes appear again and again in Bohr's essays. They provide the basis of his account of progress in physics. Here is an example from 1955; what earlier he called "the points of view best suited for the ordering of our experience" he here explicitly names a "conceptual framework," making an indirect negative reference to the Kantian conclusion:

> The main point to realize is that all knowledge presents itself within a conceptual framework adapted to account for previous experience and that any such frame may prove too narrow to comprehend new experiences. Scientific research in many domains of knowledge has indeed time and again proved the necessity of abandoning or remoulding points of view which, because of their fruitfulness and apparently unrestricted applicability, were regarded as indispensable for rational explanation. (Bohr 1958, 67–8)

This "remoulding" of "points of view" requires an analysis of "the presuppositions for unambiguous use of the concepts employed in the account of experience," a phrase which echoes throughout his writings. In the Introduction to the 1958 *Atomic Physics and Human Knowledge*, he expresses this wish:

> It is hoped that the improved terminology used to present the situation in quantum physics has made the general argument more easily accessible. In its application

to problems of broader scope, emphasis is laid especially on *the presuppositions for unambiguous use of the concepts employed in the account of experience*. The gist of the argument is that for objective description and harmonious comprehension it is necessary in almost every field of knowledge to pay attention to the circumstances under which evidence is obtained. (2; my emphasis)

Of course Bohr held not only that this was true of the advance of knowledge in general, but also in particular in the case of the advent of quantum mechanics which "led to the disclosure of hitherto disregarded presuppositions for the unambiguous use of even the most elementary concepts on which the description of natural phenomena rests" (18–19). Moreover he also maintained that in this respect the quantum revolution merely followed "the great example of relativity theory which, just through the disclosure of unsuspected presuppositions for the unambiguous use of all physical concepts, opened new possibilities for the comprehension of apparently irreconcilable phenomena" (19).

Like Bohr, Lewis (1926, 237) also emphatically emphasized that the growth of scientific knowledge requires the development of new concepts, which allowed the possibility of interpreting the given in terms of novel logical structures:

That the introduction of concepts which are novel and not generally shared may be of the highest importance is something illustrated by almost every major advance in science. Such an initial concept, whether new or old, is a definite logical structure. It sets up precise relations of certain elements of thought. And that structure—or the combination of a few such conceptual structures—may give rise to logical consequences as elaborate as mathematics or the game of chess.

Indeed, before we set out upon any systematic investigation, we must have such initial concepts in our minds. It does not matter how we get them; we can always change them for any reason, or for no reason if its suits our whim. The *real* reason why we *do* use certain concepts is, of course, practical. (Emphases in the original)

The precise logical relations among the conceptual elements in a scheme by which the given in experience is interpreted determine what is held as "a priori" within that conceptual interpretation of experience, but we have no prior assurance that any given conceptual structure will provide a unified and harmonious interpretation of experience, nor which conceptual structure will serve our practical interests most successfully. Our "free choice" of concepts has to meet the brute facts of the given:

But it is between these two [i.e., the concepts and the given], in the determination of those concepts which the mind brings to experience as the instruments of its interpretation, that a large part of the problem of fixing the truths of science and our common-sense knowledge has its place. Wherever such criteria as comprehensiveness and simplicity, or serviceability for the control of nature, or conformity to human bent and human ways of acting play their part in the determination of such conceptual instruments, there is a pragmatic element in knowledge. (237)

Because we never know in advance whether any conceptual interpretation will in fact serve these pragmatic criteria, the growth of knowledge is essentially a dialectic of trial and error.[3] Here is how Lewis described the process of conquering "the world in the name of intelligibility":

> The business of learning and the process by which mind has conquered the world in the name of intelligibility is not a process in which we have passively absorbed something which experience has presented to us. It is much more truly a process of trial and error in which we have attempted to impose upon experience one inter-pretation or conceptual pattern after another and, guided by our practical success or failure, have settled down to that mode of constructing it which accords best with our purposes and interests of action. (240–1)

As empiricists, pragmatists insist that epistemic justification within the framework of any given conceptual scheme is always a matter for experience to decide. But the conceptual element itself that makes possible the experience of a particular world is not dictated by or derived from the given which is thus interpreted.

The conceptual scheme is a "free" choice in that sense, even though it is only empiri-cal success in leading us to anticipate the results of our actions that can justify a choice. This is what produces the dialectical advance of human knowledge: "The growth of knowledge is a process of trial and error, in which we frame the content of the given now in one set of concepts, now in another, and are governed in our final decision by our relative success-by the degree to which our most vital needs and interests are satis-fied" (Lewis 1926, 242).

Where Lewis talks of a process of "trial and error," a process Bohr certainly knew very intimately from the development of quantum theory, Bohr was fond of referring to "the never ending struggle for the proper relation between content and frame." Bohr (1998, 170) advocates an "attitude"

> that may be characterized by its striving for harmonious comprehension of ever more aspects of our situation in the never ending struggle for proper relation between content and frame, recognizing that no experience is definable without a logical frame and that any disharmony apparent in such a relationship can be removed only by appropriate widening of the conceptual framework.

Of the many different pragmatic criteria, Bohr's favorites seem to have been "har-mony" and "unity."

In an argument familiar to today's philosophers of science under the name of "empirical underdetermination of theory choice," Lewis's (1926, 241) account of theory change in science argues that it is pragmatic criteria, not empirical ones, that account for conceptual change: "this mode of successful interpretation may not be dictated unambiguously by the content of experience itself." Again the example of the theory of relativity in the previous decade provided both Bohr and Lewis with an example of a theory being selected for pragmatic criteria. In 1926 Lewis writes:

> Since ... whatever happens at some distant point is known to us only by the passage of an effect through space and time, we cannot measure space without some assumption about time, or time without assumptions about space and the laws of matter which govern clocks, and so on. Therefore at the bottom of our interpretation of events in the physical universe there must be some fundamental assumptions, or definitions and criteria, to which empirical evidence cannot simply say yes or no. One set of assumptions -the relativity ones– means a reduction in the number of independent laws but a reorganization of common sense; the other set obviates this change in current notions about space and time but condemns us to forgo the simplification in fundamental principles. The determinable empirical issues, such as the perturbations of Mercury and the bending of light rays, are—so we may venture to think—by themselves not decisive. If there were no other issue, we should find some way to accommodate these recalcitrant facts to the old categories. The really final issues are pragmatic ones such as the comprehensiveness of laws and economy in unverifiable assumption. (241)

This passage bears a striking resemblance to Bohr's comments on the same historical development written a few years later in his 1934 "Introductory Survey" to *Atomic Theory and the Description of Nature*:

> It was Einstein's elucidation of the limitation which the finite velocity of propagation of all force effects, including those of radiation, imposes upon the possibilities of observation, and, therefore, upon the application of the space-time concepts, that first led us to a more liberal attitude towards these concepts, an attitude which found its most striking expression in the recognition of the relativity of the concept of simultaneity. As we know, Einstein, adopting this attitude, succeeded in tracing significant new relationships also outside the domain to which the electromagnetic theory properly applies, and in his general theory of relativity, in which the effects of gravitation no longer occupy a special position among physical phenomena, he has approached, to a quite unexpected degree, the unity in the description of nature which is the ideal of the classical physical theories. (3)

When Bohr speaks of "a more liberal attitude towards these concepts" he recognizes that Einstein retained the "space-time concepts" for reporting what an observer will experience relative to a reference frame, that is, their "sense meaning"—in Lewis's terminology—remains fixed in human experience, but changed their "linguistic meaning" by using them in a new theoretical structure. This in turn changes what we regard as a priori, and what we regard as contingent empirical fact.

As pointed out above, Bohr (1963, 6–7) often introduces his complementarity viewpoint by emphasizing what the relativity revolution and the quantum revolution have in common: both change the presuppositions for the use of our most elementary physical concepts in a way which leads us to describe phenomena "which cannot be comprehended in the pictorial conceptions adapted to the account of more limited fields of physical experience".

> Notwithstanding all difference in the typical situations to which the notions of relativity and complementarity apply, they present in epistemological respects far-reaching similarities. Indeed, in both cases we are concerned with the exploration of harmonies which cannot be comprehended in the pictorial conceptions adapted to the account of more limited fields of physical experience. Still, the decisive point is that in neither case does the appropriate widening of our conceptual framework imply any appeal to the observing subject, which would hinder unambiguous communication of experience. In relativistic argumentation, such objectivity is secured by due regard to the dependence of the phenomena on the reference frame of the observer, while in complementary description all subjectivity is avoided by proper attention to the circumstances required for the welldefined use of elementary physical concepts.

And Lewis (1926) makes effectively the same points: experience widens and concepts must often change to comprehend hitherto unknown phenomena, even while the names we give these concepts remain the same. But the engine that drives conceptual change is not motivated by empirical factors alone:

> With other fundamental concepts, it is much the same. Words such as "life," "matter," and "cause," and so on have been used since thought began, but the *meanings* of them have continuously altered ... To be sure, the telescope and microscope and the scientific laboratory have played an important part. As time goes on, the body of familiar experience widens. But that hardly accounts for all the changed interpretation which history reveals. Not sense observation alone, but accord with human bent and need must be considered. The motive to control external nature and direct our own destiny was always there. Old principles have been abandoned not only when they disagreed with newly discovered fact, but when they proved unnecessarily complex and bungling, or when they failed to emphasize distinctions which men felt to be important. (242)

Pragmatists insist that the justification for our concepts lies in their value for human interests and action, not in some supposed role of representing how the world is apart from any experience or observation of it.

Nevertheless, Lewis (1926) does not shy away from talking of "truth," although it is only truth relativized to the particular conceptual scheme by which we interpret that world. Since the choice of that scheme is subject to pragmatic values so is truth:

> When things so fundamental as the categories of space and time, the laws of celestial mechanics, and the principles of physics are discovered to depend in part upon pragmatic choice; when history reveals continuous alteration in our basic concepts, and an alteration which keeps step with changing interests; and when we recognize that without interpretation it is not a world at all that is presented to

us, but only, so to speak, the raw material of a world; then may it not plausibly be
urged that, throughout the realm of fact, what is flatly given in experience does not
completely determine truth—does not unambiguously fix the conceptual interpretation which shall portray it? (241)

Clearly here is a dynamic notion of "truth" in striking contrast to the correspondence
theory of truth that assumes a static truth, which must mirror how reality *is* quite
apart from any human interaction with the world: "Truth here is not fixed, because
interpretation is not fixed, but is left for trial and error to determine." Classical empiricism erred in thinking that experience itself "produce[d] the concepts in our minds"
by which we describe it, and transcendentalism was misled in assuming that "concepts
evoke the experience which fits them." The dynamics of this changing "truth" are pragmatic and as such are subject to the dialectic of "trial and error":

> The given experience does not produce the concepts in our minds.... Nor do the
> concepts evoke the experience which fits them, or limit it to their pattern. Rather
> the growth of knowledge is a process of trial and error, in which we frame the
> content of the given now in one set of concepts, now in another, and are governed
> in our final decision by our relative success-by the degree to which our most vital
> needs and interests are satisfied. (242–3)

But if pragmatic success is used to define truth, and success is a matter of degree, it
follows that truth also will be a matter of degree:

> The criteria of its success are accommodation to our bent and service of our inter-
> ests. More adequate or simpler interpretation will mean practically truer. Old
> truth will pass away when old concepts are abandoned. New truth arises when new
> interpretations are adopted. Attempted modes of understanding may, of course,
> completely fail and prove flatly false. But where there is more than one interpreta-
> tion which can frame the given, "truer" will mean only "better." And after all, even
> flat falsity can only mean a practical breakdown which has proved complete. (243)

Lewis admits that "speaking thus of 'new truth' and 'old truth' and of pragmatically
'truer' and 'falser' ... is a little to be regretted," but he is "following a usage which the
literature of pragmatism has made familiar." Nevertheless he realized that this sounded
sufficiently paradoxical to warrant clarification:

> Most of the paradoxes and many of the difficulties of the pragmatic point of view
> cluster about this notion that the truth can change. When we see precisely what
> it is that happens when old modes of interpretation are discarded in favor of new
> and more successful ones, all these paradoxes will, I think, be found to disappear.
> What is it that is new in such a case? The given, brute-fact experience which sets
> the problem of interpretation is not new. And the concepts in terms of which the
> interpretation, whether old or new, is phrased are -remembering Plato- such that
> the truth about them is eternal. Obviously what is new is the *application* of the

concept, or system of concepts, to experience of just this sort. The concepts are *newly chosen* for interpretation of the given data. (243–4; emphases in the original)

While this may provide an intelligible response to the paradoxical sound of "old truth" and "new truth," Lewis is under no illusions that "the problem of interpretation" presents the scientist with an easy task. Confronted with rival conceptual schemes, evolving human interests, and an ever-expanding body of experience the scientist must weave a careful choreography of concepts to capture the brutally given by heeding "the conditions of application of the concepts to this new body of total relevant experience":

Historically the situation is likely to be slightly more complex; the body of data to be interpreted itself undergoes some alteration. It is possible that old systems of thought should be rejected and replaced by new, simply through reflection and realization of the superior convenience of the novel mode. In fact, this has sometimes happened. But in the more typical case, such change does not take place without the added spur of newly discovered phenomena which complicate the problem of interpretation. The several factors which must be considered are, then: (1) the two sets of concepts, old and new, (2) the expanding bounds of experience in which what is novel had come to light, (3) the conditions of application of the concepts to this new body of total relevant experience. (244)

And Lewis's reference to the "conditions of application of the concepts" leads us back to Bohr's (1958, 19) concern with "hitherto disregarded presuppositions for the unambiguous use of even the most elementary concepts on which the description of natural phenomena rests."

4. Sense meaning and the classical concepts

Our considerations thus far establish that it is reasonable to describe Bohr's epistemological views as "pragmatist." But what can we learn from this? Can it help us to understand one of Bohr's most controversial contentions: that the classical concepts must be retained for describing measurement phenomena? Perhaps Lewis can lend Bohr a hand here. Bohr's essays are filled with references to "definition," "well-defined concepts," and "unambiguous communication." These terms are usually connected with his talk about the need to preserve "classical concepts" and "objective description," but the connection is by no means obvious.[4] Bohr was a physicist, neither a logician nor an epistemologist, and in spite of his insistence on paying attention to the presuppositions for the unambiguous use of concepts, he espoused no clear theory of meaning. Lewis was no physicist, but he was a perceptive logician and epistemologist who interested himself in matters of definition, meaning, and the proper use of concepts. Their talents were, in a word, "complementary."

Lewis developed his most detailed analysis of meaning in his 1946 *Analysis of Knowledge and Valuation*. Although he was a great herald of the significance of *Principia Mathematica*, he held that the extensional logic developed therein could not express the nature of scientific reasoning, and thus was inadequate for epistemological purposes. As A. J. Reck (1968, 10) notes in his analysis of Lewis's epistemology:

> The emphasis upon symbolic logic as an exclusively extensional logic has accompanied the thesis of the name-relation as the sole mode of meaning. Lewis' dissatisfaction with the doctrine of material implication advanced by an extensional symbolic logic has already been noted; his elaboration of an intensional logic of strict implication, with its employment of the logical modalities of necessity, possibility, impossibility, led him into the depths of epistemology. It is not surprising, then, that his theory of meaning is not limited to the narrow name-relation. Reviving scholastic and traditional conceptions, it offers four modes of meaning, of which extension is only one.

Both terms (concepts) and statements (propositions or "expressions") have meaning in four distinct modes. The "denotation" of a term is "the class of *actual* things to which the term applies," its extension in this actual universe, while the "comprehension" of a term is the class of "*possible* or consistently thinkable things to which the term would be applicable," its extension in the universe of all logically possible worlds (Lewis 1946, 39; emphases in the original). Both of these modes of meaning are forms of naming, expressing a name-to-the-thing-named relation. But the remaining two modes of meaning present *criteria for application* of the term or expression. The "signification" of a term is "that property in things the presence of which indicates that the term correctly applies," its objective correlate as a property of objects in the world (ibid.). However, it is the fourth mode of meaning, "connotation," or "intension," which is the goal of definition as ordinarily understood. Lewis originally defines "intension" generally as "the conjunction of all the terms each of which must be applicable to anything to which the given term would be correctly applicable" (ibid.). However, this definition refers to *linguistic* intension that must be distinguished from *sense meaning*, which is a form of intension or part of the connotation of the term as giving a *criterion in mind* for applying a term or expression:

> What we indicate by this phrase *sense meaning* is intention as a *criterion in mind*, by reference to which one is able to apply or refuse to apply the expression in question in the case of presented, or imagined, things or situations. One who should be able to apply or refuse to apply an expression correctly under all imaginable circumstances, would grasp its sense meaning perfectly. (133)

Both signification and intension give us criteria of application as modes of meaning; the former indicating the criteria as properties of things in the world and the latter as that which the language user has "in mind."

The assumption that the language user has to have certain sensory criteria "in mind" in order to be able to use an expression correctly is common to all empiricistic epistemologies, so Lewis (1946, 135) recognizes that the need to rely on sense meaning is by no means unique to pragmatists:

The notion of what we here call sense meaning has been emphasized by pragmatists, operationalists and holders of empiricistic theories in general ... attribution of meaning in this sense requires only two things; (1) that determination of applicability or non-applicability of a term, or truth or falsity of a statement, be possible by way of sense presentable characters, and (2) that *what* such characters will, if presented, evidence applicability or truth should be fixed in advance of the particular experience, in the determination of the meaning in question ... [That] there must be some criterion in mind, in grasping their meaning, which would determine applicability or truth by way of sense presented characters, is too obvious a thesis to be denied by anyone. Whoever approaches an empirical situation with intent to apply or refuse to apply an expression, or assert something as evidenced or its falsity as evidenced, must—if he knows what he means—be somehow prepared to accept or reject what he finds as falling under or confirming what he thus intends. (Emphasis in the original)

Following a well-known line of reasoning Lewis argues that linguistic meaning alone is ultimately "circular"; only sense meaning allows us to break out of the prison of words and set our epistemological feet on the firm ground of experience:

If we think of the meaning of words and expressions *only* as something specifiable in terms of other words and other expressions, then it must strike us that *all* use of language has somewhat the character of such an abstract deductive system ... .Linguistic meaning must be in this sense eventually circular, and can be identified only with the network of relations in which any expression in question stands in relation to others. (140; emphases in the original)

The attempt to get beyond words puts us in "a predicament characterizing the attempt to *express* meaning":

We must express meaning by the use of words; but if meaning altogether should end in words, then words altogether would express nothing. The "language system" as a whole would "have no interpretation," and there would be no such fact as the meaning of language.... . Meaning as language-pattern abstracts altogether from that function of language by which it empirically applies. Precisely what is omitted in such conception of meaning as linguistic is intension as criterion of classification and as determinant of those characters by which the sensibly presented may be recognized and made intelligible. (140)

Lewis argued that the linguistic orientation of his contemporaries, the positivists, had caused an overemphasis on linguistic meaning at the expense of sense meaning, leading to epistemological problems:

Because many logicians have of late been somewhat preoccupied with language, intension as linguistic (or "syntactic") meaning has been overemphasized, and sense meaning has been relatively neglected ... .But there are motives of some importance for epistemology which lead to their distinction; and on the whole it is sense meaning which is the more important for investigation of knowledge, and should be emphasized. (133)

Because the sensory part of sense meaning requires sensory imagery, Lewis (1946, 134) has to confront the nominalists' objection that we cannot imagine *general* terms, and he thinks that it is the "persistence of such nominalism which, in large measure, is responsible for the current tendency to identify meaning with linguistic meaning exclusively." Here he takes his cue from Kant's notion of a schematism or rule for application of an expression:

> The valid answer was indicated by Kant. A sense meaning when precise and explicit, is a schema; a rule and a prescribed routine and an imagined result of it which will determine applicability of the expression in question. We cannot adequately imagine a chiliagon, but we can easily imagine counting the sides of a polygon and getting 1000 as a result. We cannot imagine a triangle in general, but we can easily imagine following the periphery of a figure with the eye or a finger and discovering it to be a closed figure with three angles. (134)

It is because of sense meaning that Lewis emphatically rejects the conception that "analytic truth is primarily a verbal matter" (138). The sense meaning of the expressions occurring in analytic statements imply a "test-schematism" which transcends linguistic tautologies:

> When the intensional meaning of a term is taken as a test-schematism and an anticipated result of it, or when it is conceived in any other manner as a criterion of application which can be entertained in advance of the particular occasions of its application, we have meaning in a sense in which it is essentially independent of the use of language. (138–9)

Thus the "predicament" that sense meaning cannot be "put into words" can be circumvented by expressing *test-schema* for application into words:

> Sense meaning, however, as the criterion of application, can be expressed in words and the relations of words, or exhibited by exhibiting words and the relations of words; indeed, we may say that this is what words do express, but it cannot be literally put into words. Such patterns of linguistic relation can only serve as a kind of map for location of the empirical item meant in terms of sense experience. If there were no meanings in the mode of sense meaning, then there would be no meanings at all. (140–1)

The truth value of a statement depends on which mode of meaning applies to the relevant expression in the statement, and that in turn is a function of the context in which the statement appears:

> One statement in which an expression occurs may depend, for its truth or falsity, upon one mode of meaning for that expression, and another statement

may depend on another mode of its meaning. But such difference is due to the different context in which the expression stands, in two such cases, and does not argue any different meaning of the expression itself. That an expression is thus modified or limited by its context does not signify an alteration of its meaning. (77)

So Lewis holds that the truth value of a statement requires understanding the whole context in which the statement is employed.

Since Lewis (1946) defines "sense meaning" as criteria in mind, it is of course a subjective criterion in the mind of the language user, and as such inaccessible to all but that subject. So we can never determine that different subjects are employing the same mental criteria, but what can be determined empirically, of course, is their behavior and ability to communicate:

Assuming that it is essential to the meaningfulness of a linguistic expression that one who uses it should have in mind some criterion of application, still it is not clear that two persons who use the same language and "understand each other" by means of it necessarily have in mind the same thing. It *is* essential, if the purposes for which language exists are to be met, that they signalize the same objective realities by the same language.... . Hence use of the same expressions in the presence of the same objective facts does not necessarily imply the same sense-criteria of application. So far as the schematism of application involves overt behavior, common meaning presumes community of such test-routine; but so far as it is some observed resultant which must determine the decision, it is not evident that use of the same term on the same occasions indicates that the same sense meanings are in mind. (143; emphasis in the original)

The privacy of sense meaning as subjective criteria is counterbalanced by the public empirical nature of behavior, including especially linguistic behavior, as embodied in a community of language users who are able to communicate successfully by means of that language.

What may be thus in another's mind we can only argue to from the observed similarity of his behavior, including his use of language, and from his observable similarity to us in other respects. Such a conclusion may be regarded as inductively justified if we are content to class as probable a conclusion from observed similarities to similarity in a respect which can never be directly observed. (143)

Pragmatism is committed to analyzing knowledge as a guide to action and here that action implies the presence of a community of communicating language users employing the same empirical test routines.

In developing his theory of empirical knowledge Lewis (1946) distinguishes three kinds of statements. "Expressive statements" are "formulations of what is presently given in experience" (182). These are familiar as the "sense data" reports or "protocol sentences" of the positivists, and we cannot be mistaken regarding them. As Reck (1968, 19) observes, "As statements, of course, they can be true or false, but only

liars can make false expressive statements." Nonterminating or objective judgments
are "judgments which assert objective reality; some state of affairs as actual" (Lewis
1946, 184). These are familiar as the overwhelming majority of statements found in
daily common-sense discourse, statements made in the "object language" of physical-
ism, and none of them are ever certain. But Lewis's most original contribution lies in
his theory of "terminating judgments," which serve as the bridge between expressive
statements and objective judgments. They have the form " 'S being given, if A then
E,' where 'A' represents some mode of action taken to be possible, 'E' some expected
consequence in experience, and 'S' the sensory cue." S, A, and E must all be expressive
statements, and as such indubitable:

> Immediate apprehensions of data of sensation are infallible. Expressive state-
> ments refer directly to this data, but may be true or false. Taking the sense data
> as cues, the organism engages in action, anticipating experience of further data
> therefrom. Terminating judgments bridge the stages of the organism's reception
> of the initial data, of the organism's motor responses to these data, and of the
> organism's reception of the final data upon completion of the action. These final
> data either measure up to or fail to measure up to anticipations. Terminating
> judgments always find their cue in what is given; but they also always state some-
> thing which is verifiable only by a course of action resulting in sensory imme-
> diacy. (Reck 1968, 19–20)

Thus Lewis's theory of empirical knowledge returns us to the etymological heart of
pragmatism: the significance of cognition lies in its role as a guide to *action*.

How does Lewis, the pragmatist epistemologist, help shed light on Bohr, the quantum physicist? For the physicist the laboratory, the arena of empirical observation and
measurement, is where the "action" is at. And that action is performing experiments,
bringing about interactions between laboratory apparatus and "observed" objects.
More generally it is where the physicist explores the phenomena that theories are
intended to explain, from whence the sensory cues emanate. "Observations," or more
specifically "measurements," are inter*actions*.

Complaints about Bohr's lack of an account of measurement are common
throughout the literature. Most of these complaints originate from trying to interpret quantum theory from the "top down," that is, from the formalism to the world,
and thus have to confront the issue of the "reduction" or "collapse" of the wave function. Bohr never approached the issue this way; for him the problem was always
a matter of how to use concepts unambiguously in objectively describing certain
laboratory *phenomena* as instances of making a "measurement," that is, from the
empirical "bottom up."

As we have seen Lewis's account of sense meaning presupposes a community of
language users who are able to communicate successfully. Bohr (1958, 71) begins his
defense for retaining the classical concepts with the same presumption of a community
of successfully communicating scientists who can attain intersubjective agreement on
the results of an experiment:

In this context, we must recognize above all that, even when the phenomena transcend the scope of classical physical theories, the account of the experimental arrangement and the recording of observations must be given in plain language, suitably supplemented by technical physical terminology. This is a clear logical demand, since the very word "experiment" refers to a situation where we can tell others what we have done and what we have learned.

A conceptual framework permitting unambiguous description of observed phenomena is therefore a necessary condition for scientific practice; it is the avenue of escape from the subjectivity of individual experiences to the "objective description" essential to scientific knowledge.

Since the description of the results of the phenomenon of an observational interaction requires specification of the experimental arrangement in which the observation is made, it follows that different experimental arrangements present different phenomena to be described. As we know, in Bohr's (1958) vocabulary, the description of a measurement "phenomenon" must include a description of the entire experimental arrangement on pain of falling into "ambiguity."

> As soon as we are dealing, however, with phenomena like individual atomic processes which, due to their very nature, are essentially determined by the interaction between the objects in question and the measuring instruments necessary for the definition of the experimental arrangements, we are, therefore, forced to examine more closely the question of what kind of knowledge can be obtained concerning the objects. (25)

Because the arrangements of laboratory equipment for a well-defined position measurement preclude those for a well-defined momentum measurement and vice versa, the two values cannot be predicated simultaneously of the same object.

> Any unambiguous use of the concepts of space and time refers to an experimental arrangement involving a transfer of momentum and energy, uncontrollable in principle, to fixed scales and synchronized clocks which are required for the definition of the reference frame. Conversely, the account of phenomena which are characterized by the laws of conservation of momentum and energy involves in principle a renunciation of detailed space-time coordination. These circumstances find quantitative expression in Heisenberg's indeterminacy relations which specify the reciprocal latitude for the fixation of kinematical and dynamical variables in the definition of the state of a physical system. (Bohr 1958, 72)

Bohr understood the indeterminacy relations from the "bottom up" as the expression in the formalism of this *physical* fact about the phenomena that provide the empirical basis of quantum theory.

What has changed as a consequence of the exclusion of simultaneous observations determining values of both kinematic and dynamic observables is that while classically an observed object could be considered as for all practical purposes as detached from any interaction with the observing instruments, once action becomes quantized the observational interaction "forms an integral part of the phenomena":

> However, the fundamental difference with respect to the analysis of phenomena in classical and in quantum physics is that in the former the interaction between the objects and the measuring instruments may be neglected or compensated for, while in the latter this interaction forms an integral part of the phenomena. The essential wholeness of a proper quantum phenomenon finds indeed logical expression in the circumstance that any attempt at its well-defined subdivision would require a change in the experimental arrangement incompatible with the appearance of the phenomenon itself. (Bohr 1958, 71)

Because of this "wholeness," or what he also calls the "atomicity" or "individuality" of the interaction phenomenon at the atomic level—the consequence of the quantum postulate—the determination of one value at one time cannot be used to predict (or retrodict) the value of that observable at a later (or prior) time. Therefore, Bohr concludes that "under conditions where the quantum of action plays a decisive part and where such an interaction is therefore an integral part of the phenomena, there cannot to the same extent be ascribed a mechanically well-defined course" (98). Hence, the presuppositions for the use of the concepts of position and momentum, for example, that all physical objects have at all times properties corresponding to these observables, must be rescinded.

> In accordance with the character of the quantum mechanical formalism, such [indeterminacy] relations cannot, however, be interpreted in terms of attributes of objects referring to classical pictures, but we are here dealing with the mutually exclusive conditions for the unambiguous use of the very concepts of space and time on the one hand, and of dynamical conservation laws on the other. (72–3)

These classical mechanical state-defining properties can apply to microsystems only in *the context of the description of a particular observational phenomenon*, and paying attention to this new presupposition for using the classical concepts is how we avoid falling into "ambiguity."

So why must it be the "classical concepts" that are used to communicate unambiguously the results of a measurement? Our situation in atomic physics is "unprecedented":

> We must realize that the situation met with in modern atomic theory is entirely unprecedented in the history of physical science. Indeed, the whole conceptual structure of classical physics ... rests on the assumption, well adapted to our daily experience of physical phenomena, that it is possible to discriminate between the behaviour of material objects and the question of their observation. (Bohr 1958, 19)

The assumption that we can unambiguously describe an object independently of the means of observation by which we obtain this information is "already inherent in ordinary conventions of language" and "fully justified by all everyday experience" (25). As we have seen, the wholeness of an observational interaction invalidates that assumption. Thus the conceptual framework of quantum theory has altered the presuppositions for the unambiguous use of the concepts by which we report the result of an observation. And this changes their relations to other concepts, for example, exact position and exact momentum now become exclusive, while in the classical framework they can both be predicated simultaneously of the same object:

> In fact, the limited commutability of the symbols by which such variables are represented in the quantal formalism corresponds to the mutual exclusion of the experimental arrangements required for their unambiguous definition. In this context, we are of course not concerned with a restriction as to the accuracy of measurements, but with a limitation of the well-defined application of space-time concepts and dynamical conservation laws, entailed by the necessary distinction between measuring instruments and atomic objects. (Bohr 1963, 4)

But this alteration of the presuppositions for the unambiguous use of the classical concepts is concerned with their *linguistic* intension, their relations to other concepts, not as Lewis would say, with their *sense meaning*. However in those expressions in which the physicist is not relating one concept to another, but is describing phenomena, specifying the "experimental arrangements required for their unambiguous definition," the mode of meaning intended must be sense meaning, which remains fixed under the alteration of linguistic meaning.

Bohr (1958, 63) clearly believed that the classical concepts had evolved from the concepts used to interpret the "given" sensory input of ordinary human subjects in everyday situations, but the classical framework had proved too narrow: "The very conceptual frame, appropriate both to give account of our experience in everyday life and to formulate the whole system of laws applying to the behaviour of matter in bulk and constituting the imposing edifice of so-called classical physics, had to be essentially widened if it was to comprehend proper atomic phenomena." Although the framework is "widened," the sense meaning of the classical concepts used to describe the measurement phenomenon stays rooted to human sensory cues, because it is sense meaning that remain "unambiguous" in the description of experience. The laboratory sets the context that determines this mode of meaning: an observer applies the relevant test-schema to satisfy him or herself that the criteria for asserting that a certain result has been obtained are indeed satisfied. This can be communicated publicly and is unambiguous to the relevant community. It is not because a human experimenter as a conscious being becomes cognizant of the result that a measurement is made.[5] But it is (i) because physics is an empirical science that must be able to communicate unambiguously contingent claims about the world, in order to test them by experience; (ii) because the sense meanings of the relevant concepts lie on that foundation; and (iii) because the definition of a "measurement" as something that can be described unambiguously as recording a certain result, that it must be described unambiguously in terms of the direct sense meanings of the classical concepts: "The aim of every physical experiment—to gain knowledge under reproducible and communicable conditions—leaves us no choice but to use everyday concepts, perhaps refined by the terminology of classical physics, not only in all accounts of the construction and manipulation of the measuring instruments but also in the description of the actual experimental results" (25–6). What makes a physical system into a measuring instrument is how we describe it. It remains theoretically possible to describe a measuring instrument, like any physical object, in terms of the quantum formalism, but to do so would preclude its serving its epistemic function. The measuring instrument is a measuring instrument precisely because it is described as possessing properties "directly determinable" by experience:

> In the first place, we must recognize that a measurement can mean nothing else than the unambiguous comparison of some property of the object under investigation with a corresponding property of another system serving as a measuring instrument, and for which *this property is directly determinable according to its definition in everyday language or in the terminology of classical physics.* (Bohr 1938, 100; emphasis mine)

When we employ classical language to describe the measuring instrument we must "renounce" the attempt to describe the observational interaction in a manner that would exceed the limitations of the indeterminacy relations, but the "influence" this has on the "phenomenon itself" is not one of physically (or mentally) disturbing an otherwise well-defined state. Rather, the "influence" is on the circumstances in which the relevant concepts are well-defined:

> While within the scope of classical physics such a comparison can be obtained without interfering essentially with the behaviour of the object, this is not so in the field of quantum theory, where the interaction between the object and the measuring instruments will have an essential influence on the phenomenon itself. Above all, we must realize that this interaction cannot be sharply separated from an undisturbed behaviour of the object, since the necessity of basing the description of the properties and manipulation of the measuring instruments on purely classical ideas implies the neglect of all quantum effects in that description, and in particular the renunciation of a control of the reaction of the object on the instruments more accurate than is compatible with [the indeterminacy relations]. (100)

Thus we see that Bohr's argument for "widening" the physicist's conceptual framework while at the same time insisting that observational phenomena must be described in the classical framework loses its paradoxical character once we understand Lewis's distinction between linguistic intension and sense meaning. Moreover we note that just as Lewis argued that meaning is determined by context, the context of the laboratory, it is unambiguously communicable through empirical test-schema.

5. Conclusion

The historical record reveals no evidence at all of any possible direct connection between the epistemological views of Bohr and Lewis, but it also reveals that both thinkers had great enthusiasm for William James, who was the inspirational fount of much pragmatic philosophizing in the first part of the twentieth century. And, like James himself, both men came of age in a philosophical climate much shaped by Kant. This shared intellectual paternity permits us to say that they were spiritual brothers, at least insofar as their epistemological views were concerned. Moreover, it seems fair to say that the similarities in their views documented above justify saying this.

But Bohr was a physicist first, and by choice; he was an epistemologist only second, and by necessity. Thus, not surprisingly, he always kept at arm's length from associating his views with any philosophical school or movement, making him something of an outsider to those who made epistemology their profession. Perhaps an opportunity was lost. Pragmatist philosophers who did make epistemology their profession explored a variety of distinctions and arguments that might have been helpful in expressing the "epistemological lesson" of complementarity. We have seen one such instance with Lewis's analysis of sense meaning. It is worth remembering that Bohr approached the quantum revolution from the other side, participating and guiding its development on the basis of a wide variety of empirical phenomena, not from the formalism, which it attained only at the end, after the empirical stage had been set. It is this "sense meaning" of the language used to describe these experimental phenomena that provides the foundation, literally and metaphorically, for Bohr's theory of knowledge.

Notes

1 Unfortunately one of the frustrations of attempting to elucidate the thinking of historical figures is that often there is reasonable circumstantial evidence for a channel connecting different pools of thought but the direct evidence is either absent or ambiguous. Bohr's connection to pragmatism is a case in point. In his final interview with Thomas Kuhn, Bohr tantalizingly expresses great enthusiasm for William James and singles out his famous chapter "The Stream of Thoughts" for special admiration. However, the date of his acquaintance with James's work, which Bohr recalled to have been during his student days, was contested by his assistant Léon Rosenfeld as a mistaken memory, and that his first acquaintance with James dated to the 1930s. It is also unclear whether Bohr was familiar with only James's *Principles of* Psychology or if he also knew James's later explicitly pragmatic essays.

2 My own first teacher of logic and epistemology was Harold N. Lee, who himself had been a student of Lewis at Harvard. He related this anecdote: it happened that one year Lewis was not scheduled to teach his Kant course. A colleague noticed that he was looking somewhat out of sorts. "Is anything the matter, Prof. Lewis?" it was asked. He responded, "It is the first time in so many years I've not read the *Critique of Pure Resason.*"

3 This is a view of the growth of knowledge that contemporary philosophers of science are more likely to associate with Karl Popper than American pragmatism; therefore it

is worth noting that Lewis had advocated this view the decade before Popper's 1934 *Logik der Forschung*. But of course, as in so many other philosophical matters, in some sense Socrates was there first.

4 Surely those who expected Bohr to respond to EPR by scoring a point in physics must have been disheartened to learn that the failure of the argument rested on an "ambiguous" criterion of reality.

5 Here Bohr parts ways with subjectivist readings of the quantum revolution.

References

Bohr, N. 1934. *Atomic Theory and the Description of Nature*. Cambridge: University Press.

Bohr, N. 1958. *Essays 1932–1957 On Atomic Physics and Human Knowledge*. New York: Wiley.

Bohr, N. 1962. Interview with T. S. Kuhn, L. Rosenfeld, Aa. Petersen, and E. Rudinger, Session V: 17 Nov 1962. *Archive for the History of Quantum Physics*. Copenhagen: Niels Bohr Institute.

Bohr, N. 1963. *Essays 1958–1962 On Atomic Physics and Human Knowledge*. New York: Wiley.

Bohr, N. 1998. *The Philosophical Writings of Niels Bohr, Vol. IV: Causality and Complementarity*, edited by J. Faye and H. J. Folse. Woodbridge, CT: Ox Bow Press.

Faye, J. 1991. *Niels Bohr: His Heritage and Legacy*. Dordrecht: Kluwer Academic Publishers.

Hunter, B. 2014. "C.I. Lewis." In *Stanford Encyclopedia of Philosophy*, Spring 2014. http://plato.stanford.edu/archives/spr2014/entries/lewis-ci/.

James, W. 1890. *The Principles of Psychology*. New York: Henry Holt.

Lewis, C. I. 1923. "A pragmatic conception of the a priori." In *Human Knowledge: Classical and Contemporary Approaches*, edited by P. K. Moser and A. vander Nat, 234–44. New York: Oxford. Originally printed in *The Journal of Philosophy* 20: 169–77.

Lewis, C. I. 1926. "The pragmatic element in knowledge." In *Human Knowledge: Classical and Contemporary Approaches*, edited by P. K. Moser and A. vander Nat, 234–44. New York: Oxford. Originally printed in *University of California Publications in Philosophy* 6.

Lewis, C. I. 1929. *Mind and the World Order*. New York: Dover.

Lewis, C. I. 1946. *An Analysis of Knowledge and Valuation*. LaSalle, IL: Open Court.

Reck, A. J. 1968. *The New American Philosophers*. Baton Rouge: Louisiana State University Press.

5

Complementarity and Human Nature

Jan Faye

In this chapter I will not only defend Niels Bohr's philosophy of quantum mechanics against years of misinterpretations but also argue that it is an interpretation that makes sense in the light of Darwin's theory of human evolution. Bohr's claim that the use of classical concepts is indispensable for describing quantum phenomena seems to be grounded in an assumption that these concepts reflect our cognitive adaptation to our physical environment. However, Bohr also claimed that the use of classical concepts has to be restricted, due to the quantum of action, so their application makes sense only in relation to the context of specific experiments. The consequence is, he argued, that the quantum formalism does not represent the quantum world as it is in itself but can be used to represent the world as it appears to us given the experimental results. Such a pragmatic stance seems to stand in stark contrast to most other contemporary interpretations that take realism and representationalism for granted. A closer look at his philosophy reveals strong pragmatic positions similar to those held by the American pragmatists.

1. Classical concepts

Why did Bohr believe that ordinary concepts such as space, time, causation, and unity—and their physical equivalences which he referred to as the "classical concepts"—were indispensable for any description of physical experience including that of the quantum world? Was it because he believed much like Kant that human beings are restricted by a priori reasons to use certain basic categories the existence of which acts as a transcendental precondition for human experience, or did he have a more naturalistic and pragmatic view in mind? First of all he seems to think of the classical concepts as a kind of Kantian categories of the understanding by which human beings are able to organize their sense impressions and thereby make experience possible. Let me provide four quotations that may give a better grip on what Bohr had in mind:

> From a logical standpoint, we can by an objective description only understand a communication of experience to others by means of a language which does no

admit ambiguity as regards the perception of such communications. In classical physics, this goal was secured by the circumstance that, apart from unessential conventions of terminology, the description is based on pictures and ideas embodied in common language, *adapted to* our orientation in daily-life events. (Bohr [1953] 1998, 156–7; emphasis added)

The main point to realize is that all knowledge presents itself within a conceptual framework *adapted to* account for previous experience and that any such frame may prove too narrow to comprehend new experiences. (Bohr [1954] 1958, 67; emphasis added)

All account of physical experience is, of course, ultimately based on common language, *adapted to* orientation in our surroundings and to tracing relationships between cause and effect. (Bohr [1958] 1963, 1; emphasis added)

And finally, while speaking about relativity theory and quantum mechanics Bohr ([1958] 1963, 6–7) stated:

Indeed, in both cases we are concerned with the exploration of harmonies which cannot be comprehended in the pictorial conceptions *adapted to* the account of more limited fields of physical experience. (Emphasis added)

These quotations tell us that the classical concepts are said to be indispensable for any physical description, since these concepts are already embodied in the ordinary language of perception before they received a physical explication. We may call this Bohr's *indispensability thesis*. The classical concepts are indispensable, not because of Kant's a priori reasons, but because ordinary language constitutes the only unambiguous means by which we can communicate about our perceptions and actions. Furthermore, Bohr seems to hold that any description of physical experience must use a language that reflects those concepts and ideas that have been installed in human beings as a result of our adaptation to the physical environment. The latter may be called Bohr's *adaptation thesis*.

Thus, while defending his indispensability thesis, Bohr did not take Kant's line of thought that the categories of the understanding are necessary for human experience since they act as the transcendental conditions for human empirical knowledge. According to Kant, the categories of the understanding frame our impressions given in the intuition of space and time. However, Bohr seems to argue quite the reverse. Human experience comes first, the categories come second. He seems to say that the categories of the understanding stem from human experience, since our common language has been formed by and adapted to human experience. He holds that classical physics should be considered as an attempt to give a physical description of the world based on a precise articulation of these categories. Strangely enough Bohr does not mention Charles Darwin's theory of natural selection. But from the use of the term "adaptation" he opens the possibility for an evolutionary interpretation of his insistence on the necessary use of classical concepts in physics.

That such a construal is not totally far-fetched is testified by the fact that Bohr's longtime associate Léon Rosenfeld ([1961] 1979, 514), an ardent advocate of complementarity, presented a similar suggestion as a reaction to his self-imposed question[1]: "Is the recourse to complementarity logic so to speak an accident, another of the many imperfections of human nature, and is it not conceivable that other rational beings, some kind of Maxwellian demons, could develop a different mode of description of Nature not affected by any complementary limitations?" He rejected such a possibility by saying:

> The tentative answer I should like to submit is that there is nothing more acciden-
> tal in the emergence of complementary logic than the emergence of man him-
> self as a product of organic evolution. My point is that the concept of a thinking
> Maxwellian demon is self-contradictory. I suspect the development of a comput-
> ing and communication system like our brain demands about that complexity of
> organization which has been reached by our own species in the course of evolu-
> tion. After all, it is an empirical fact that that rational thinking has only arisen at
> the stage of mankind, presumable because it was only at this stage that a brain
> developed with a sufficient number of interconnexions between nerve cells to
> make possible this very peculiar behaviour we call rational thinking. (515)

Since the complementary logic arises from the use of classical concepts in the description of atomic phenomena, Rosenfeld seems to indicate that given human evolution, and the way it turned out, the use of classical concepts is constitutive for rational thinking and therefore that which makes talk about atomic objects intelligible.

So if human evolution was part of Bohr's thinking concerning the use of classical concepts, which I assume it was, one may present the following argument. A propon-
ent of evolutionary epistemology maintains that much of our cognitive capacity has evolved over the course of eons by natural selection and adaptation to our natural habi-
tat. Hence, our basic cognitive mechanisms were already in place when our predeces-
sors split from the common ancestors we share together with the chimpanzees, a split estimated to have taken place around 7 million years ago. Perception, memory, imagin-
ation, concept-formation, and intentional action are some of the cognitive processes we have in common. Color vision, the ability to sense unities and differences, spatial and temporal relations, movements, to categorize various kinds, and the capacity to set up strategies for action to gain imagined goals are all basic cognitive mechanisms that we find in human beings as well as in other larger brained animals. The ability among apes and monkeys to distinguish between different kinds of colors, various kinds of fruits, and diverse kinds of animals, as well as their ability to count small numbers, indicates that our capacity for forming concepts had been developed long before our nonlinguistic forebears were set on the evolutionary path that led to language users. The most sophisticated evidence tells us that Homo sapiens evolved around 200,000 years ago, but that advanced language first came about between 150,000 and 50,000 years ago. From then on the cognitive development that took place by virtue of natural selection among hominids evolved toward an increasing capacity for abstract thinking in combination with proper language skills. Thus, in order for the process to be successful the rise of ordinary language had to evolve from an adaptation of the natural concepts by which our nonlinguistic predecessors ordered their experience.

In the light of this scenario, Bohr's adaptation thesis makes a lot of sense. Ordinary language is adapted to human experience, and similarly human experience is the cognitive disposition to response to the sensory information brought to us via our sense organs, a disposition which is based on our biological adaptation to the physical environment in which our nonlinguistic predecessors lived. A naturalistic explanation of the inevitability of classical concepts is then that these concepts are already embedded in our ordinary language as reflecting our genetically inherited categories and organization of thinking. Hence physics, as an exact empirical discipline about our physical environment, had to adjust its descriptive capacity according to the way we are genetically disposed to experience the world. Therefore, as long as physics is an empirical discipline, it cannot but be confined to describing the world in terms of those concepts that we are adapted to use and by which we are genetically forced to convey perceptual information.

The evolution of Homo sapiens also gave rise to a proliferation of abstract and reflective thinking from which both mathematics and physics have benefitted a great deal. Abstract thinking goes beyond our sensory experience and creates concepts which may or may not have an origin in a real thing. But without an unambiguous connection to human experience such abstract thinking has a tendency to fool its thinkers into believing that it objectively represents something, as it is independently from our experience. In such cases the concepts of abstract thinking are reified into standing for some allegedly real entities. I think this may be the reason, as we shall see, why Bohr denied the wave function, and the quantum mechanical formalism in general, have any representational function.

2. Bohr's contextualism

If the above analysis of the arguments for Bohr's indispensability thesis is correct, his view on quantum mechanics seems to follow given the empirical fact that the quantum of action plays such a significant role in the study of atomic objects. The indispensability thesis plus the adaptation thesis, together with the discovery of Planck's constant, led Bohr to another thesis, which Dugald Murdoch (1987, 139 ff.) calls Bohr's *indefinability thesis*, according to which an object cannot meaningfully be said to possess exact values, say of position and momentum simultaneously. Literally the indefinability thesis stipulates the conditions within quantum mechanics under which it is *not* meaningful to apply the classical concepts in terms of two conjugate variables. Here I also want to refer to the reverse thesis which may be called Bohr's *definability thesis*. This specifies the conditions under which it *is* meaningful to apply classical concepts to a quantum system by restricting the application of classical concepts in quantum mechanics to a specific setting with a definite outcome.

In his analysis of various thought experiments in terms of classical concepts Bohr was able to confirm the consistency of Heisenberg's indeterminacy thesis, say, that the position of an electron could not be exactly determined at the same time as the precise determination of its momentum. As Bohr ([1929] 1934, 18) said: "Obviously, these [kinds of] facts not only set a limit to the extent the information is obtainable by measurements, but they also set the limit to the *meaning* which we may attribute to such information." So the restrictive use of the classical concepts in quantum mechanics arises not merely because measurements fail to provide information about an exact position and an exact momentum at once, but also because the classical concepts cannot be applied meaningfully to the states of objects which are physically impossible to observe. Why, one may ask? Apparently, the answer lies in the scope of *the adaptation thesis*. Natural, experience-based concepts meaningfully apply only to physical states of a system that are directly accessible to our experience. This is due to the fact that these natural notions have evolved by the adaptation of an ability to grasp the environment as we experience it. The natural notions are in particular fitted by natural selection to structure and organize our experience. And the classical concepts are just more precise extrapolations of these naturally evolved notions. By them we can grasp our sensory experience, not what is inaccessible to human observation, and their adaptive origin causes their restrictive application.

In classical physics it makes sense to attribute a definite position and a definite momentum to a system that is not under observation. It makes sense due to the fact it is always possible in principle to bring an observer in connection with the system so this person has a sensory experience of the system in a way that allows a description in terms of the classical concepts. But the discovery of Planck's constant led Bohr to believe that the quantum of action physically excludes an observer from being able to be brought into a similar situation with respect to a quantum system. Thus, concepts like the classical ones adapted to grasp our experience do not apply to such non-experiential situations, where we have no empirical possibility to determine their applicability. So the indispensability thesis states that all physical experiences have to be described in terms of the classical concepts, whereas the adaptation thesis claims that the classical concepts are properly used only within those areas of experience from which they have evolved. However, the discovery of Planck's constant creates those experientially inaccessible situations we find in quantum mechanics, *in casu* experimentally inaccessible situations, where it is impossible at the same time to assign exact values to two conjugate variables. Together with the *indispensability thesis* and the *adaptation thesis* this fact implies that the definability thesis holds for those cases where we have determinate observational results. Hence the *definability thesis* states that experience-adapted notions are applicable in quantum mechanics only if the use of these notions corresponds to a physically possible experimental arrangement that actually brings about an appropriate observation or measurement.

A consequence of Bohr's definability thesis is that the classical concepts such as position and momentum do not apply in quantum mechanics to cases where we have no empirical access. It does not make sense, according to Bohr, to apply these notions to a "free" quantum object because such an object is not one about which we can have any empirical information. As Bohr ([1951] 1998, 151) put it:

> In atomic physics, we can no longer uphold the idea of a behaviour of objects, independent of the circumstances under which the phenomena are observed. It is here not a question of a practical limitation of the accuracy of measurements, but of an

aspect of the laws of nature, associated with the quantum of action, which sets a lower limit to the interaction between the objects and the measuring instruments.

Information about a system is something we can acquire only when we physically interact with the system. But unlike the situation in classical physics the situation in quantum mechanics is such that we cannot control or determine this interaction. Such a determination would be possible only if it had made sense to assign a "free" quantum system with an exact value of position and momentum.

Bohr's main conclusion was that the attribution of dynamical and kinematic properties to a quantum system should be considered to be complementary, in order to escape from being inconsistent. Their correct attribution depends on the experimental context that established the physical condition for having the relevant experience to which one of the classical concepts applies. Bohr's discussion with Einstein convinced him that no experiment, which allows an exact value attribution of one of the conjugate variables to the system, permits also an experiment determining the exact value of the other conjugate variable.

Indeed Bohr's ([1938] 1958) naturalistic epistemology had ontological implications. This is obvious from the following long quotation:

> In this respect we must, on the one hand, realize that the aim of every physical experiment—to gain knowledge under reproducible and communicable conditions—leaves us no choice but to use everyday concepts, perhaps refined by the terminology of classical physics, not only in all accounts of the construction and manipulation of the measuring instruments but also in the description of the actual experimental results. On the other hand, it is equally important to understand that just this circumstance implies that no result of an experiment concerning a phenomenon which, in principle, lies outside the range of classical physics can be interpreted as giving information about independent properties of the objects, but is inherently connected with a definite situation in the description of which the measuring instruments interacting with the objects also enter essentially. (25–6)

First Bohr mentions the indispensability thesis; he then argues that the definability thesis has the consequence that the attribution of properties, like position and momentum, is not to "independent properties." In classical mechanics we assumed that any system has a definite position and a definite momentum independently of any other system because the values of these properties are always empirically accessible in principle. But with respect to the ontology of quantum mechanics the situation is different. Here information about these values is not always accessible, and when it is accessible the information comes about in an uncontrollable interaction with the measuring instrument. Hence Bohr holds that the information is not about independent properties. Instead it is about relational properties, that is, properties which the system has only in relation to a specific experimental context.

This form of contextualism, according to which the attribution of certain properties to a system is meaningful only in relation to a certain outcome of an experimental arrangement, highlights Bohr's comparison between quantum mechanics and the theory of relativity. He believed that both theories should be interpreted in a way that is in conflict with well-established assumptions of common sense and classical physics. Relativity had revealed that "length" and "duration" do not designate intrinsic and independent properties. The observed value of these properties depends on the frame of reference that determines the perspective under which an observer can ascribe a value to the system. Similarly, quantum mechanics can be understood to be showing that attributes like "position" and "momentum" are also frame dependent. The experimental arrangement is here acting as the frame of reference. Every physical experiment has to relate to a spatially and temporally fixed coordinate system in order for the physicists to be able to identify and determine measurable properties. It is only in relation to a spatially and temporally fixed frame of reference that physicists can define a body as moving or at rest. Hence it is only with respect to a frame of reference established by observation that position and momentum are well-defined. The experiment itself has to be described in classical terms for us "to gain knowledge of the quantum objects under reproducible and communicable conditions."

Seen from an evolutionary perspective it is not strange that some of the basic ways of how we experience and conceptualize our environment turn out to be strongly perspectival. Our capacity for experiencing our surroundings is indeed not adapted to ultrahigh velocities or microscopic objects. We are adapted by natural selection to perceive medium-sized things which have an immediate influence on our survival and daily life such as finding food, shelter, or a mating partner. But ultrafast microscopic objects have no such impact. Nevertheless, science is what it is today because we have been able to use the very same sensory experience that evolution has given us to pick up information about our environment on matters that reach beyond our immediate environment.

3. Beyond classical concepts

Although Bohr believed that the classical concepts and ordinary language are necessary for understanding the experimental and observational practice of science, there is another part of the scientific enterprise that reaches above observation and experimentation. This part consists of theory construction and model building. From a Darwinian point of view this cognitive activity is due to a later phase of our evolution in which reflective and abstract thinking supplies various forms of embodied and experiential thinking. The capacity for abstract thinking helps human beings in foreseeing the future and preserving memory of the past, creating arts, crafts, and technologies, and organizing a civil society. But the same capacity seems also to lead our thinking astray in believing that real things corresponds to our abstract thoughts and symbols. A typical victim of such a common inclination was John Bell (1988, 128) as he declared with respect to quantum mechanics: "no one can understand this theory until he is willing to think of ψ as a real objective field rather than just a 'probability amplitude.' Even though it propagates not in 3-space but in 3N-space."

Some of the first philosophers who saw the drawbacks of such a strong realist tendency of reification were the American pragmatists. They argued that scientific concepts and theories should be evaluated according to how effective they were in explaining and predicting observational phenomena in opposition to how accurately they described reality itself. Although Bohr believed in the existence of atomic objects, he held, as we shall see, a similar anti-realist and instrumentalist view about the wave function and scientific theories.

It is well-known that Bohr himself established the first predictively successful model of the hydrogen atom by incorporating Planck's constant into the atomic structure. Moreover, it was first with Heisenberg and Schrödinger's contributions that the physics community achieved a consistent theory of atomic objects which made some important corrections to Bohr's original model. However, both Bohr and Heisenberg stressed that the theory of quantum mechanics was a result of the use of Bohr's correspondence principle. This prescription required that a consistent quantum mechanics should provide radiation frequencies for the transition between energy states of high quantum numbers that would correspond with similar frequencies predicted by classical physics.

To a certain extent the correspondence principle clashes with Thomas Kuhn's later claims that succeeding paradigms are incommensurable and that scientific observations are theory-laden. Obviously the correspondence principle is inconsistent with strong versions of the incommensurability thesis. But when frameworks, or paradigms, change, there is bound to be *some degree* of variation in meaning, and Kuhn served a good purpose in making philosophers sensitive to that fact. Bohr seemed to believe that it would be impossible to compare the predictions of two theories, unless the basic concepts in terms of which they make their predictions share much of the same content. In his opinion such a comparison between the predictions of two theories, or models, is possible because all physical experience has to be described in terms of ordinary language and the classical concepts. So already with the early introduction of correspondence principle we see Bohr's implicit exemplification of the indispensability thesis and the adaptation thesis.

But what about the theory of quantum mechanics itself? How did Bohr react to its very abstract formalism? Originally Heisenberg's formulation was based on the idea that so-called quantum jumps could be characterized as a transition probability that required that both position and momentum were represented by their own set of Fourier coefficients with indices for the initial state and the final state. Heisenberg therefore discarded as fictive most of the frequencies that appear in the analysis of the electron's orbit as a classical Fourier-series. Instead he used only those frequencies in his analysis that can be observed as radiation frequencies. Hence Heisenberg's construction of matrix mechanics did stay as close as possible to the assumption that a consistent theory should take care of only observable properties. From this goal of his construction, no invitation follows to interpret matrix mechanics as a literal representation of a quantum system.

The situation seems to be a bit different when we turn to Schrödinger's formulation. Unlike Heisenberg, Schrödinger designed wave mechanics based on his explicit hope that a quantum system could be represented by a wave function, or a state vector, that figures in his wave equation. Schrödinger himself attempted to give the wave function a realist interpretation. But a general realist interpretation of his formulation seems to have been undermined as soon as Born proposed his probability interpretation associating the wave function with a probability amplitude. Again, Born's interpretation cleared the ground for an instrumentalist interpretation of the formalism. Now the wave function could be regarded as an instrumental device for calculating probabilities whose values depend on our knowledge of previous measurements. The theory can be used for bookkeeping, but we have no further evidence for advocating a more representationalist interpretation of the quantum formalism.

This also seems to be Bohr's ([1954] 1958) view. He considered the wave equation as merely a symbolic means for calculation. He says, for instance, that the purpose of scientific theories "is not to disclose the real essence of phenomena but only to track down, so far as it is possible, relations between the manifold aspects of experience" (71). As we have seen, Heisenberg created matrix mechanics based on exactly such an approach. In other places Bohr ([1951] 1998, 152) distances himself from any representationalist view of theories: "The ingenious formalism of quantum mechanics, which abandons pictorial representation and aims directly at a statistical account of quantum processes." Similarly, he says, "The formalism thus defies pictorial representation and aims directly at prediction of observations appearing under well-defined conditions" (Bohr [1956] 1998, 172). Bohr's interpretation of quantum mechanics rests on the same skeptical attitude toward scientific theories as representing our knowledge of the inner structure the world just as the pragmatist view of knowledge did.

Indeed, this pragmatic approach to scientific theories had consequences for Bohr's view of the quantum state. First of all, the wave function should not be interpreted in any pictorial sense, or rather *cannot* be so interpreted because it is a function in a multidimensional Hilbert space. The state vector has only a symbolic and calculative function. It doesn't represent anything; it doesn't represent some kind of novel "quantum reality" but "the entire formalism is to be considered as a tool for deriving predictions of definite and statistical character" (Bohr [1948] 1998, 144). One of Bohr's reasons for considering the wave function to be symbolic is that Schrödinger's wave function is defined in terms of imaginary numbers. Real numbers can be associated with the recording of measurement values in ordinary space and time, whereas imaginary numbers function only in abstract vector spaces and have no counterparts in measurement. However, Bohr had two more equally important reasons to regard the wave function as symbolic: its mathematical interpretation is formulated in an n-dimensional configuration space that usually contains more than three dimensions, and the deterministic evolution of the wave function is not limited by the speed of light.

Thus in Bohr's opinion there was no possibility for an interpretation of the quantum formalism that postulates an ontological structure based on mathematical features of the formalism. Even among more realist-minded philosophers the difficulties of turning a mathematical representation into an ontological description have been acknowledged. With a reference to Tim Maudlin, Mauro Dorato and Frederico Laudisa (2014) mention that "mathematical representations of physical phenomena are not a clear guide to ontology, since they often do not guarantee even isomorphic relations between themselves and the latter. Furthermore, for obvious algorithmic reasons they must greatly simplify and idealize the target they are a vehicle for, and so they are not necessarily similar to what they are supposed to denote." Maudlin (2013, 152–3) himself points out that "mathematical objects acquire algebraic and numerical properties that the physical objects do not have; there are purely gauge degrees of freedom in the mathematics." However, Bohr was certainly not a mathematical realist. He would deny that mathematics has any representational function by itself. We may, I would say, use mathematics to express our (possibly true) beliefs about the world; just as we can use ordinary language to express other (possibly true) beliefs about the world.

Mathematics, according to Bohr, does not express a certain kind of knowledge of reality; it is a language just like the natural language with its own syntactical rules. Both mathematics and our ordinary language are used for communication about our experience. This is revealed by the following remark: "Rather than a separate branch of knowledge, pure mathematics may be considered as a refinement of general language, supplementing it with appropriate tools to represent relations for which ordinary verbal expression is unprecise and cumbersome" (Bohr [1962] 1998, 190). The purpose of mathematics is the same as that of ordinary language but where the latter has evolved as a tool for smooth communication about our qualitatively grasped experiences, the former has developed as a tool for precise communication about quantitatively grasped experiences. Again, if this is a correct interpretation of Bohr's thoughts, we realize how close his ideas on mathematics were to a naturalistic and pragmatic point of view.

4. The measurement problem

In Bohr's view there is no measurement problem. Many physicists and philosophers claim there is a problem. But if you, like Bohr, do not take the wave function to designate an objective property of the system but to be a manual for thinking about the system, then the problem becomes a figment of the imagination. The claim that there is a problem simply presupposes that scientific realism is correct, and that scientific theories tell us how the world really is. So their accusation against Bohr and the Copenhagen interpretation builds more on a misunderstanding of the nature of scientific explanation than on substantial arguments.

This is how a realist may see the problem: The wave function in Schrödinger's equation represents the quantum state of an isolated system of one or more particles and the equation itself gives a deterministic description of the evolution of the wave function. A quantum state is supposed to consist of a superposition of eigenstates of a certain observables and when a measurement is performed on the system the superposition undergoes a "collapse" and we get a particular result corresponding to one of the possible eigenvalues of the observable. If you assume that the wave function represents something real, you would have to conclude that the mathematical reduction from a superposition to a single eigenstate represents a real physical process. But Bohr never talked about the wave function collapse since he didn't assume that the state vector plays a representational role. In fact he mentioned the wave function or the state function only in very few places. Nor did he believe that the quantum formalism gives us an expression for our knowledge. According to Bohr, the quantum formalism is the device by which we can communicate the statistical distribution of the outcomes of possible measurements based on actual measurements.

Since Bohr never spoke about the collapse of the wave function, and rarely mentioned the wave function or the state vector, but talked about the symbolic function of the quantum formalism, it cannot come as a surprise that he never addressed the alleged paradox of Schrödinger's cat. But we might ask, how would he have responded if somebody had pressed him?

Here is the situation: If we describe this entire system of the cat in the box in quantum mechanical terms, the cat is supposed to be, during the time in which the box is cut off from its environment, in a state of superposition between being dead and being alive. But when the box is opened the observer will see either a dead cat or a live cat because everything we believe about feline biology tells us that being alive and being dead are mutually exclusive. This fact is not because of the epistemic fact that no one has ever seen this, Bohr would say, but because ontologically such a state is physically impossible for a cat to be in. Since the wave function does not signify an objective property of the system, then the mathematical superposition does not represent a real physical state. Bohr would probably have added that it does not make much sense to apply quantum mechanics in order to describe biological states like "alive" or "dead" since neither of these terms is unambiguously defined outside biology.

The paradox, as Schrödinger saw it, is that if the cat is dead when the box is opened, or if the cat is alive when the box is opened, at what time did the original state of superposition collapse into a dead-cat mode or into a live-cat mode? Did it happen when the cat was poisoned or did it happen when the physicist opened the box and became aware of the state of the cat? Since Bohr explicitly denied that observation could affect the outcome of an experiment and explicitly states that the wave function is a nonrepresentational tool for calculation, the point of Schrödinger's thought experiment was lost on him. If the cat was dead, when somebody opens the box, it was already dead at the time the quantum trigger was pulled. It is just that simple. The poor cat would have acted as a classical measurement instrument and its state of being alive or dead would indicate whether or not the radioactive decay had occurred.

Ironically, Bohr was accused for inviting the observer into our understanding of quantum mechanics by philosophers like Karl Popper and Mario Bunge, but he had no such intentions. Once he said that it is the observer who always chooses the experimental setup, and therefore also which kind of observable that can be measured, but from this moment it is nature herself that determines which value the outcome will have. No human subject has any influence on the actual outcome. So for Bohr the measurement outcome was an objective fact due to a physical interaction between the object and the measuring instrument.

5. Realism and representationalism

Was Bohr really an instrumentalist? I once characterized Bohr as an objective anti-realist. (Faye 1991). By this expression I wanted to say that Bohr believed in the reality of atomic objects but abstained from holding a representational view of scientific theories. So, in my opinion, Bohr was a realist with respect to the existence of atomic objects, but an anti-realist with respect to the representational structure of quantum mechanics. But if the quantum mechanical formalism does not represent the world as it is, what then had Bohr in mind about its function? Well, it is not only an instrument for prediction; it is also an unambiguous means of communication about our experience concerning the quantum objects. The formalism provides us with the syntactical rules that determine the logical order and structures by which inferences and predictions become possible, whereas the physical interpretation of the formalism gives us the semantics of the mathematical symbols. However, such an interpretation is not due to free stipulations but is based on an already well-established linguistic practice which is formed by our common-sense experience.

The resemblance of Bohr's view on language to Ludwig Wittgenstein's theory of language as it is exposed in *Philosophical Investigations* is quite compelling. Earlier Edward MacKinnon (1985) has noticed a similar comparison. Wittgenstein accepted that the natural language contains many ambiguous expressions, which arise from the development of language. The way of handling ambiguities is to understand the context in which they appear. The meaning of words and sentences in a language is determined by how they are used, and in usage there are ambiguities. In one context an utterance can mean one thing, in another the same utterance can mean something else. The problem is that when we are presented with a word, we are not simultaneously presented with its application, and words appear to us as being ambiguous. Words can— and are—given to us as being something that can be used in a variety of ways, like a brick that can be used as a building block, a doorstop, or a murder weapon. However, it is the use in a context that assures us what the application of a word is meant to have. Thus, the context in which the word is used makes it less ambiguous by determining its meaning and therefore its reference.

Thus, Wittgenstein's theory of meaning is not challenged by ambiguities in ordinary language, because the ambiguities of ordinary language are necessarily parts of its *use*. But what about the language of physics? Wittgenstein ([1953] 2001) did not oppose the view that it is possible to construct an ideal language, such as mathematics, that consists only of propositions that are either true or false regardless of the context in order to avoid ambiguous expressions.

> We want to establish an order in our knowledge of the use of language: an order with a particular end in view; one out of many possible orders; not *the* order. To this end we shall constantly be giving prominence to distinctions which our ordinary forms of language easily make us overlook. This may make it look as if we saw it as our task to reform language.

Such a reform for particular practical purposes, an improvement in our terminology designed to prevent misunderstandings in practice, is perfectly possible. But these are not the cases we have to do with. The confusions which occupy us arise when language is like an engine idling, not when it is doing work. (§ 132; emphasis in the original)

According to Wittgenstein, such an ideal language is not impossible but this is not what philosophy should be occupied with: "the task of philosophy is not to create a new, ideal language, but to clarify the use of our language, the existing language" (Wittgenstein 1974, 115). This seems to be exactly what Bohr argued with respect to understanding the use of a classical terminology and the meaning of quantum mechanics.

Working as a philosopher the physicist should clarify the ambiguities of the existing language of physics in relation to the mathematical formalism per se. Mathematics is, according to Bohr, a language by which we can describe our experience unambiguously. He sometimes talked about mathematics as a tool of description. Like the natural language it does not apply to anything unless it is put into a definite use. The language of physics is exactly a formal language in the Wittgensteinian sense that has been developed to avoid those ambiguities we find in the use of the natural language. The use of mathematics gives us the possibility of an objective description. But the use of mathematics must be empirically relevant; it has to be given a physical interpretation in terms of what we can see and measure. "From a logical standpoint, we can by an objective description only understand a communication of experience to others by means of a language which does not admit ambiguity as regards the perception of such communications" (Bohr [1953] 1998, 156–7). Thus, the use of mathematics in classical physics provides the physicists with the possibility of clear and unambiguous descriptions but in order for such descriptions to be empirically relevant their application must be grounded in our experience. As Bohr continued, "In classical physics, this goal was secured by the circumstance that, apart from unessential conventions of terminology, the description is based on pictures and ideas embodied in common language, adapted to our orientation in daily-life events."

However, Bohr thought that the discovery of the quantum of action renders the language of classical physics ambiguous when applied to observations in quantum mechanics due to the fact that it is no longer able to recover a visual description of the events: "The exploration of new fields of physical experience has, however, revealed unsuspected limitations of such approach and has demanded a radical revision of the foundation for the unambiguous application of our most elementary concepts, like space and time, and cause and effect" (Bohr [1953] 1998, 157). The way to restore an unambiguous communication, which at the same time is empirically relevant, is to tie the application of the words of classical physics to what we actually can observe in the context of particular experiments.

Indeed Bohr does not deny that there exists an external world independently of our description, and we can describe what we perceive so our assumptions represent these external facts. But it just is not enough to establish a physical interpretation of any formalism. The meaning emerges from the linguistic practice of physics by handling experiments while enabling us to tell others what we have done. Communication and rational thought are only possible between people who share a common ground, and the common ground is the linguistic practice which physicists eventually have articulated by using classical concepts in response to their perceptual experience. This is why Bohr ([1929] 1934, 16) maintained that "it would be a misconception to believe that the difficulties of the atomic theory may be evaded by eventually replacing the concepts of classical physics by new conceptual forms."

In his posthumous book *On Certainty* Wittgenstein argued that such a linguistic practice is possible because certain basic propositions, even though they seem to have an empirical content, specify a certain conceptual framework within which empirical statements can make sense. This analysis catches out quite well some of the same insights Bohr had when he insisted on the classical description of physical experiments. A common empirical statement is normally meaningful only within a given context that, inter alia, consists of certain background assumptions that support this statement. Thus our general practice of empirical study and our explanatory justification of empirical evidence produce assumptions which we in a given situation hold for true, and which help us to justify further assumptions. But certain assumptions, says Wittgenstein, underlie our practice itself and their certainty is not derived from a singular practice of investigation and justification, but from those propositions that are embedded as fixed hinges in every possible context in which we use language to express empirical knowledge. We cannot imagine situations where there is room for doubt about their truth as a possible alternative while we act rationally. These kinds of propositions are not unquestionable by virtue of their truth. In other words, their necessity for any investigation and reasoning builds not on their truth, but on that they act as a kind of epistemic hinges that allow such practices.

Bohr contested any pictorial interpretation of quantum mechanics by calling it a symbolic or an abstract representation. Much like Wittgenstein, but quite independently of him, Bohr believed that the physical meaning of the mathematical formalism is determined by its use in the description of observations in particular contexts. A mathematical formalism cannot be given an independent interpretation apart from its use in connection with our experience of physical objects. The application of the abstract formalism to something in the world cannot be reached by conventional rules of interpretation or by an act of baptism. Rather, it is important for us to be able to report to others what is happening, or has happened, at a given situation, in order to convey a certain intention, or to reach a certain goal. If a physicist wants to inform another physicist about the results of a certain prediction in comparison with the outcomes of an experiment, the only way to make certain that the same terms are used "in a clear and unambiguous way" is to rely on the observational context in which these terms have an already settled meaning.

I think such an account explains why Bohr was not a representationalist with respect to the theory of quantum mechanics. Both the natural language and the mathematics of physics get their meaning by the way we use particular words or utterances, phrased in these languages, to express our experience outside and inside physics. Representationalists assume that it makes sense to interpret a language by stipulating certain conventional rules that connect words to the world, even though we have no empirical access to the object to which the terms allegedly refer. But if meaning of our terms lies in the act of communication, as both Bohr and Wittgenstein seemed to argue, then there are no such link between language, mathematics, and the world. Rather, in daily life as well as in physics, the connection between language and the world goes through bodily interactions and experiments. The experiment sets the conditions for what we can experience and therefore also the conditions for the use of those classical concepts in quantum mechanics that can be regarded as the explication of innate concepts emerging from the human adaptation to the physical environment.

Conclusion

Biological evolution has brought cognitive powers to Homo sapiens that we do not find in any other animal. We may find sensory modalities in other animals, which in every sense are more sensitive or extend further than similar modalities in human beings. Some animals see better, hear better, and smell better. They are also able to sense things like magnetic lines of force and sonar signals for which we have no sensory modalities at all. But no other species has gained a reflective power of reasoning like human beings. Abstract thinking and the use of language and mathematics go hand in hand. For that reason we alone have been able to create social structures and acquire knowledge of things that are not present before our eyes. We naturally believe in the existence of those things we can see with the naked eye. Of course those beliefs may be mistaken, if, for example, we are hallucinating or being tricked by a master illusionist. In dreaming we believe in the reality of the things we see as long as we are "in" the dream; when we awake, we realize any beliefs we formed while dreaming are deceptive. As long as we have experimental evidence of things we cannot directly see, we feel justified in believing that these invisible things exist, especially if we can use them, while taking them to be and behave in certain ways, to construct new artifacts or improve existing technologies. But from an evolutionary perspective it is much trickier to argue convincingly that by virtue of scientific theories, which contain a lot of abstract mathematics without any empirical grounding, we possess a capacity of representing the world as it really is. If the only evidence we have for justifying a representational view of reality builds on a nonrepresentational presentation of reality, namely, our actions and sensory experience, it seems to be a mere dogma that mathematical theories as such can be interpreted realistically.

I think some of these reflections were the philosophical assumptions behind Bohr's interpretation of quantum mechanics. His view was based on a naturalistic and a pragmatic approach to science. In the light of Darwin's discovery of human evolution, I think that Bohr's interpretation emerges as one of the most plausible interpretations today. It represents an interpretation that is most consistent with what we know about human origins and the evolution of our cognitive capabilities. By nature we are born realists. There is a world outside of ourselves, and it is more or less as we experience it. This kind of common-sense realism is included as a part of our innate cognitive understanding of the world. But it is the same instinct that drives many physicists to interpret their theories realistically. The disease which infects those physicists is not realism as such but representationalism, the view that "knowing" something means being able to "picture" what something looks like when nobody is looking at it, a canvas of reality painted by a ghost spectator. The realist instinct explains quite naturally their realistic tendencies. Yet, there are good reasons to believe that their realistic interpretation causes them to postulate a reality that we are completely unable to have knowledge about.

The reality that these realists have in mind looks more reminiscent of God's vision of the world than our own point of view. Physics is, to use an expression of Kant, unable to describe the world as it is in itself, but only as it appears to us. This was also Bohr's view. Anything else would mean that we are able to "bootstrap" human cognition beyond its evolutionary origins and describe reality independently of our epistemic situation. Naturally, such an epistemic goal is unfeasible. We are part of reality. We cannot describe the world from a "nowhere." It must always be done from a "somewhere." As being part of reality we must describe reality in virtue of the concepts by which our biology has adapted us to understand it.

Note

1 I would like to thank Kristian Camilleri for having directed my attention to these remarks.

References

Bell, J. S. 1988. *Speakable and Unspeakable in Quantum Mechanics*. Cambridge: Cambridge University Press.

Bohr, N. [1929] 1934. "Introduction." In *Atomic Theory and the Description of Nature*. Cambridge: Cambridge University Press.

Bohr, N. [1938] 1958. "Natural philosophy and human cultures." In *Atomic Physics and Human Knowledge*, N. Bohr. New York: J. Wiley and Sons.

Bohr, N. [1948] 1998. "On the notion of causality and complementarity." In *Niels Bohr's Philosophical Writings Volume IV, Causality and Complementarity*, edited by J. Faye and H. J. Folse. Woodbridge, CT: Ox Bow Press.

Bohr, N. [1951] 1998. "Medical research and natural philosophy." In *Niels Bohr's Philosophical Writings Volume IV, Causality and Complementarity*, edited by J. Faye and H. J. Folse. Woodbridge, CT: Ox Bow Press.

Bohr, N. [1953] 1998. "Physical sciences and the study of religion." In *Niels Bohr's Philosophical Writings Volume IV, Causality and Complementarity*, edited by J. Faye and H. J. Folse. Woodbridge, CT: Ox Bow Press.

Bohr, N. [1954] 1958. "The unity of knowledge." In *Atomic Physics and Human Knowledge*, N. Bohr. New York: J. Wiley and Sons.

Bohr, N. [1956] 1998. "Physical science and man's position." In *Niels Bohr's Philosophical Writings Volume IV, Causality and Complementarity*, edited by J. Faye and H. Folse. Woodbridge, CT: Ox Bow Press.

Bohr, N. [1958] 1963. "Quantum physics and philosophy—causality and complementarity." In N. Bohr *Essays 1958–1962 on Atomic Physics and Human Knowledge*. New York: J. Wiley and Sons.

Bohr, N. [1962] 1998. "Address delivered at the Second International Germanist Congress." In *Niels Bohr's Philosophical Writings Volume IV, Causality and Complementarity*, edited by J. Faye and H. J. Folse. Woodbridge, CT: Ox Bow Press.

Dorato, M., and F. Laudisa. 2014. "Realism and instrumentalism about the wave function. How should we choose?" In *Protective Measurements and Quantum Reality: Toward a New Understanding of Quantum Mechanics*, edited by G. Shan. Cambridge: Cambridge University Press.

Faye, J. 1991. *Niels Bohr: His Heritage and Legacy. An Anti-realist View of Quantum Mechanics*. Dordrecht: Kluwer Academic Publishers.

MacKinnon, E. 1985. "Bohr on the foundations of quantum mechanics." In *Niels Bohr: A Centenary Volume*, edited by A. P. French and P. Kennedy, 101–20. Cambridge, MA: Harvard University Press.

Maudlin, T. 2013. "The nature of the quantum state." In *The Wave Function. Essays on the Metaphysics of Quantum Mechanics*, edited by A. Ney and D. Z. Albert. Oxford: Oxford University Press.

Murdoch, D. 1987. *Niels Bohr's Philosophy of Physics*. Cambridge: Cambridge University Press.

Rosenfeld, L. [1961] 1979. "Foundations of quantum theory and complementarity." In *Selected Papers of Léon Rosenfeld*, edited by R. S. Cohen and J. J. Stachel, 503–16. Dordrecht: D. Reidel Publishing Company.

Wittgenstein, L. [1953] 2001. *Philosophical Investigation*. Oxford: Blackwell.

Wittgenstein, L. 1974. *Philosophical Grammar*. Oxford. Blackwell.

6

Bohr's Relational Holism and the Classical-Quantum Interaction

Mauro Dorato

1. Introduction: A conflict in Bohr's philosophy?

Bohr's philosophy of quantum mechanics has often been charged for what is alleg-edly one of its major shortcomings, namely, the advocacy of an unambiguous classical/quantum distinction (let me refer to this view with the label "the distinction thesis"). As is well known, such a distinction is needed to defend Bohr's view that any commu-nicable measurement outcome must presuppose a classically describable instrument, with respect to which any reference to the quantum of action can be neglected (Bohr 1958, 4). Critics have then often insisted on the fact that the distinction in question is hopelessly vague (Bell 1987, 29) or at least strongly contextual (Ghirardi 2004), so that Bohr's interpretation of quantum mechanics suffers from the same vagueness and adhoc-ness. The resulting problem is, allegedly, a renunciation to describe the dynami-cal interaction between system and apparatus in a physically precise, theoretically based, and noncontextual way, and therefore to offer a much-needed solution to the measurement problem.

In my chapter I will present and critically discuss the main strategies that Bohr used and could have used to defend from this charge his interpretation of quantum mechan-ics. In particular, in the first part, I will reassess the main arguments that Bohr used to advocate the indispensability of a classical framework to refer to quantum phenomena by trying to look at them from a new angle. I will then go on to discussing the nature of the indispensable *link* between classical measurement apparatuses and observed system that he also advocated. Typically, this link has been interpreted as a mere neo-positivistic appeal to the fact that it is *meaningless* to talk about state-dependent prop-erties of quantum entities independently of a measurement apparatus (Redhead 1987, 49–51; and Beller and Fine 1994). On the contrary, other authoritative scholars have rejected this minimalistic reading by stressing the fact that Bohr's view implies the presence of a holistic nonseparability between quantum system and classical appara-tuses (Bohm 1951; Folse 1985; Faye 1991; Whitaker 2004, 1324). Howard (2004) and Tanona (2004, 691) explicitly interpret Bohr's relational view of a measurement by invoking the notion of *entanglement*. And it is clear that in this latter case, Bohr should have offered an explicit and well-articulated theory of measurement, a challenge that has been accepted among others also by Zinkernagel (2016) who, by relying on Landau and Lifshitz's (1981, 2–26) brief treatment, insisted that the quantum system interacts with only a *part* of a classical apparatus. In order to evaluate this discussion and give due emphasis to Bohr's holistic understanding of a quantum "phenomenon" (see Bohr 1958 among other sources), it is important to distinguish among different senses of Bohr's "holism" and consider which of these can be reconciled with "the distinction thesis" that he also explicitly advocated.

Let me add at the outset that in the following I will *not* try to argue that Bohr's interpretation of quantum mechanics is free from all conceptual difficulties. Nevertheless, I will try to correct some frequent misunderstandings that many of his critics have fallen prey to. After the 1935 confrontation with Einstein, it is well known that philosophers and physicists attributed to Bohr a definite victory over Einstein's criticism. But since the late 1960s surge of interest in the foundations of physics caused by Bell's theorem and his sympathy for alternative formulations of quantum mechanics (Bohmian mechanics and dynamical reduction models), Bohr has become been regarded as responsible—and not just by philosophers—for having "brainwashed a whole generation of physicists into thinking that the job was done 50 years ago" (Gell-Mann 1976, 29). It is about time to achieve a more balanced picture of Bohr's contribution to the philosophy of quantum mechanics. The conclusion that Bohr's interpretation of the formalism is untenable can *only* be established by giving his arguments as much force as possible, which is what I will try to do in the following by remaining as faithful as possible to his published work.

2. Bohr's recourse to classical concepts

Various misunderstandings of Bohr's (1949, 233) philosophy of quantum mechanics have certainly been favored by the obscurities of his prose that he himself later acknowledged,[1] as well as by overly polemical remarks due to some of his brilliant opponents. Here is one example:

> Rather than being disturbed by the ambiguity in principle ... Bohr seemed to take satisfaction in it. He seemed to revel in contradictions, for example between "wave" and "particle," that seem to appear in any attempt to go beyond the pragmatic level. Not to resolve these contradictions and ambiguities, but rather to reconcile us to them, he put forward a philosophy, which he called "complementarity." (Bell 1987, 189)

A possible source of an important "contradiction" that to my knowledge has not received sufficient attention is given by the conflict between the sharp separation thesis and Bohr's holism, that is, his defense of the inseparability of the quantum systems from classical describable measurement apparatuses. As evidence for the former thesis, consider the following quotation: "The essentially new feature in the analysis of quantum phenomena is, however, the introduction of a fundamental distinction between the measuring apparatus and the objects under investigation" (Bohr 1963, 3–4). Even though, as we will see, Bohr does *not* identify "the quantum" with "objects under investigation," it is well known that for him the "measuring apparatus" must be described with the language of classical physics, while the object of investigation is typically an atomic particle, not another classical object. Evidence for Bohr's full endorsement of the holistic nature of any measurement interaction (the latter thesis) is provided a little further down the same text: "While, within the scope of classical physics, the interaction between object and apparatus can be neglected or, if necessary, compensated for, in quantum physics this interaction thus forms an inseparable part of the phenomenon" (4).

This problem is in part a side effect of a perduring lack of consensus on what his use of "classical physics" really amounts to. In order to make some progress on this question, it is important to distinguish among at least six related reasons that motivated Bohr's reliance on classical concepts. Even though some of these motivations have been already amply investigated, the perspective that I am stressing here is hopefully able to shed some additional light on each of them as well as on their logical relationship. My conciseness in this section of the chapter can be justified in view of what comes next. Schematically, classical physics is needed

1. To defend the intersubjective validity of physical knowledge.
2. To provide a solid epistemic evidence for, and assign meaning to, claims about non-directly observable but *mind-independent*, real physical entities, that are codified in an otherwise abstract symbolism.
3. By using the Principle of Correspondence to defend the continuity of quantum theory with classical physics in a period characterized by radical scientific revolutions.
4. To be able to rely on a physical realm to which Heisenberg's Indeterminacy Principle does *not* apply.
5. Following Einstein's (1954) distinction,[2] to formulate quantum mechanics as a *theory of principle* based on classical physics, and not as a constructive theory, in the same sense in which the special theory of relativity was formulated as a theory of principles (Bohr 1958, 5).[3] This was meant to guarantee the completeness of the theory on the basis of the practical impossibility to use the wave function to describe classical apparatuses and atomic particles.
6. To replace states of superpositions between object and apparatus by mixed states in which both have classical, definite properties, to be specified as a function of the particular measurement setup (Howard 1994).

Let me expand in some more details on the five points above, with the exception of 6, which will be developed in the last section of the chapter.

1. Bohr's insistence on the indispensability of classical physics is first and foremost justified by the need of communicating in a *nonambiguous way* the results of scientific research and therefore, in particular, of experiments. Such a nonambiguity, on which Bohr insists so much, is guaranteed by our reliance on a *shared language*. This point is of crucial importance and it should not be downplayed.[4] As I understand it, "shared

language" in Bohr (1958, 3) is a technical term, since it refers to *plain, ordinary language, supplemented and enriched by physical concepts derived from classical physics.* The role of a shared language is explicitly motivated by Bohr's vigorous and explicit refusal of a solipsistic or subjectivistic interpretation of quantum physics that we would lapse into if this theory implied some sort of relativity of experimental outcomes to the observation of *single* individuals or to mental observations of human beings in general.[5] In a word, given the indispensability of a shared language in the report of holistic quantum phenomena, the consequent elimination of any reference to subjective mental states is blocked. The fact that language is shared necessarily leads to the view that a measurement is an "irreversible registration," as Wheeler (1998, 337–8) often put it *a propos of* Bohr.[6]

2. The necessity of relying on a shared language takes us to the strictly related second point, *evidence.* This second point does not just express the trivial fact that any piece of evidence in favor of physical theories, quantum mechanics included, depends on empirical reports expressed by a shared language in technical sense above. More importantly, there are at least *two* remarks that need to be stressed:

2.1 The first relates to Faye's convincing suggestion that Bohr was an *entity realist* (Faye 1991).[7] The process of irreversible amplification due to the interaction between quantum systems and apparatuses described by a shared language is in fact essentially *causal.* As a result of an inference to the best explanation, it follows that the macroscopic effects observed in any classically describable experimental device justify our belief in the reality of quantum entities regarded as causes of what we observe in the laboratories (Hacking 1983).

2.2 In order to qualify what I stated in point 1 above, it is necessary to elaborate on what can be found in Bohr's work. His belief in the unrevisability of the description of the world afforded by the "shared language" vis-à-vis changing scientific worldviews does *not* entail that the new physical theories proposed in the twentieth century do *not* put stress on the categories of ordinary language as they have been refined by classical physics (space, time, particle, wave etc.). It only entails that any imaginable evidential link between physical hypotheses and experiments presupposes that the so-called "manifest image of the world" (Sellars 1962)—as it is expressed in part by our reliance on a shared language—cannot provide a *completely illusory* representation of the world around us. *If we had to accept this conclusion, we would have no reason to believe in physics to begin with.* In fact, our belief in a physical realm that goes beyond our unaided senses ultimately depends on evidence coming from our senses plus the referential power of "the shared language," which also entails our belief in the reality of causal relations.

By using a distinction coming from an apparently unrelated philosophical corner, we could say that Bohr is not a *revisionist* philosopher of physics but rather a *descriptivist* one (for this important distinction, see Strawson [1959]).[8] Strawson's main idea, that Bohr would have certainly endorsed, can be thus summarized: "there is a shared and universal conceptual scheme which we human beings have, and know that we have, and for which no *justification* in terms of more fundamental concepts or claims can be given" (Snowdon 2009, 32; my emphasis). On the contrary, *whenever there is some free room for different interpretations of physical theories,* revisionist philosophers favor worldviews in which the main tenets of the manifest image sense are overthrown. An example of a revisionist metaphysics is configuration space realism (see papers in Ney and Albert 2013).[9]

The task of the philosopher of physics for Bohr is rather to describe the way in which the new quantum physics categorizes the world in interaction with the categories provided by the "shared language." To put it in a nutshell by using other words, Bohr is against "quantum fundamentalism" (Rugh and Zinkernagel 2011; Zinkernagel 2016), which, in my way of putting it, is the view that quantum physics grounds conceptually and epistemically—but not ontologically, given Bohr's entity realism—classical physics.

3. In his search for the new quantum mechanics, Bohr wanted it to be the case that it implied some kind of limitation but not a complete overthrow of the classical concepts. Consequently, according to him quantum theory had to be regarded as a *rational generalization* of classical physics that applies to physical phenomena in which h is non-negligible (Bokulich and Bokulich 2004; Faye 2014). In the same sense, general relativity is a generalization of Newtonian gravitation that applies when gravitational fields are strong, and special relativity is a generalization of Newtonian mechanics that applies when the relative velocity of bodies approaches c. Likewise, as is well-known, general relativity, with its manifolds with variable curvature, generalizes special relativity, which is valid in infinitesimal regions of a Riemannian manifold, which can be considered to all effects Euclidean and therefore flat.

It is sometimes argued that Bohr's philosophy of quantum mechanics was mainly inspired by Einstein's operationalist analysis of the concept of simultaneity. This claim, if correct, would make Bohr into a full-blown neopositivist (see below). While the influence of Einstein's verificationist analysis on the physicists of Bohr's generation and on Bohr can hardly be overstated, reference to the 1905 paper in discussions about the epistemic status of quantum mechanics is much more frequent in Heisenberg and in Born (1971). Despite their different views about the theory, Bohr agreed wholeheartedly with Einstein's (1920, 102) methodological approach to the construal of a new physical theory, which insisted on the importance to recover the previous theory from the successive one: "No fairer destiny could be allotted to any physical theory, than that it should of itself point out the way to the introduction of a more comprehensive theory, in which it lives on as a limiting case." In our case, Einstein's "pointing out" refers to the idea that any physical theory has in itself the germs for a generalization: it is in this sense that classical physics was for both Einstein and Bohr an indispensable heuristic means. The validity and truth of classical physics is only *restricted* to some values of the variables intervening in the physical laws and not to all. For Bohr, the fact that classical mechanics cannot be extended to all possible values, as was thought beforehand, did not mean that classical physics had to be abandoned, since the very concepts of "generalization" and "limiting case" imply—against unrealistic philosophies of science based on incommensurability (Faye 2014) or pessimistic meta-induction—a full commensurability between the quantum and the classical epistemic and ontic domain. In a word, the epistemic reliability of classical physics was the main justification for an extension to the quantum realm via the rational generalization principle.

Furthermore, "mixed treatments" of quantum domains with classical theories[10] that were fundamental in establishing the new mechanics (Bohr's 1913 model of the atom) already implied that the latter had to be regarded in any case as approximately true, otherwise it would have been heuristically useless in predicting new phenomena: "The problem with which physicists were confronted was therefore to develop a *rational generalization* of classical physics, which would permit the harmonious incorporation of the quantum of action" (Bohr 1963, 2; my emphasis). We should realize that the importance of correspondence rules was particularly felt in a period in which, given the revolutionary changes happening in physics, a principle of continuity had to be reaffirmed in order to guarantee the rationality of scientific progress.

4. The epistemic, foundational role of classical physics depends also on the fact that a classical measurement apparatus is not subject to Heisenberg's indeterminacy relations. Unlike atomic objects, a classical object has a well-defined position in space-time and *at the same time* obeys the causal laws of classical mechanics (conservation of momentum). According to the distinction thesis, complementary magnitudes of the apparatus are always simultaneously sharp. However, as a consequence of Bohr's complementarity principle, the experimental arrangements suitable for locating an atomic object in space and time, and those that are fit for a determination of momentum-energy are mutually exclusive. Consequently, we cannot use *the same apparatus* to measure complementary magnitudes. Clearly, if an apparatus did not have a precise position before and during measurement (as confirmed by our senses), we could not measure the precise location where an electron hits a photographic plate. At the same time, the fact that, as we can see, a position-measurement apparatus always has *also* a definite momentum, that is, it does not move, is what makes the measurement of the position of the particles possible. However, if the diaphragm suspended with springs did not have a precise position and momentum before, during, and after a measurement, we could not use the law of conservation of momentum to calculate the momentum of the particle that goes through it. In the next section, we will see whether Bohr's response to Einstein's thought experiment during the Solvay conference is not a withdrawal of the distinction thesis, since Bohr applies Heisenberg's relations also to the macroscopic, classical screen suspended on springs, thereby *apparently* abandoning this thesis.[11] In fact this is an aspect of Bohr's holism.

5. In order to provide a deeper understanding of Bohr's reliance on classical physics, it is illuminating to recall Einstein's (1954, 227–32) distinction between *principle* physical theories and *constructive* physical theories and suggest that Bohr conceived of quantum mechanics as a *principle* theory—as Einstein did with special relativity—and not as a *constructive* theory (see also Bub 2000). The distinction is best explained with Einstein's (1954, 228) own words: "constructive theories attempt to build up a picture of the more complex phenomena out of the materials of a relatively simple formal scheme from which they start out." A constructive theory for Einstein is the kinetic theory of gases, which explains thermal, macroscopic processes in terms of underlying molecular motions. On the contrary, "the elements which form their basis and starting-point [i.e., of principle theories] are not hypothetically constructed but empirically discovered ones, general characteristics of natural processes, principles that give rise to mathematically formulated criteria which the separate processes or

the theoretical representations of them have to satisfy" (228). His example of a princi-ple theory is *thermodynamics*, which "seeks by analytical means to deduce necessary conditions, which separate events have to satisfy, from the universally experienced fact that perpetual motion is impossible."

Einstein regarded special relativity as a principle theory based on two very general postulates, namely, the principle of relativity and the independence of the speed of light from the motion of the source. These postulates act as meta-laws (Lange 2013a, b), that is, as very general constraints that all separate processes ("separate events") governed by all lower-level physical laws must obey. As plausibly argued by Bub (2000), in Bohr's philosophy of quantum mechanics the two general principles are the quantum pos-tulate and the principle of correspondence, whose role is analogous to Einstein's two postulates. Qua axioms of the theory, they exclude any attempt to further explain them by constructing them causally or dynamically, in the precise sense that Einstein gave to "constructions." Of course, the two postulates of quantum mechanics do *not* function as meta-laws in exactly the same sense as the principle of relativity. Yet, when they were formulated, they fixed structural constraints about how to calculate transition prob-abilities (78). Furthermore, as argued in a later paper written together with Clifton and Halvorson, by changing somewhat Bohr's two principles in order to introduce three information-theoretic axioms, it is possible to make the analogy with Einstein's "theory of theory" approach even stronger. These three principles in fact "constrain the law-like behavior of physical systems" (Clifton, Bub, and Halvorson 2003).

Even though Bohr was very probably unaware of the *Times* 1919 article republished in (1954), it is rather striking that—in order to justify his philosophy of quantum mechanics—he refers explicitly to Einstein's purely *kinematic* foundations of the spe-cial theory of relativity, which, in Einstein's intention, was called for by the shaky foun-dations of electrodynamics, torn at that time between Maxwell's "continuity" approach implied by the use of differential equations and Einstein's own *Lichtquantum* hypoth-esis. In his 1949 piece written in honor of Einstein, Bohr wrote:

> Notwithstanding all differences between the physical problems which have given rise to the development of relativity theory and quantum theory, respectively, a comparison of purely logical aspects of relativistic and complementary argumen-tation reveals striking similarities *as regards the renunciation of the absolute signifi-cance of conventional physical attributes of objects.* Also, *the neglect of the atomic constitution of the measuring instruments themselves, in the account of actual expe-rience, is equally characteristic of the applications of relativity and quantum theory.* (Bohr 1949, 236; my emphasis)

Notice that in the first part of the italicized quotation, Bohr is directing our attention to the historical process of relativization of physical quantities that were previously regarded as absolute ("the renunciation of the absolute significance of conventional physical attributes of objects"). Exactly as in the special theory, spatial and tempo-ral intervals "taken by themselves" become relative to an inertial frame, in quantum mechanics it is the possession of definite physical properties that becomes relative to experimental situations.[12]

The second part of Bohr's passage pursues the analogy with relativity one step forward, since it makes explicit the nonconstructive character of special relativity. Unlike what happens in Lorentz's *dynamical* theory of the inner constitution of the rulers, contractions of classical objects and temporal dilations of clocks don't need to be explained bottom-up from a constructive theory, since they can be "explained" geometrically from the structural constraints of a four-dimensional space-time that codifies the two axioms of the theory. Contractions don't need to be explained from a dynamic theory because they are non-invariant (Dorato and Felline 2010; Dorato 2014).

In the same spirit, according to Bohr, the two axioms/principles from which quantum mechanics depend (quantum postulate and the principle of complementarity) render futile any attempt to *construct* classical objects out of underlying quantum entities with their laws. If classical objects could be explained bottom up by a theory T that fully reduced them to quantum properties, we would lack any evidence for T to begin with, since evidence must ultimately be expressed in classical language for the reasons already given.

There are however some differences between the epistemic and methodological foundations of the two theories of which Bohr was certainly aware: quantum physics is committed to a greater amount of "relationality" and holism than the special theory of relativity. In the latter theory there is an *invariant* that plays the crucial role: the four-dimensional metric with its light cone structure supersedes completely the relational nature of space and time "separately considered" (Minkowski 1952, 75).

In Bohr's philosophy of physics, the relativistic invariant metric might be matched by the indispensability of a description provided by classical physics (the "experimental invariant"). This analogy, however, does not extend very far: the necessary presence of a classical experimental setup does not avoid complementary descriptions: apparatuses capable of revealing the wave-like feature of an atomic object are incompatible with setups that are necessary to reveal its particle-like property. Consequently, the relational nature of a quantum property is by all means much more relevant than a classical one. Despite the fact that any experiment takes place in a particular frame, the fact that an observer O in frame F measures length L in relativity holds for all possible observers.

In any case, the principle/constructive distinction has a double function. On the one hand, it illustrates in a clear way that Bohr drew inspiration from Einstein's early methodological attitude. On the other hand, it is also very useful to account for their later divergence about the completeness of quantum mechanics, a fact that Bub (2000) and other scholars have neglected. In 1905 Einstein is *provisionally* construing special relativity as a principle theory because of the poorly understood, dual nature of the underlying electromagnetic components of rulers and rigid bodies (Brown 2005, 72–3). He was convinced that in physics one should first proceed with principle theories as the case of thermodynamics and only later attempt to hypothesize constructive theories, which provide a deeper understanding of physical phenomena (Howard 2015, section 6). To the extent that Einstein regarded quantum mechanics as a theory of principle that unified a lot of empirical phenomena without offering a deeper comprehension,[13] we can understand his distance from Bohr, who thought that no deeper explanation was forthcoming because no explanation was needed.

In fairness to Bohr (1949, 223; my emphasis), the fact that he regarded quantum mechanics as a principle theory did not entail the impossibility of any bottom-up explanation of the properties of classical bodies: "Although, of course, the existence of the quantum of action is ultimately responsible for the properties of the materials of which the measuring instruments are built and on which the functioning of the recording devices depends, this circumstance is *not relevant* for the problems of the adequacy and completeness of the quantum-mechanical description in its aspects here discussed." Against Zinkernagel (2015), we should note that passages like this seem to imply that the quantum realm is ontically more fundamental than the classical realm, especially in view of the fact that Bohr was an entity realist. Of course, as noted above, an ontic primacy intended in this sense is compatible with the epistemic primacy of the classical physics illustrated above.

The irrelevance in question is based on a very *general* fact concerning any kind of measurement in physics: the need for approximations and idealization that in the case of quantum mechanics is due to the scale of phenomena:

> The smallness of the quantum of action compared with the actions involved in usual experience, including the arranging and handling of physical apparatus, is as essential in atomic physics as is the enormous number of atoms composing the world in the general theory of relativity which, as often pointed out, demands that dimensions of apparatus for measuring angles can be made small compared with the radius of curvature of space. (Bohr 1949, 238)

In addition to what we have presented in this section, we might add that it is the scale of the phenomena to be measured that renders a classical description unavoidable.

3. Bohr's relational holism and the classical-quantum interaction

We can now return to the question that I introduced at the beginning of the chapter: is there a conflict between the distinction thesis and the holistic character with which Bohr characterized any quantum phenomenon? On the one hand, the previous section has presented the main reasons that drove Bohr to rely on the domain of classical physics as something *ontologically* distinct from that of quantum mechanics but necessary to gain empirical evidence about atomic objects. As his critics have insisted, however, this thesis, which for his approach is absolutely central, presupposes as a necessary condition the existence of a clear-cut, noncontextual criterion to distinguish the quantum from the classical realms.[14] On the other hand, however, Bohr (1949, 210) insists on the "*impossibility of any sharp separation between the behavior of atomic objects and the interaction with the measuring instruments which serve to define the conditions under which the phenomena appear*" (emphasis in the original).

A defender of Bohr's interpretation could remark that the contradiction here is only apparent, as it is based on an ambiguity that is merely semantical: "lack of sharp separation" is compatible with "distinctness." When Bohr talks about the "impossibility of a sharp *separation*" he is referring to a relational link between quantum objects and classical apparatus, or even more radically, to a form of holism or nonseparability that has been empirically proved after Bell's inequalities, and that EPR interpreted as a type of nonlocality that had to be avoided. Without some further assumption, the friends of Bohr insist, this notion of "nonsharp separation" is compatible with Bohr's assumption that the classical and the quantum domains possess radically incompatible properties and are therefore *ontically distinct*.

However, Bohr's enemies note, where and how can we draw a clear-cut, nonambiguous distinction between the two ontological levels if in the *physical* process of measurement these two levels *interact*? Suppose that the inseparability of the parts of a holistic "phenomenon" in Bohr's sense implies also for Bohr an entanglement of the atomic objects with a classical apparatus (see Howard 1994, 2004). If one assumes at the same time the linearity of the evolution equation and the *completeness* of the theory, as Bohr certainly did, one has then the problem of explaining the observed definiteness of the outcomes that are revealed by classical apparatuses. To avoid this problem, Zinkernagel follows Landau and Lifshitz's (1981) approach to the measurement problem by (i) denying the existence of an entanglement relation between the atomic entity and the *whole* classical apparatus, and by (ii) claiming that the former interacts only with a *part* of the latter (Zinkernagel 2016, 11). However, the problem represents itself also with this suggestion: how are we to distinguish the "part" of a classical object from the whole of it? How big must the part be for it to follow the laws of classical mechanics? Is there a precise, theory-derived criterion that enables us to answer these questions? At this point Bohr's foes could insist that the claim that the distinction between part of a classical object and the whole is contextual counts as a purely pragmatic solution to the problem. As such, isn't it a just a way to sweep the dust under the carpet?

In order to answer these important questions, we need to dig deeper into the meaning of "nonseparable" in Bohr's account of a "phenomenon," and consequently into his theory of measurement, no matter how weak it is. In fact, without such a clarification, it is not clear whether Bohr's "relational holism" is really compatible with the existence—on which his interpretation depends—of two clearly distinct descriptions of the world, one referring to the quantum and the other to the classical realm.

First of all, all interpreters seem to agree that by talking about a "*finite, uncontrollable interaction*" between system and apparatus, Bohr assigns to the interaction a *physical* meaning. The interaction may *cause* an increase of information, but by virtue of this very statement one has to assume that here we are dealing with a physical, nonfictitious cause (i.e., the interaction). There is ample evidence for this claim in his texts: in his famous reply to EPR's paper, for example, he discusses the consequences of measuring precisely the position of an atomic particle with a double-slit diaphragm and a photographical plate bolted to a common support.[15] The interaction of the electron with the slit is *finite* because of the discrete, indivisible nature of the quantum of action. The fact that it is *uncontrollable* means that in these experimental conditions it is impossible to come to know the momentum of the electron because it cannot be *manipulated* at will: the result of this impossibility is unpredictability and indeterminism. While Bohr is not always explicit about the nature of this uncontrollability, it is a consequence of a practical impossibility due to the smallness of the atomic particles, as well as to the fact that their dynamical behavior is analogous to the jump of an electron from an orbit to the other when hit by photons: in his response piece in the Schilpp volume, he refers in this respect to spontaneous emission processes and radioactive decays, attributing these examples to Einstein's theory of radiation published in 1917 (Bohr 1949, 202). All of this terminology shows that the measurement interaction for Bohr is a *physical, irreversible* process.

If this is to be taken for granted, there is still some important disagreement as to whether this interaction makes room for a genuine, even though qualified, "collapse" of the wave function (Zinkernagel 2016), or it rather only causes an increase of our information about the system (Faye 1991). In the latter case, the wave function is merely an instrumental, bookkeeping device, devoid of any ontological significance, while in the former the fact that the wave function symbolizes something, stands for something "out there," needs to be further discussed.

Bohr's *contextual* view of measurement does not help per se to solve this dispute. It is in fact also noncontroversial that the particular complementary property of an atomic entity revealed by an experiment *depends* on the particular setting that is appropriate to measure that property (either spatiotemporal properties or dynamical properties). However, there seem to be at least three senses of dependence that need to be disentangled, since they are logically independent: beyond the problem of establishing which of these senses (maybe more than one) is defended by Bohr, I am interested in asking which of them renders his whole philosophical approach least amenable to criticism.

The first sense of dependence is also the weakest, since it is compatible with assuming that also *before* a measurement any quantum system possesses a definite value of both position and momentum. These values however are unknowable in principle, due to the uncontrollable *disturbance* that the macroscopic apparatus causes in the measured system. Whether Bohr defended this position before 1935 is still controversial, but after EPR's he certainly rejected it (see also Faye 2014, 13).[16] After all, talking of "disturbance" implies that there is a definite property that existed before the measurement and that it is disturbed by it (Beller and Fine 1994), or at least it does not exclude it. It is not plausible to believe that Bohr did not see this logical possibility.

In other words, this first sense of dependence is weak because it holds also in classical physics: the result of an experiment on a physical system—the experimental answer—depends on the kind of question we ask and is in any case influenced by the apparatus. However, in the classical case it is *in principle* possible to discount the effect of the apparatus on the measured system, so that it is much more reasonable to assume that experiments reveal preexisting values of any physical magnitudes (even if it might be de facto rather difficult to determine them). As is well known, according to Bohr's complementarity principle the story is indeed quite different. Bohr never tires of insisting on the fact that, as a consequence of Heisenberg's principle of indeterminacy and the mathematical formalism, asking a "position-question" to an electron, and therefore providing a *spatiotemporal description*, precludes in principle asking a "momentum-question", and therefore giving a *causal description* of the process, and conversely. The two experimental settings are mutually exclusive, even though for an exhaustive description of an atomic object we need both properties.

The second, stronger sense of dependence, which is nearer to Bohr's position after 1935, suggests that there is *no* preexisting value that the apparatus measures and the coming into existence of a definite value is due to the measurement interaction. This is the sense of measurement that is assumed by EPR-type of arguments: since it is not possible that a measurement in a region of space causes an instantaneous coming into existence of the value of the distant particle, quantum theory is incomplete. Interestingly not only does John Bell—the staunchest opponent of Bohr—attribute him this second sense of "dependence," but he even agrees with him that this is the right view to take about measurement in quantum mechanics: "the word [measurement] very strongly suggest the ascertaining of some pre-existing property ... Quantum experiments are not just like that, *as we learnt especially from Bohr*. The results have to be regarded as the joint product of "system" and "apparatus," the complete experimental set-up" (quoted in Whitaker 1989, 180; my emphasis).

An obvious objection to the distinction between the two senses might come from neopositivist corners. For a neopositivist, in fact, there is no difference between the claim that there are definite magnitudes before measurement that cannot be measured due to a "disturbance" (first sense), and the claim that it is the measurement that causes the acquisition of a definite position or momentum previously undetermined (second sense). In both cases it would be meaningless to talk about magnitudes without a measurement, and in any case one cannot measure simultaneously position and momentum. Interestingly, *to the extent that* in Bohr's approach to quantum measurement there has been a transition from the first sense of dependence of measurement results from apparatuses to the second, there would be an additional reason not to count him as a neopositivist. Since without a verificationist approach one cannot move from the impossibility of simultaneous measurement to the impossibility of a common value, the fact that Bohr took so seriously EPR's argument is another piece of evidence that he was not a neopositivist. Conversely, as remarked by Whitaker (2004), the impossibility of common values conjoined with the projection postulate implies the impossibility of measuring complementary magnitudes.[17]

Given this understanding of "dependence on measurement," how did Bohr conceive of an atomic entity? If it were exclusively a particle or a wave, we could not explain the experiments: in the former case, it would not generate interference patterns; in the latter, it would not localize in the photographic plate. Likewise, for Bohr, it could not be both a particle *and* a wave in the sense of de Broglie's pilot wave theory, where the particles' velocities are fixed by guiding waves (Bacciagaluppi and Valentini, 2009, 28). More plausibly, for him it is neither a particle nor a wave until it interacts with a particular apparatus. One way to cash this out would be to argue that an atomic particle has *two dispositional properties* to be both a particle and a wave, depending on "the experimental question we ask," that is, in the dispositionalist language, on the particular *stimulus*. Certainly this position seems closer to Heisenberg's (1958) thought but it is interesting to go over the reason why Bohr did not consider it, independently of well-known personal enmity with the German physicist.

We saw that two properties are complementary if and only if they are *mutually exclusive* and *jointly exhaustive* (see Murdoch 1987). Kinematical and dynamical properties of atomic entities are *mutually exclusive* because they cannot be attributed to the same system at the same time via a single measurement, but they are also jointly exhaustive because in classical terms, they are both necessary for a complete knowledge of the dynamical properties of the entity.

However, from a *dispositionalistic* point of view, before the experiment we can attribute the *same* particle *two irreducible and intrinsic dispositions* (powers) for manifesting a particle-like behavior *and* for a wave-like behavior, depending on the experiment. When one disposition is inhibited, the other is selected as a result of the interaction with the chosen experimental setup, in such a way that they two can never be manifested at the same time by a single experiment. The presence of two dispositional properties is the reason why complementary descriptions about spatiotemporal trajectories and causal properties of the atomic particle cannot be used simultaneously. Analogously, for the trajectory of the particle (its position) and the interference pattern: if we know through which slit the particle enters, the wave-like disposition is inhibited by the experimental setup. Conversely, if we observe the manifestation of the wave-like disposition (the interference effect), we cannot know where the particle went because the particle-like disposition is inhibited. In both cases, the possession of both dispositional properties before the measurement interaction does not imply the possession of definite values for the respective magnitudes. If the possession of definite values is regarded as the categorical basis of the dispositional properties, the two dispositional properties are to be regarded, as they should, as irreducible to the categorical basis. In a word, an electron is neither a particle nor a wave, but the relevant dispositional properties coexist in the same entity.[18]

In order to make the reasons for Bohr's refusal of this interpretation explicit, it is sufficient to point out that the holistic link between the atomic particle and the apparatus stressed by Bohr involves the *manifestation* of the dispositions—*the measurement event*—and not the two dispositions *before* measurement. In this sense, the dispositional approach—if it is not a merely verbal redescription of the holistic interaction between system and apparatus[19]—would not help to clarify the nature of this interaction. Possibly, this approach might make this link even less perspicuous by suggesting that an introduction of dispositions that get selected by the classical apparatus can by itself solve the measurement problem (as suggested by Suárez 2004).

The third, more controversial sense of dependence calls into play the "nonseparability" or "indivisibility" of the measurement apparatus from the measured entity. It is important to remark at the outset that the fact that before measurement the measured system has no definite value (second sense of dependence) per se is compatible with the fact that before measurement quantum entities and classical instruments have an independent reality: the coming into existence of the definiteness of the conjugate variable depends on the apparatus, but the latter has properties that don't depend on the former. In other words, according to the second sense, the dependence is asymmetric.

This remark provides evidence for the fact that if indeed Bohr understood the "impossibility of any sharp separation between the behavior of atomic objects and the interaction with the measuring instruments which serve to define the conditions under which the phenomena appear" (Bohr 1949, 210) as a form of holism, he was referring to a third sense of dependence, which implies the other two but is not implied by them. As a matter of fact, in 1927, at the Como lecture, he wrote "the quantum postulate implies that any observation of atomic phenomena will involve an interaction with the agency of observation not to be neglected. Accordingly, an independent reality in the ordinary physical sense can neither be ascribed to the phenomena nor to the agencies of observation" (quoted in Zinkernagel 2016, 11). In a passage like this, it is clear that the dependence we are trying to clarify is *symmetric*: not only does the reality of the atomic phenomenon depend on the apparatus, but also the converse is the case. Before trying to understand the nature of this third sense of dependence in Bohr's philosophy, we should ask whether it is reasonable to attribute it to Bohr.

In some cases, admittedly Bohr's ([1956] 1998, 168) texts allow multiple interpretations: "quantum phenomena find logical expression in the circumstance that any attempt at a well-defined subdivision [between particle and apparatus] would require a change in the experimental arrangement that precludes the appearance of the phenomenon itself." Here the change may be regarded as compatible with an asymmetric form of dependence, and therefore with the first two senses of dependence. This is probably why the neopositivist reading has it that according to Bohr it is meaningless to attribute a definite property to a quantum system unless we specify a particular measurement context, which functions as a frame of reference in the special theory of relativity: "To regard Bohr as endorsing a nonlocal or nonseparable conception of reality strains his carefully tailored language of measurement and his picture of the operational presuppositions on physical magnitudes posed by conditions of measurement" (Beller and Fine 1994, 24).

This neopositivist interpretation neglects the fact that the need to have a "well-defined subdivision" that Bohr is referring to in the quotation above *presupposes* that there is something to be divided, which in its turn implies that instruments and atomic particle form a one, a whole. This whole, however, cannot be studied without introducing some separation, whether conventional or not: given the holistic nature of what Bohr (1985, 141) calls a phenomenon, "one must therefore cut out a partial system somewhere from the world, and one must make 'statements' or 'observations' just about this partial system."

Let us then take for granted that in Bohr's writings, ambiguous as they sometimes may be, there are important hints toward a stronger sense of holistic dependence such that, due to the measurement interaction, not only does the atomic particle lack a definite property, but also a part or the whole of the classical instrument does, because the description we give of it depends on the whole experimental condition. When we measure the exact momentum of an incoming particle by using a mobile screen suspended with springs (see Bohr 1949), we include *this* classical apparatus in the description of the holistic phenomenon in Bohr's sense. The fact that the mobile spring is subject to Heisenberg's relation implies that *it is treated as a quantum system*, independently of its size. In fact, as Howard has pointed out, the distinction between quantum and classical according to Bohr does not coincide with that between atomic system and apparatus. The fact that a measuring instrument can be described quantum mechanically implies that only some of its properties can be regarded as classical (i.e., those that are correlated, e.g., the momentum of the particle before measurement and the momentum of the spring after measurement; see Howard [1994, 214]), but this implies, in virtue of Heisenberg's uncertainty relations, that the position of the particle before and that of the mobile spring after measurement are entangled, because they both lack definite properties. For these reasons a measurement of the particle position in this particular situation is impossible.

Of course, talking of a third sense of dependence in terms of nonseparability or entanglement generates the problem of explaining our observations of a definite world within a theory that relies on a linear equation and is thought to be complete. However, since for Bohr there is clearly *no* room for a collapse of the wave function regarded as a physical process causally describable in space-time, how do we justify the transition from a dynamic indivisibility between system and apparatus to the possession of a definite quantity on the part of the former? Here I will focus only on two explanatory strategies. The first is to explain the transition from the quantum to the classical as the passage from an entangled pure state to a mixture (Howard 1994, 203). The second appeals to a view of measurement regarded as an irreversible physical process, due to Landau and Lifshitz (1981) and revived among others by Zinkernagel (2016).

1. As acknowledged by Howard himself, his thought-provoking suggestion to regard Bohr's holism as a form of entanglement between the two systems is not grounded in his texts. Independently of questions of hermeneutic faithfulness, which here can be in part neglected, according to Bohr the third sense of dependence (the measurement-induced nonseparability between a property of the atomic system and that of the classical apparatus) is never such that the two systems are jointly in a pure state. For Bohr there is *no* measurement problem exactly because according to him there is no situation in which *both* properties of the system and the apparatuses before, during, and after measurement are indeterminate, despite their symmetric dependence and despite the fact that for Bohr a classical object *can* be attributed a quantum state (as in the case of the screen suspended with movable springs).

There are at least two reasons for this claim. The first has been explored in various ways in the previous section: to summarize, in order to confer some meaning to our experimental practice, we must assume that at the end of the measurement interactions there is a classical object endowed with definite classical properties. The second reason stresses the idea that quantum mechanics is a theory of principle linked to some intrinsic, insurmountable limitations. If there existed an underlying causal description capable to explain bottom-up the definiteness of measurement outcomes, this description in any case could not be given in a spatiotemporal language. To the extent that our form of understanding is limited to the intuitions of phenomena in space and time as Kant had it, this description would not add to our comprehension of quantum theory. From this conclusion, however, the tension between Bohr's need of a condition of sharp distinction between the classical and the quantum and his belief in the nonseparate reality of atomic system and apparatuses—a nonseparability that does not prohibit treating a classical object as obeying Heisenberg's relations and therefore to quantum mechanics—has not been completely removed.

2. The second approach claims that Bohr's holism is to be interpreted as an entanglement between particles and, crucially, only *a part* of the classical, measuring apparatus (Zinkernagel 2016). The entanglement of the particle with a part of the classical apparatus calls into play Bohr's (1985, 141) appeal to the necessity of a "cut" between

entangled stuff: "one must therefore cut out a partial system somewhere from the world, and one must make 'statements' or 'observations' just about this partial system."

Zinkernagel draws inspiration from passages like these and, as we will see, from Landau and Lifshitz's (1981) treatment. He claims that the evolution of the wave function describes a real process in the world that cannot be given a purely epistemic meaning. At the same time, however, it cannot be subject to more fundamental explanations in terms of a constructive theory: "We can ... say that, for Bohr, the collapse is not physical in the sense of a physical wave (or something else) collapsing at a point. But it is a description—in fact the best or most complete description of something happening, namely the formation of a measurement record" (Zinkernagel 2016, 14). In other words, not only is the wave function an abstract object defined in the mathematical model, but it also refers to an irreversible process of amplification in the physical world. This real process is the outcome of the entanglement of the atomic system with the part of the classical apparatus that is involved in producing a well-defined record. This interpretation of the measurement interaction can at least in part be traced back to Bohr (1954, 73), with the obvious proviso that the term "entanglement" does not explicitly occur in his work. Measurement results are due to "suitable amplifications devices with irreversible functioning such as, for example, permanent marks on a photographic plate."[20]

However, this solution, despite its ingenuity, raises some difficulties: (i) given the existence of an entanglement, doesn't Bohr need some sort of collapse theory, even if limited to the dynamic interaction between some part of the apparatus and the atomic systems? And if he does not need such a theory, as Zinkernagel claims by following faithfully Bohr's texts, (ii) doesn't he still need a more precise account for the obtaining of a definite record in the part of the classic system that is entangled with the atomic particle? If these questions were left unanswered, typical objections concerning how to draw the boundary or a distinction between "a part" (which can enter in an entangled state) and a whole of a classical object (which cannot) would stick up their ugly heads again. It is at this stage that the notions of approximation and idealization might play an essential role.[21]

In order to rebut the well-known charge that Bohr does not justify in a noncontextual way the cut between the classical and the quantum it seems necessary to exploit the importance of position in any form of measurement that has been often signaled also by Bell: the classical limit involves the obtaining of a well-defined spatial trajectory in a bubble chamber, or as mentioned in the above quotation, "the formation of permanent marks." The answer to the question "how do we identify the part that is entangled with the system?" may then be provided by an appeal to an *approximation* of a quantum system with a classical system. So classicality is given by the existence of a well-defined trajectory. This is essentially Landau and Lifshitz's (1981) "Bohrian" account of measurement, which begins by noting that in order to describe the trajectories of atomic particles and attribute them dynamical properties, one needs apparatuses that can be regarded as classical "to a sufficient degree of accuracy" (2). The two Russian physicists note that such apparatuses need not be macroscopic, but the deliberate vagueness of "sufficient degree of accuracy" allows us to count as apparatuses also microscopic instruments, which, I take it, is the source of Zinkernagel's appeal to the entanglement between particles and "parts of a classical object." In our case, the formula "sufficient degree of accuracy" refers to the track left by ionized molecules in a Wilson chamber; as noted by Landau and Lifshitz, the track is classical to the extent that it is very large with respect to the electrons, so that the latter becomes a quasi-classical object. However, they continue, the track measures the electron path with a very low degree of accuracy, which is the necessary price to pay for having definite outcomes and well-defined paths. In conclusion, the readings of the apparatus are described by quasi-classical functions and the fact that the apparatus is classical means that "all of these possible values have always definite value" (22).

In order to comment on these neglected pages—to which, however, Zinkernagel makes explicit reference[22]—we should note that measures that are realized with a "sufficient degree of accuracy" are used throughout physics and not just in the context of quantum measurements. *The degree of accuracy cannot but depend on the aim at hand*, and it is for this reason that the boundary between classical and quantum must depend on the context. According to Bohr, this is compatible with the fact that, *for any given experimental context, the distinction between classical and quantum is always sharp and nonarbitrary.*[23] *Contextualism would imply vagueness only if we tried to distinguish the classical from the quantum across all experimental contexts.*

These italicized sentences go back to the main issue raised in this chapter: a nonarbitrary, nonmovable, distinction between classical and quantum is needed to cut definite outcomes out of an entangled mess (holism). However, contextualism and holism are called into play to take into account the principle of complementarity and the variability of the experimental setup. This alleged conflict is precisely the Achilles's heel that his enemies have been focused on. Is Bohr's contextualism really leading to a pernicious vagueness?

The conceptual situation that we are facing in this case is very similar to the famous sorite paradox. Even though the concept of "being bald" is vague—in certain cases we don't know how to distinguish between a bald person and a hairy one, and if we pluck one hair at a time from a hairy person, there is no moment at which she becomes bald—at the *extremes* of the spectrum there is still a clear-cut distinction between baldness and hairiness. Likewise in the quantum/classical case: this fact, for Bohr, is enough. If in any experiment we can find nonarbitrary borders at the extreme of the spectrum—say in terms of spatiotemporal paths—the distinction remains sharp and the vagueness of the borders in "intermediate" cases can be avoided.

In conclusion, whether the opposition between distinction and contextuality implies a violation of criteria of physical precision, as Bell has it, is of course dependent on one's own general philosophical intuitions of physics, or one's *stances* (van Frassen 2002). As such, as of now the problem cannot be settled only on the basis of philosophical or conceptual arguments. Until, of course, theoretical and experimental progress forces us to revise these stances. However, this experimental progress so far is lacking, and the alternative theories are not producing new predictions. At best, nonstandard interpretations or theories are based only on a superior explanatory power. This epistemic virtue, while so far not sufficient to consider them superior also on empirical grounds, is certainly sufficient to encourage their practitioners to further pursue new avenues: without Einstein's *philosophical* worry on nonlocality, we would not have quantum computing and quantum cryptography.

Notes

Thanks to the editors of the volume for their attentive reading of a previous draft of the chapter.

1 "Rereading these passages, I am deeply aware of the inefficiency of expression which must have made it very difficult to appreciate the trend of the argumentation aiming to bring out the essential ambiguity involved in a reference to physical attributes of objects when dealing with phenomena where no sharp distinction can be made between the behaviour of the objects themselves and their interaction with the measuring instruments." Here Bohr (1935) is referring to his reply to EPR.

2 In referring to this article, I will be quoting from the 1954 reprint.

3 This point has been already stressed by Bub (2000), but used to defend an anti-realist philosophy of quantum mechanics.

4 See Petersen (1968).

5 In some popular expositions we still read that quantum physics has vindicated the centrality of human beings in the physical world (due the role of the concept of observation).

6 As is well known, Wheeler spent quite some time with Bohr in Copenhagen, and is therefore a reliable guide to report faithfully his thought.

7 Faye's hypothesis that Bohr was an anti-realist about quantum theory is more controversial since, given that independently of other readings of Bohr (Zinkernagel 2016), it is not clear whether entity realism can be defended *without* theory realism.

8 The relevance for Bohr of an unrevisable conceptual scheme has been also stressed by MacKinnon (1982) and Murdoch (1987) and is of course at the basis of the Kantian readings of Bohr.

9 Of course, whether one can go revisionist or not does not depend only on one's philosophical taste, but is almost always strongly constrained by physical theories.

10 This is a common phenomenon in the history of twentieth-century physics: semiclassical quantum gravity mixes *quantum* field theory with *classical* general relativity, and is regarded as an approximation to a full theory of quantum gravity.

11 The localizability of macro-objects in space-time is important for an additional reason, which is linked to the fact that space and time are *principia individuationis*: the localization of macroscopic objects in space is sufficient to ascribe them an identity. On the contrary, quantum particles are indistinguishable and come to possess an identity only when they are localized by a classical apparatus.

12 Rovelli's (1996) relational interpretation of quantum mechanics extends Bohr's approach to any physical system, thereby avoiding any reference to a classical apparatus. For an evaluation of Rovelli's attempt, see Dorato (2016).

13 Not surprisingly, Bub (2000) ignores this aspect of Einstein's attitude toward quantum mechanics as principle theory, because he is mainly interested in using it to defend his informational approach to quantum mechanics (Bub 1997).

14 Furthermore, how can we come to know the wave function of the universe if nothing classical can interact with it from the outside?

15 The fact that this is an ideal experiment, as Bohr recognizes, is irrelevant for the point he wants to make.

16 By referring to EPR, in his 1949 work Bohr wrote "there is in a case like that just considered no question of a mechanical disturbance of the system under investigation

during the last critical stage of the measuring procedure (235). The "disturbance argument" was suggested by Heisenberg in 1930, in his optical, "microscope argument."

17 "Once one has agreed to accept the [projection] postulate, it is easy to close the gap and to move from the impossibility of common values to that of simultaneous measurement. A measurement of momentum should leave the system in an eigenfunction of momentum; a measurement of position should leave it in an eigenfunction of position. Since there can be no common eigenfunctions, there can be no simultaneous measurements" (Whitaker 2004, 1313).

18 One could adopt the same approach by invoking the existence of just one dispositional property. I am not going to explore this alternative here.

19 In Dorato (2007) I argue that it is not.

20 The irreversibility called into play here is rather different from the temporal evolution in decoherent approaches to quantum mechanics, given that the latter involve, as noted by Schlosshauer and Camilleri (2011, 5), improper mixtures, and therefore quantum states, while in Bohr's account the irreversibility concerns classical states.

21 The notion of approximation or idealization is indeed crucial in Bohr's philosophy (Tanona 2004).

22 I must thank Nino Zanghì for his suggestion to look at Landau's treatment as one of the best defenses of Bohr's interpretation.

23 For Bohr, even though not for Heisenberg, the sharp distinction thesis did not imply a movable shift between the classical and the quantum. For him, there is a distinction between the contextuality of the boundary classical/quantum and its "changeability." He accepted the former but rejected the latter, as recorded by Heisenberg in a letter he wrote to Heelan (see Schlosshauer and Camilleri 2011, 4).

References

Bacciagaluppi G., and Valentini A. 2009. *Quantum Theory at the Crossroads: Reconsidering the 1927 Solvay Conference.* Cambridge: Cambridge University Press.

Bell, J. S. 1987. *Speakable and Unspeakable in Quantum Mechanics.* Cambridge: Cambridge University Press.

Beller, M., and A. Fine. 1994. "Bohr's response to EPR." In *Niels Bohr and Contemporary Philosophy*, edited by J. Faye and H. J. Folse, 1–31. Boston Studies in the Philosophy of Science, Vol. 153. Berlin: Springer.

Bohm, D. 1951. *Quantum Theory.* Englewood Cliffs, NJ: Prentice Hall.

Bohr, N. 1949. "Discussion with Einstein on epistemological problems in atomic physics." In *Albert Einstein: Philosopher-Scientist*, edited by P. A. Schilpp, 200–41. Evanston: The Library of Living Philosophers. Reprinted in N. Bohr (1958), *Atomic Physics and Human Knowledge*, 32–66.

Bohr, N. 1954. "Unity of knowledge." In *Atomic Physics and Human Knowledge*, N. Bohr (1958), 67–82.

Bohr, N. [1956] 1998. "Mathematics and natural philosophy." In *The Philosophical Writings of Niels Bohr*, Vol IV: *Causality and Complementarity*, edited by J. Faye and H. Folse, 164–9. Woodbridge, CT: Ox Bow Press.

Bohr, N. 1958. *Atomic Physics and Human Knowledge.* London, New York: John Wiley and Sons. Reprinted as *The Philosophical Writings of Niels Bohr*, Vol. II. Woodbridge, CT: Ox Bow Press.

Bohr, N. 1963. *Essays 1958–1962: Atomic Physics and Human Knowledge*, London: John Wiley and Sons. Reprinted as *The Philosophical Writings of Niels Bohr*, Vol. III. Woodbridge, CT: Ox Bow Press.

Bohr, N. 1985. *Collected Works, Vol. 6. Foundations of Quantum Mechanics I (1926–1932)*, edited by Jørgen Kalckar. Amsterdam: North-Holland.

Bokulich, P., and A. Bokulich. 2004. "Niels Bohr's generalization of classical mechanics." *Foundations of Physics* 35: 347–71.

Born, M. 1971. *The Born Einstein Letters. Correspondence between Einstein and Max and Edwig Born, with Commentaries by Max Born from 1916 to 1955*, translated by I. Boim. London: MacMillan.

Brown, H. 2005. *Physical Relativity*. Oxford: Oxford University Press.

Bub, J. 1997. *Interpreting the Quantum World*. Cambridge: Cambridge University Press.

Bub, J. 2000. "Quantum mechanics as a principle theory." *Studies in History and Philosophy of Modern Physics* 31: 75–94.

Clifton, R., J. Bub, and H. Halvorson. 2003. "Characterizing quantum theory in terms of information theoretic constraints." *Foundations of Physics* 33: 1561.

Dorato, M. 2007. "Dispositions, relational properties and the quantum world." In *Dispositions and Causal Powers*, edited by M. Kistler and B. Gnassonou, 249–70. Oldercroft: Ashgate.

Dorato, M. 2014. "Causal *versus* structural explanations in scientific revolutions." *Synthese* 1–21.

Dorato, M. 2016. "Rovelli's relational interpretation of quantum mechanics, antimonism and quantum becoming." In *The Metaphysics of Relation*, edited by D. Yates and A. Marmodoro, 290–324. Oxford: Oxford University Press.

Dorato, M., and L. Felline. 2010. "Structural explanations in Minkowski spacetime: which account of models?" In *Space, Time, and Spacetime—Physical and Philosophical Implications of Minkowski's Unification of Space and Time*, edited by V. Petkov, 189–203. Springer: Berlin, Heidelberg.

Einstein, A. 1920. *Relativity: The Special and General Theory. A Popular Exposition*, Translated by R. W. Lawson. London: Methuen.

Einstein, A. 1954. "What is the theory of relativity?" *Ideas and Opinions*. New York: Bonanza Books, 227–324; originally published as "Time, Space, and Gravitation." *Times* (London), 28 November 1919, 13–14.

Faye, J. 1991. *Niels Bohr: His Heritage and Legacy. An Antirealist View of Quantum Mechanics*. Dordrecht: Kluwer Academic Publishers.

Faye, J. 2014. "Copenhagen interpretation of quantum mechanics." *The Stanford Encyclopedia of Philosophy* (Fall 2014 Edition), edited by Edward N. Zalta. http://plato.stanford.edu/archives/fall2014/entries/qm-copenhagen/.

Folse, H. J. 1985. *The Philosophy of Niels Bohr: The Framework of Complementarity*. Amsterdam: North-Holland.

Gell-Mann, M. 1976. "What are the building blocks of matter?" In *The Nature of the Physical Universe: Nobel Conference*, edited by H. Douglas and O. Prewitt, 27–45. New York: John Wiley & Sons.

Ghirardi, G. C. 2004. *Sneaking a Look at God's Card: Unravelling the Mysteries of Quantum Mechanics*. Princeton, NJ: Princeton University Press.

Hacking, I. 1983. *Representing and Intervening*. Cambridge: Cambridge University Press.

Heisenberg, W. 1958. *Physics and Philosophy*. New York: Harper and Collins.

Howard, D. 1994. "What makes a classical concept classical? Toward a reconstruction of Niels Bohr's philosophy of physics." In *Niels Bohr and Contemporary Philosophy*, edited by J. Faye and H. Folse, 201–29, Boston Studies in the Philosophy of Science, Vol. 158. Dordrecht: Kluwer Academic Publishers.

Howard, D. 2004. "Who invented the 'Copenhagen interpretation?' A study in mythology." *Philosophy of Science* 71: 669–82.

Howard, D. 2015. "Einstein's philosophy of science." *The Stanford Encyclopedia of Philosophy* (Winter 2015 Edition), Edited by Edward N. Zalta. http://plato.stanford.edu/archives/win2015/entries/einstein-philscience/.

Landau, L., and E. Lifshitz. 1981. *Quantum Mechanics: Non-relativistic Theory*, translated by J. B. Sykes and J. B. Bell. New York: Pergamon Press.

Lange, M. 2013a. "How to explain the Lorentz transformations." In *Metaphysics and Science*, edited by S. Mumford and M. Tugby, 73–98. Oxford: Oxford University Press.

Lange, M. 2013b. "What makes a scientific explanation distinctively mathematical?" *British Journal for the Philosophy of Science* 64: 485–511.

MacKinnon, E. 1982. *Scientific Explanation and Atomic Physics*. Chicago: Chicago University Press.

Minkowski, H. 1952. "Space and time." In *The Principle of Relativity: A Collection of Original Memoirs on the Special and General Theory of Relativity*, H. A. Lorentz, A. Einstein, H. Minkowski, and H. Weyl, 75–91. New York: Dover.

Murdoch, D. 1987. *Niels Bohr's Philosophy of Physics*. Cambridge: Cambridge University Press.

Ney, A., and D. Albert (eds). 2013. *The Wave Function*. Oxford: Oxford University Press.

Petersen, Aa. 1968. *Quantum Physics and the Philosophical Tradition*. Cambridge: MIT Press.

Redhead, M. 1987. *Incompleteness, Nonlocality and Realism*. Oxford: Oxford University Press.

Rovelli, C. 1996. "Relational quantum mechanics." *International Journal of Theoretical Physics* 35: 1637. http://arxiv.org/pdf/quant-ph/9609002v2.pdf.

Rugh, S., and H. Zinkernagel. 2011. "Weyl's principle, cosmic time and quantum fundamentalism." *Explanation, Prediction, and Confirmation*, edited by Dennis Dieks et al., 411–24. Berlin: Springer.

Schlosshauer, M., and K. Camilleri. 2011. "What classicality? Decoherence and Bohr's classical concepts." *Advances in Quantum Theory: AIP Conference Proceedings* 1037: 26–35.

Sellars, W. 1962, "Philosophy and the Scientific Image of Man." In *Frontiers of Science and Philosophy*, edited by Robert Colodny. Pittsburgh: University of Pittsburgh Press, 35–78.

Snowdon, P. 2009. "Peter Frederick Strawson." In *The Stanford Encyclopedia of Philosophy* (Fall 2009 Edition), edited by Edward N. Zalta. http://plato.stanford.edu/archives/fall2009/entries/strawson/.

Strawson, P. 1959. *Individuals*. London: Methuen.

Suárez, M. 2004. "Selections, dispositions and the problem of measurement." *The British Journal for the Philosophy of Science* 55: 219–55.

Tanona, S. 2004, "Idealization and Formalism in Bohr's Approach to Quantum Theory." *Philosophy of Science* 71: 683–95.

van Frassen, B. 2002. *The Empirical Stance*. New Haven: Yale University Press.

Wheeler, J. A., with K. Ford. 1998. *Geons, Black Holes and Quantum Foam*. London: W.W. Norton.

Whitaker, A. 1989. *Einstein, Bohr and the Quantum Dilemma*. Cambridge: Cambridge University Press.

Whitaker, M. A. B. 2004. "The EPR paper and Bohr's response: A reassessment." *Foundations of Physics* 34: 1305–40.

Zinkernagel, H. 2015. "Are we living in a quantum world?" In *One hundred years of the Bohr atom*. Proceedings from a Conference, Scientia Danica. Series M: Mathematica et Physica, Vol. 1, edited by F. Aaserud and H. Kragh, 419–34. Copenhagen: Royal Danish Academy of Sciences and Letters.

Zinkernagel, H. 2016. "Niels Bohr on the wave function and the classical/quantum divide." *Studies in History and Philosophy of Modern Physics* 53: 9–19.

7

Complementarity as a Route to Inferentialism

Stefano Osnaghi

1. Introduction

Several commentators have pointed out that Bohr's inquiry into the foundations of quantum mechanics is, in many respects, a reflection on language (Petersen 1968; 1985; Stapp 1972; Bub 1977; MacKinnon 1985; Murdoch 1987; Faye 1991; Katsumori 2011). Semantic analysis is crucially involved in Bohr's argument for the completeness of quantum mechanics; especially in the formulation he gave of it after 1935, when, under the pressure of Einstein's challenges, he was led to clarify the nature and implications of complementarity. Being faced with the untenability of the epistemic construal of the uncertainty relations, Bohr (1998, 100) shifted the emphasis of his interpretation toward semantics: "The statistical character of the uncertainty relations in no way originates from any failure of measurements to discriminate within a certain latitude between classically describable states of the object, but rather expresses an essential limitation of applicability of classical ideas to the analysis of quantum phenomena." This move, in turn, led him to sharpen the pragmatist element that was latent in his conception of meaning, as a means to ground his instrumentalist interpretation of quantum probabilities (Stapp 1972; Murdoch 1987; Faye 1991).

Retrospectively, Bohr's (1958a, 88) view that "every analysis of the conditions of human knowledge must rest on considerations of the character and scope of our means of communication" appears to be remarkably in tune with the philosophical trend that was emerging in the 1930s from the work of Wittgenstein and Carnap. At the time, however, none of the physicists who were wrestling with the interpretive puzzles of quantum mechanics seemed to realize that the contemporary inquiry into the constitution of meaning might be relevant to the problem of formally establishing the objectivity of the quantum account of phenomena. To be sure, the failure to recognize the semantic, in fact *meta*semantic, implications of Bohr's reflection—a circumstance which can be traced to the very "semantic blindness" that, according to Alberto Coffa (1991), affected, at the time, so many philosophers, including those who pioneered the semantic approach—resulted in a formidable obstacle to the appraisal of complementarity. In particular, the way complementarity was supposed to offer a synthesis of two prima facie conflicting aspects of Bohr's approach, namely, the fundamental and irreplaceable role that it assigned to ordinary *concepts* on the one hand, and the call for a revision of the classical *conceptual framework* on the other, remained to a large extent obscure.

As a matter of fact, Bohr was not exactly innocent vis-à-vis the misunderstandings that surrounded his philosophical *passe-partout*. Bohr's ambiguities can be partly explained, if not removed, by retracing the dialectical process that resulted in the mature formulation of his argument for the completeness of quantum mechanics. One striking aspect of this process is the gap between the objections that were raised against complementarity and the way Bohr chose to reply to them. Revealingly, Bohr's responses often appeared, to his interlocutors, even more puzzling than the arguments they were intended to clarify. Indeed the first thesis that I will put forth and scrutinize in the present chapter is that Bohr's responses *were* wanting—or at least so they legitimately appeared to those who took the standard "denotational" model of semantic explanation (see, e.g., Schroeder-Heister 2014) for granted.

The second question that I will address is whether the endorsement of a nonstandard theory of conceptual content may help make sense of Bohr's approach. Assuming that this is the case, a subordinate question is whether Bohr was explicitly or implicitly committed to such a theory. While the bulk of the chapter is devoted to substantiating a positive answer to the former question, my assessment of the latter will be less clear-cut. What makes the issue delicate in this case, besides the familiar difficulties encountered by any attempt to read Bohr through the lens of traditional philosophical doctrines, is the fact that we are concerned with questions that the philosophy of his time was only beginning to thematize. This makes the task of reconstructing Bohr's standpoint especially intricate. For not only are we faced with Bohr's own personal elaboration of the concepts he borrowed from the philosophical tradition, but we must also take note of the fact that the concepts themselves were originally worked out in intellectual contexts in which issues relating to meaning were, to a large extent, implicit.

Bohr's classic debates with Einstein and Schrödinger provide a number of interesting hints to address the preceding questions. In the present chapter, I will, however, mainly focus on a much less studied series of documents, which refer to a debate that took place in the 1950s between the Copenhagen group, John Wheeler, and Hugh Everett (who was at the time a student of Wheeler's). These documents are relevant because they show that, being faced with charges of failing to provide a satisfactory account of measurement, Bohr explicitly refused to endorse any strategy aimed at dealing with *that* problem. Instead, to his interlocutors' frustration, he systematically rejected the objections—which depicted his view as facing a dilemma between either endorsing the incompleteness of quantum mechanics or solving the measurement problem—as simply misguided. A natural hypothesis to explain the gulf between Bohr and his critics, I will argue, is that they were committed to different views of meaning.

Claiming that complementarity presupposes a nonstandard theory of conceptual content implies providing at least a sketch of how such a theory should look. I undertake this task in the last section, where I suggest that the inferentialist approach put forth by Robert Brandom (1994) presents some interesting analogies with Bohr's method and might provide a suitable framework for making sense of Bohr's instrumentalist interpretation of quantum probabilities.

2. Ordinary concepts and quantum probabilities

Unlike other physicists of his time, who appeared to be convinced that a consistent account of the unconventional statistical features displayed by quantum phenomena required new concepts in the first place, Bohr (1958a, 64) systematically opposed the idea of relinquishing ordinary concepts, arguing instead that "the appropriate physical interpretation of the symbolic quantum-mechanical formalism amounts only to predictions, of determinate or statistical character, pertaining to individual phenomena appearing under conditions defined by classical physical concepts" (see also Bohr 1998, 144).

Bohr's conservative stance with regard to ordinary concepts was a major source of puzzlement and impatience for his critics. Bohr appeared to be convinced that it is "only within the framework of ordinary concepts" that "we can speak of well-defined experiences" and he considered it to be "[un]likely that the fundamental concepts of the classical theories will ever become superfluous for the description of physical experience" (Bohr 1958a, 39; 1934, 16). Accordingly, he famously proclaimed that "*however far the phenomena transcend the scope of classical physical explanation, the account of all evidence must be expressed in classical terms*" (Bohr 1998, 87; emphasis in the original; see also Bohr 1958a, 67). The grounds for these assertions, however, were not clear. As Schrödinger wrote to him: "There must be quite definite and clear grounds, why you repeatedly declare that one must interpret observations classically, which lie absolutely in their essence.... It must belong to your deepest conviction—and I cannot understand on what you base it" (Schrödinger, letter to N. Bohr, October 13, 1935, quoted in Moore 1989, 313). Indeed, even granting that the "basing of quantum mechanics upon classical physics was a necessary provisional step," why should not we "proceed to something more fundamental," enabling us to treat quantum mechanics "in its own right as a fundamental theory without any dependence on classical physics, and to derive classical physics from it"? (Everett, letter to Aa. Petersen, May 31, 1957, quoted in Osnaghi, Freitas, and Freire 2009, 217).

In 1956, John Wheeler discussed with Bohr and his group the possibility of deducing the characteristics of our conceptual and pragmatic apparatus from a suitable physical model, in line with his tenet that "the kind of physics that occurs does not adjust itself to the available words; the words evolve in accordance with the kind of physics that goes on" (Wheeler, letter to A. Stern, May 25, 1956, quoted in Osnaghi, Freitas, and Freire 2009, 118). Wheeler suggested that Everett's "model philosophy" could in principle explain the necessity of relying upon classical concepts in terms of human adaptation to the environment. He was persuaded that, since human practices (including communication and experimentation) are an outgrowth of (the complex physical interactions underlying) biological selection, it must be possible to describe them in terms of some process occurring within Everett's "model universe": "Thinking, experimentation and communication—or psychophysical duplicates thereof—are all taken by Everett as going on within the model universe" (ibid.).

In reporting to Everett how these ideas had been received in Copenhagen, Wheeler noted that while "Aage Petersen had a tendency to insist that small interaction, small $e^2/\hbar c$, was essential for a world in which one could use normal words," he had "argued that the world came first." And he went on to explain:

> It could have small or large $e^2/\hbar c$, but grant only complex systems, and evolution, and you have systems that must find a way to communicate with each other to give mutual assistance in the struggle for existence; in the struggle for survival words would necessarily be invented to deal with a large $e^2/\hbar c$. You don't first give a list of words and then ask what systems are compatible with them; instead, the system comes first, and the words second. (Wheeler, letter to H. Everett, May 22, 1956, quoted in Osnaghi, Freitas, and Freire 2009, 118)

From the available documents, it is pretty clear that Bohr regarded Wheeler's endeavor as misguided, insofar as it overlooked the fact that any claim based on a physical model—however sophisticated and encompassing—*presupposes* a conceptual frame-work. This applies in particular to any purported account of the process through which our conceptual apparatus is constituted. What Bohr may have thought of the attempt to derive the concepts, which we need for making sense of a physical model, *from* the model itself is apparent from the notes that Wheeler took during his discussion with Petersen. The view according to which language is to be regarded as "second" was, Petersen said, "very contrary to Bohr" (quoted in Osnaghi, Freitas, and Freire 2009, 118).[1]

In a paper published a few years later, Léon Rosenfeld (1965, 222) explained: "No formalization can be complete, but must leave undefined some "primitive" concepts and take for granted without further analysis certain relations between these concepts, which are adopted as "axioms": the concrete meaning of these primitive concepts and axioms can only be conveyed in a "metalanguage" foreign to the formalism of the the-ory." These remarks help understand why, according to Bohr (1934, 16), "it would be a misconception to believe that the difficulties of the atomic theory may be evaded by eventually replacing the concepts of classical physics by new conceptual forms" (see also Bohr 1998, 87). An analysis of this "misconception" can be found in a letter of 1959, in which Rosenfeld indirectly responded to Wheeler's claim that Everett's theory did "not require for its formulation any reference to classical concepts" and was "con-ceptually self-contained" (Wheeler 1957, 463–4):

> Everett's work ... suffers from the fundamental misunderstanding which affects all the attempts at "axiomatizing" any part of physics. The "axiomatizers" do not realize that every physical theory must necessarily make use of concepts which cannot, in principle, be further analysed, since they describe the relationship between the physical system which is the object of study and the means of obser-vation by which we study it: these concepts are those by which we give information about the experimental arrangement, enabling anyone (in principle) to repeat the

experiment. It is clear that in the last resort we must here appeal to common experience as a basis for common understanding. To try (as Everett does) to include the
experimental arrangement into theoretical formalism is perfectly hopeless, since
this can only shift, but never remove, this essential use of unanalysed concepts
which alone makes the theory intelligible and communicable. (Rosenfeld, letter
to S. M. Bergmann, December 21, 1959, quoted in Osnaghi, Freitas, and Freire
2009, 117)[2]

Indeed, Bohr's (1963, 3) *pragmatic* argument for granting a primitive role to ordinary "conceptual forms" was grounded in the remark that objectivity requires
communicability:

The decisive point is to recognise that the description of the experimental arrangement and the recording of observations must be given in plain language, suitably
refined by the usual physical terminology. This is a simple logical demand, since
by the word "experiment" we can only mean a procedure regarding which we are
able to communicate to others what we have done and what we have learnt ... The
description of atomic phenomena has in these respects a perfectly objective character, in the sense that no explicit reference is made to any individual observer and
that therefore ... no ambiguity is involved in the communication of information.
(See also Chevalley 1995)

Quite independently of the justification of Bohr's conceptual conservatism, there
remains the question of whether the ensuing "instrumentalist" interpretation of the
quantum formalism is viable. Specifically it must be verified whether such an interpretation manages to avoid the difficulties that prevent a straightforward semantic account
of the probabilistic inferences licensed by quantum mechanics. Such difficulties can be
illustrated as follows. Assuming that all the (*incompatible*) quantum observables pertaining to a given system must in any circumstance be assigned a definite value (viz.,
the value that would be revealed by the appropriate measurement), there exist cases in
which a (probabilistic) conclusion follows from a (probabilistic) premise even though
no model (i.e., no set of possible worlds) can be exhibited, in which the conclusion
holds whenever the premise does (Greenberger et al. 1990). In Bohr's (1998, 144) characteristically semantic phrasing: "No unambiguous interpretation of [the indetermin-
acy] relations can be given in words suited to describe a situation in which physical
attributes are objectified in a classical way."
   How can one deal with this problem? In line with the gist of Bohr's approach, which
seeks to avoid the difficulties arising from the unrestricted application of classical concepts to quantum systems by bringing the experimental context explicitly into play,
one can endorse a "conditional" interpretation of quantum probabilities, according to
which the probabilities refer to the results that can be obtained *if the appropriate measurement is performed*. If one further postulates that only in *some* experimental contexts, namely, the contexts in which a certain property is measured, can the question
of whether a given quantum system has that property meaningfully be asked, then,
provided that the simultaneous occurrence of some experimental contexts is precluded, it

becomes possible to devise models which can account for all the statistical correlations that are actually recordable (van Fraassen 1991b, chapter 5). As Bohr (1998, 80) puts it: "*[I]t is only the mutual exclusion of any two experimental procedures*, permitting the unambiguous definition of complementary physical quantities, which provides room for new physical laws, the coexistence of which might at first sight appear irreconcilable with the basic principles of science" (my emphasis). In other words, if the experimental setups of incompatible observables are mutually exclusive, we are not compelled to assign a truth-value to sentences which attribute values to *pairs* of incompatible observables. Since the troubles with the account of quantum probabilities result from attempting simultaneously to attribute a value to incompatible observables, the preceding condition ensures that "the classical propositional calculus continues to hold for all statements which have a truth-value" (Murdoch 1987, 153).

Though somewhat unorthodox, the idea that observables might not have a specific value outside the context of measurement is not necessarily problematic: as Bas van Fraassen (1991b, 109) notes, "we attribute colours to liquids; but when the liquid is vaporized, the question of what colour it is then has no answer." Moreover, "that a given parameter should sometimes have no value at all may look like a violation of the logical principle of Excluded Middle. But it is not" (110). One can introduce a three-valued semantics or even a syntactic prescription that rules out sentences attributing a value to an observable O in a context designed to measure an observable incompatible with O as ill-formed, but less radical solutions are available (see van Fraassen and Hooker 1976; Bub 1979).

Insofar as the conditional interpretation of quantum probabilities is assumed, therefore, nothing seems to preclude a standard model-theoretic account of the related inferences. Consider, for example, a specific state vector. A model for the probability distributions it assigns to the relevant family of incompatible observables might be constructed as follows. First, in each possible world exactly one measurement context should be realized. Second, within the class of worlds characterized by the occurrence of a particular measurement context, the ratio of worlds in which a particular result obtains should reflect the statistical distribution predicted by the state vector for that result.

So far so good. However, a problem arises as soon as one undertakes to include the measurement context itself within the physical description. Suppose we have an experimental apparatus which can be used to measure a whole family of incompatible observables and suppose we want to label each possible configuration of the apparatus based on the observable it allows one to measure. The task is one of characterizing *states of affairs*, which quantum mechanics performs by associating a suitable state vector with each of them. So, for example, whenever the setup happens to be in a configuration apt to measure O_α, it will be attributed the state vector $|m_\alpha\rangle$. What can we say about $|m_\alpha\rangle$? Surely it should assign probability 1 to the situation in which one of the values of O_α *has been recorded*. The issue, however, is how the observable M of which $|m_\alpha\rangle$ is an eigenvector can itself be formally characterized, given that a value of M will typically refer to, say, a spot on a detection screen, but the same material record can evidently *signify* different things—hence, in particular, denote different results—in different contexts. In classical physics the problem is formally solved by requiring the existence of a functional link between the measured value and the degree of freedom that "represents" the measurement result. But this is precisely what quantum mechanics cannot provide if the result has to qualify as a result at all. Instead, it looks as if, in order to capture the *semantic* dimension of a measurement outcome, one must specify the infinite degrees of freedom that fix the context of its occurrence.[3]

Unless one manages to answer Bell's (1990, 34) infamous question: "What exactly qualifies some system to play the role of 'measurer'?," therefore, the preceding analysis seems to preclude the kind of semantic account of quantum inferential patterns that results from endorsing the instrumentalist interpretation. A number of proposed "improvements" of Bohr's interpretation have sought to deal with this problem by eliminating any explicit reference to measurement in favor of a more fundamental objective description, in which the measuring apparatuses (as well as the observers) would be part of a "closed system inside of which the contextual selections are determined by purely physical factors" (van Fraassen 1991a, 499).

In the next section I will scrutinize Bohr's attitude vis-à-vis some of these proposals. The fact that, as we will see, he did not hesitate to point out their flaws and inherent circularity suggests, I think, that he regarded the problem of establishing the *soundness* of his approach as largely irrelevant. The impossibility to "think of quantum theory as a putative autonomous description of the world in neutral physical terms" (van Fraassen 1991b, 284) did not seem to bother him. Nor did he feel pressed to show how "a theory which speaks *only* of the results of external interventions on the quantum system" could be derived from "one in which that system is attributed *intrinsic properties*" (Bell 1990, 38). If this analysis is correct, far from being conceived as a sophisticated means to save the "representational" order of semantic explanation instantiated by Tarskian semantics, complementarity should be understood as an attempt to take note of its intrinsic limitations.

3. The controversy over the status of the "agency of measurement"

It is generally taken for granted that Bohr was committed to the idea that, "in order to obtain a means of *interpreting* the [quantum] wave function, we must ... at the outset postulate a classical level in terms of which the definite results of a measurement can be realized" (Bohm 1951, 626, my emphasis). A corollary of the preceding view is that "without an appeal to a classical level, quantum theory would have no meaning." Hence, in particular, "classical concepts cannot be regarded as limiting forms of quantum concepts" and "quantum theory presupposes the classical level and the general correctness of classical concepts in describing this level; it does not deduce classical concepts as limiting cases of quantum concepts" (624–5). Everett's (1973, 111) rephrasing of this claim "the deduction of classical phenomena from quantum theory is ... impossible simply because no meaningful statements can be made without pre-existing classical apparatus to serve as a reference frame" is perhaps less precise, but it is equally correct. Indeed, as we saw, Bohr (1963, 3–4) considered it to be necessary to account "for the functions of the measuring instruments in purely classical terms, excluding in principle any regard to the quantum of action" which, he added, compels us to introduce "*a fundamental distinction between the measuring apparatus and the objects under investigation*" (emphasis in the original).

One would accordingly expect that Bohr would renounce the idea that quantum mechanics is an all-encompassing theory, giving up, in particular, the possibility of providing a quantum account of the measuring apparatus. This expectation is responsible for the entrenched view according to which Bohr advocated "the inapplicability of quantum theory to macrophysics" and held that the "classical world [was] physically distinct from the microsystems" and constituted "a portion of the world" to be treated "in a different manner than the rest of it" (Rovelli 1996, 1671). Analyses of this sort are widespread in the literature on the foundations of quantum mechanics. Thus, for example, Steven Weinberg (2005, 31) argues that the "Copenhagen interpretation describes what happens when an observer makes a measurement, but the observer and the act of measurement are themselves treated classically"; and Wojciech Zurek (2003, 716) contends that "Bohr's solution was to draw a border between the quantum and the classical and to keep certain objects—particularly the measuring devices as well as the observers—on the classical side"—a strategy which implied "suspending" the "principle of superposition ... 'by decree' in the classical domain."

Bohr's alleged commitment to the view that "macrosystems are relatively immune to quantum effects" and that "even in principle quantum mechanics cannot describe the process of measurement itself" was one of Everett's main motivations for developing the alternative approach which resulted in his "relative state" formulation. Everett regarded Bohr's approach not only as "hopelessly incomplete," but also as "somewhat repugnant, since it [led] to an artificial dichotomy of the universe into ordinary phenomena, and measurements."[4] In his PhD dissertation, Everett (1957) offered a more sophisticated characterization of the "orthodox" approach, which he referred to as the "external observation formulation." According to it, "quantum theory was devised to describe only situations in which an observer (or at least, the measuring environment) is involved, while leaving that part out of the description" (van Fraassen 1991b, 273). The "external observation" formulation implies no rigid distinction between two physical domains: "Any system can be part of the system under study, but the line between system under study and (ultimate) measuring apparatus must be drawn somewhere" (van Fraassen 1972, 332). The fact that "a measurement is an interaction *incompletely described*, by leaving out something or other" merely means that, in any experimental context, the "ultimate" measurement interaction, as well as the "external" observer who performs the measurement, is not described by the unitary quantum model of the phenomenon under study (van Fraassen 1991b, 273; emphasis in the original).

Insofar as the "external observation" formulation "leaves out" something crucial from the physical description, the search for a "generalization" which would allow one "to deal with a situation where several observers are at work, and ... to include the observers themselves in the system that is to receive mathematical analysis" (Wheeler, letter to N. Bohr, April 24, 1956, quoted in Osnaghi, Freitas, and Freire 2009, 119) seems entirely legitimate. To Bohr, however, these purported improvements appeared completely beside the point, given that he could see no sense in which the account of phenomena provided by the instrumentalist interpretation of quantum probabilities needed to be completed. In a letter to Everett, Petersen was peremptory: "I don't think that you can find anything in Bohr's papers which conforms with [*sic*] what you call the external observation interpretation" (Petersen, letter to H. Everett, April 24, 1957, quoted in Osnaghi, Freitas, and Freire 2009, 117).

The root of the misunderstanding is arguably to be sought in Bohr's frequent allusions to a generic *physical* argument to the effect that the description of microscopic phenomena will not be affected if we *assume* that the measuring apparatus behaves classically:

> As all measurements ... concern bodies sufficiently heavy to permit the quantum to be neglected in their description, there is, strictly speaking, no new observational problem in atomic physics. The amplification of atomic effects, which makes it possible to base the account on measurable quantities and which gives the phenomena a peculiar closed character, only emphasizes the irreversibility characteristic of the very concept of observation. (Bohr 1958a, 89; see also ibid., 98; 1963, 3)

The same point is made in a couple of letters written in response to Everett's criticisms:

> There is no arbitrary distinction between the use of classical concepts and the formalism since the large mass of the apparatus compared with that of the individual atomic object permits that neglect of quantum effects which is demanded for the account of the experimental arrangement. (Petersen, letter to H. Everett, April 24, 1957, quoted in Osnaghi, Freitas, and Freire 2009, 117)

> It is extremely fundamental that this cutoff [of the "measuring chain"] is made after the measuring result has been recorded in a permanent way, so that it no longer can be essentially changed if it is observed on its turn (i.e. if the chain is set forth). This recording has to be more or less irreversible and can only take place in a macrophysical (recording) system. This macrophysical character of the later part of the measuring chain is decisive for the measuring process. I do not think that it can be left out of consideration in its description. (Groenewold, letter to H. Everett and J. A. Wheeler, April 11, 1957, quoted in Osnaghi, Freitas, and Freire 2009, 119)

The problem with the preceding arguments is that they lend themselves to be interpreted as implying that quantum mechanics delivers (hence, a fortiori, *can* deliver) the kind of semantic structure ("classical level") purportedly required to make the instrumentalist interpretation viable. Conflating the semantic and the physical levels results in the deceptive expectation that a more accurate account of the measurement process might prove helpful in constructing a consistent model for quantum probabilities. Bohr was accordingly criticized for not having provided such an account.

"One of the fundamental motivations of [my] paper," Everett wrote to Petersen, "is the question of *how can it be* that mac[roscopic] measurements are 'irreversible', the answer to which is contained in my theory, but *is* a serious lacuna in the other theory" (emphasis in the original). And he continued:

> You talk of the massiveness of macrosystems allowing one to neglect further quantum effects (in discussions of breaking the measuring chain), but never give any justification for this flatly asserted dogma. Is it an independent postulate? It most

certainly does not follow from wave mechanics ... In fact, by the very formulation of your viewpoint you are totally incapable of any justification and must make it an independent postulate—that macrosystems are relatively immune to quantum effects. You vigorously state that when apparatus can be used as measuring apparatus then one cannot simultaneously give consideration to quantum effects—but proceed blithely to apply [the uncertainty relations] to such devices, tacitly admitting quantum effects. (Everett, letter to Aa. Petersen, May 31, 1957, quoted in Osnaghi, Freitas, and Freire 2009, 217)

Furthermore, Everett claimed that while the Copenhagen interpretation took "the fundamental irreversibility of the measuring process" to be what "allows the destruction of phase relations and make possible the probability interpretation of quantum mechanics," there was "nowhere to be found any consistent explanation of this 'irreversibility.'" And he concluded: "Another independent postulate?" (ibid.).

Bohr did not think an "independent postulate" was required (see Teller 1981). This was made clear by Rosenfeld in a paper of 1965, in which he endorsed the quantum theory of measurement put forth by Daneri, Loinger, and Prosperi (1962) in the framework of the "thermodynamic approach" prompted by Günther Ludwig.[5] As Rosenfeld explained in a couple of letters written a few years later (in the context of the controversy generated by Wigner's analysis of the measurement problem):

The crux of the problem which worries Wigner so much is that the reduction rule appears to be in contradistinction with the time evolution described by Schrödinger's equation. The answer, which was of course well known to Bohr, but has been made formally clear by the Italians [Daneri et al.], is that the reduction rule is not an independent axiom, but essentially a thermodynamic effect, and accordingly, only valid to the thermodynamic approximation.

Rosenfeld concluded that the "recording process" was "entirely describable by quantum mechanics" provided that one took the macroscopic character of the measuring instruments into account. He also pointed out that "the reduction rule is nothing else than a formal way of expressing the idealized result of the registration," without which "the phenomenon is not well defined" (since "the measurement is our only way of attaching a meaning to the mathematical symbols of the theory, by associating such symbols with some direct observation (position of a pointer, spot on a photographic plate, and so on)") (Léon Rosenfeld, letters to F. Belinfante, July 24 and August 24, 1972, quoted in Osnaghi 2008, 167–8).

While it is tempting to understand the preceding claims as (wrongly) implying that a suitable quantum model of measurement can contribute to establishing the soundness of the instrumentalist approach, this is unlikely to have been what Rosenfeld had in mind. As vividly illustrated by Wigner's (1961) infamous thought experiment, one might always imagine that a hypothetical meta-observer could "undo" the putative measurements performed by a "quantum" observer. Therefore the "irreversibility for all practical purposes" that can be derived from unitary quantum models of measurement does not deliver the kind of objective criterion to identify measurement contexts that would fix the issue discussed in the previous section. Bohr (1998, 104) seems to allude to this in a passage of his Warsaw lecture:

> In the system to which the quantum mechanical formalism is applied, it is of course possible to include any intermediate auxiliary agency employed in the measuring process. Since, however, all those properties of such agencies which, according to the aim of the measurements have to be compared with the corresponding properties of the object, must be described on classical lines, their quantum mechanical treatment will for this purpose be essentially equivalent with a classical description. The question of eventually including such agencies within the system under investigation is thus purely a matter of practical convenience, just as in classical physical measurements; and such displacements of the section between object and measuring instruments can therefore never involve any arbitrariness in the description of a phenomenon and its quantum mechanical treatment.
>
> The only significant point is that in each case some ultimate measuring instruments, like the scales and clocks which determine the frame of space-time coordination—on which, in the last resort, even the definitions of momentum and energy quantities rest—must always be described entirely on classical lines, and consequently kept outside the system subject to quantum mechanical treatment.

Here Bohr points out that the quantum model of a measuring apparatus must, *as a matter of empirical adequacy*, reproduce the classical features which allow the apparatus to qualify as a *measuring* apparatus. However, this somewhat tautological remark does not remove the need for "some ultimate measuring instrument" to be "described *entirely* on classical lines" (in a way which *prevents* it from being considered as part of the system under study) if the claim that the quantum model of the apparatus and its classical description are essentially equivalent is to be meaningfully expressible at all.[6]

I think the preceding passage shows that Bohr was not so much concerned with the *possibility* of describing the measuring apparatus as a quantum system as he was persuaded that—insofar as issues of consistency were at stake—pursuing such a strategy was not only pointless, but also fatally misleading. This worry was made explicit by Rosenfeld in his correspondence:

> The fact, emphasized by Everett, that it is actually possible to set up a wave-function for the experimental apparatus and Hamiltonian for the interaction between system and apparatus is perfectly trivial, but also terribly treacherous; in fact, it did mislead Everett to the conception that it might be possible to describe apparatus + atomic object as a closed system. This, however, is an illusion: the formalism used to achieve this must of necessity contain parameters such as external fields, masses, etc. which are precisely the representatives of the uneliminable residues of unanalysed concepts. (Rosenfeld, letter to Saul M. Bergmann, December 21, 1959, quoted in Osnaghi, Freitas, and Freire 2009, 117)

More generally, Bohr's concerns emerge from his collaborators' attitude toward von Neumann's "quantum theory of measurement." In an unpublished report of 1957 about Louis de Broglie's book *La théorie de la mesure en mécanique ondulatoire*, for example, Rosenfeld complained that von Neumann's treatise, "though excellent in other respects, ha[d] contributed by its unhappy presentation of the question of measurement in quantum theory to create unnecessary confusion and raise spurious problems" (quoted in Osnaghi, Freitas, and Freire 2009, 99). According to Rosenfeld:

> Bohr's considerations were never intended to give a "theory of measurement in quantum theory," and to describe them in this way is misleading, since a proper theory of measurement would be the same in classical and quantal physics, the peculiar features of measurements on quantal systems arising not from the measuring process as such, but from the limitations imposed upon the use of classical concepts in quantum theory. By wrongly shifting the emphasis on the measuring process, one obscures the true significance of the argument and runs into difficulties, which have their source not in the actual situation, but merely in the inadequacy of the point of view from which one attempts to describe it. This error of method has its origin in v. Neumanns book Foundations of Quantum Mechanics. (118)

Rosenfeld also made some sarcastic remarks on the efforts made by a group of physicists "to develop their own 'theory of measurement' in opposition to what they believed to be the 'orthodox' theory of measurement, as presented by v. Neumann." According to Rosenfeld, "these reformers ... involved themselves in a double misunderstanding, criticizing a distorted and largely irrelevant rendering of Bohr's argument by v. Neumann, and trying to replace it by a 'theory' of their own, based on quite untenable assumptions" (ibid.).

Similar views were expressed in the abovementioned discussions between Wheeler and Petersen. In his notes, Wheeler attributes to Petersen the following remarks: "Von N[eumann]+Wig[ner] all nonsense; their stuff beside the point; ... Von N[eumann]+Wig[ner]—mess up by including [the] meas[uring] tool in [the observed] system ... Silly to say apparatus has Ψ-function" (Wheeler, Notes taken in Copenhagen, 1956, quoted in Osnaghi, Freitas, and Freire 2009, 118). When considering the "paradox outlined by Everett," Petersen said, one should keep the "distinction between Bohr way & the 2 postulate [i.e., von Neumann's] way to do q[uantum] mech[anics]" in mind and take note of the fact that "QM description of measuring tool prevents its use as a meas[uring] tool" (ibid.).

It is thus relatively clear what Bohr thought the "ultimate" measurement is *not*: it is neither a physical interaction nor a sort of extraphysical process involving an "external" observer. Far more opaque is the way he thought such a notion *should* be understood. A hint may perhaps be found in the following passage taken from a letter in which Alexander Stern reported to Wheeler and Everett Bohr's response to the criticisms contained in the first draft of Everett's dissertation:

> One cannot follow through, nor can one trace to the interaction between the apparatus and the atomic system under observation. It is not an "uncontrollable interaction," a phrase often used in the literature. Rather, it is an INDEFINABLE

interaction. Such a connotation would be more in accord with the fact of the irreversibility, the wholeness of the quantum phenomenon as embodied in the experimental arrangement. (Stern, letter to J. A. Wheeler, May 20, 1956, quoted in Osnaghi, Freitas, and Freire 2009, 115)

4. From the "external" observer to metasemantics

If, as Bohr clearly implies, the use of ordinary concepts to describe quantum phenomena cannot be justified based on a *physical* analysis, what kind of justification has he in mind? Michel Bitbol (1998, 9–10) has offered the following Kant-inspired reconstruction of Bohr's "pragmatic" attitude with regard to this question:

> Objects operate in our experience as anticipative frameworks ... But they are by no means the most general anticipative frameworks one may conceive ... [A] substitute for the objects qua anticipative structures ... can be afforded by the concept of a reproducible global experimental situation ... However, ... if the previous kind of operationalistic anticipative framework is to be efficient at all, it must be grounded on a reliable procedure for ascertaining that (experimental) situations are effectively reproduced. Of course, this procedure could itself amount to describing and performing a second-order experiment, whose anticipated outcome is precisely the instrumental set-up of the first-order experiment. But the regress has to be stopped somewhere. It is at this point that the object-organization of experience and discourse rises again. Indeed, predicating a property of an object is a way of implying the class of situations in which the appearances arising from the dispositional content of this property are observed. As Kant claimed repeatedly, referring to objects and properties is not tantamount to stepping back in "cosmic exile" (that is in no worldly situation at all), thus talking about things as they are in themselves; it only means that one endorses implicitly the sort of situation which is common to every sentient and rational being inhabiting the environment of mankind. Describing an experimental set-up in terms of reidentifiable objects possessing properties is therefore a natural way of stopping the regress of explicitly stated situations and anticipations, by means of their implicit use.

Bitbol concludes that, although "the familiar object-like organization of the surrounding world is ... only one among the many structures which are able to afford communicable anticipations," it is "designed to be the last-order one." For him, "Bohr's insistence on everyday language and concepts to describe the experimental apparatuses, and Wittgenstein's remark in *On certainty* that 'no such proposition as there are physical objects can be formulated' are two ways of expressing this special limiting status of the object-like organization" (ibid.).

Read through a semantic lens, the preceding reconstruction might be taken to amount to a pragmatic "proof" of the indispensability of the extensional model of conceptual relations. Bohr's reasoning would, in other terms, involve the following steps: (i) It is an empirical fact that we can rely on the information encoded in the quantum symbolism in order to draw effective inferences. (ii) Reliable information can only be conveyed by means of ordinary concepts. (iii) The way ordinary concepts are used is adequately described (for the limited scope of an idealized experimental language) by denotational semantics. Therefore, (iv) the inferences derived from the quantum symbolism must ultimately be explainable by denotational semantics. The problem with this line of reasoning is that, while it assumes, *as a matter of empirical evidence*, that ordinary concepts are crucially involved in communication, it *also* takes a particular theory of concept-use for granted. Given the tautological flavor of the resulting argument, it is hardly surprising that Bohr's approach was so often taken to rest on blatantly dogmatic assumptions.

In contradistinction to the preceding reconstruction, I submit that the kind of theory of conceptual content that we should endorse in the light of (our manifest capability to describe) quantum phenomena is precisely what Bohr regarded as the issue at stake. This interpretation is instrumental in explaining why, for all his professed commitment to ordinary *concepts*, Bohr (1958a, 67) contended that a *conceptual framework* may always "prove too narrow to comprehend new experiences" and why, in particular, he claimed that the classical framework needed to be "widened" in view of quantum phenomena (82). On the face of it, such a revisionary claim—which seems to clash with Bohr's firm belief that, "in spite of their limitation, we can by no means dispense with [the] accustomed forms of perception [which] colour our whole language and in terms of which all experience must ultimately be expressed" (Bohr 1934, 5)—sounds distinctively un-Kantian (Folse 1978; Kaiser 1992). Yet, one can easily discern, in Bohr's method, the characteristic pattern of "transcendental" explanation, which goes *from* (the description of) experience *to* the elucidation of the conceptual conditions of possibility for (the description of) experience. Consider, for instance, the following passage:

> [A] consistent application even of the most elementary concepts indispensable for the description of our daily experience, is based on assumptions initially unnoticed, the explicit consideration of which is, however, essential if we wish to obtain a classification of more extended domains of experience ... The analysis of new experiences is liable to disclose again and again the unrecognized presuppositions for an unambiguous use of our most simple concepts. (Bohr 1998, 83–4)

Here Bohr is not simply urging a more thorough analysis of the truth conditions of sentences involving some particular empirical concepts. In other words, he is not merely concerned with identifying the axioms which can capture the empirical content of a particular theory. Rather, he is prompting an inquiry into the *structural* constraints which govern the way concepts in general are combined in empirical descriptions that we recognize as objective. As he put it, the "novel" situation with which quantum physics faced us "demanded a renewed analysis of the presuppositions for the application of concepts used for orientation in our surroundings" (Bohr 1958a, 98; see also Bohr 1998, 83–4). Such an analysis might eventually enable us to "utilize all the classical concepts by giving them a suitable quantum-theoretical reinterpretation" (Bohr 1934, 8). "Investigating as accurately as we can the conditions for the use of our words" (Bohr

1958b; quoted in Honner 1982, 23) is indeed what Bohr takes to be the fundamental task of scientific inquiry. For, as an oft-cited slogan has it: "It is wrong to think that the task of physics is to find out how nature is. Quantum physics is about what we *can* say about nature" (Petersen 1985, 305; my emphasis).

This *expressive* endeavor is arguably the most original feature of Bohr's approach.[7] As Petersen explains in a letter: "The aim of [Bohr's] analysis is only *to make explicit* what the formalism implies about the application of the elementary physical concepts." (Petersen, letter to H. Everett, April 24, 1957, quoted in Osnaghi, Freitas, and Freire 2009, 115; my emphasis). Provided that the role played by a concept in inferential practice is viewed as the most relevant aspect for characterizing its "application," this sort of "making explicit" has some affinities with the function that Wilfrid Sellars (1953) assigned to modal statements such as "A causally necessitates B": namely, that of "making explicit, in the form of assertible rules, commitments that had hitherto remained implicit in inferential practices" (Brandom 2000, 56).

With this in mind, Bohr's (1958a, 68) notion of a "conceptual framework," by which term he refers to "an unambiguous logical representation of relations between experiences," may be understood as the sort of formal structure which enables us to translate the inferential practices encoded in a physical symbolism into a set of *claims*. In other words, according to this reading, Bohrian conceptual frameworks would be attributed a role similar to that which Bob Brandom, following a tradition that he traces back to the Frege of the *Begriffsschrift*, assigns to logic. Brandom takes the semantic task of logical vocabulary to be that of making explicit the implicit inferential commitments that articulate claims or assertions. Thus, for example, the *conditional* "*if* A, *then* B" expresses the fact that by asserting "A" one is committed to assert "B." By using logical vocabulary, the implicit commitment is thus turned into an explicit assertion, whose significance within a social practice with the structure of a *game of giving and asking for reasons* depends on its capacity to "both serve and stand in need of reasons" and to play the role of both "premises and conclusions in inferences" (Brandom 2000, 189). This, as Brandom notes, is reminiscent of Sellars's "Socratic method," understood as

> a way of bringing our practices under rational control by expressing them explicitly in a form in which they can be confronted with objections and alternatives, a form in which they can be exhibited as the conclusions of inferences seeking to justify them on the basis of premises advanced as reasons, and as premises in further inferences exploring the consequences of accepting them. (56)

Bohr's instrumentalist interpretation of the quantum geometric symbolism can be viewed as an implementation of this method. On the one hand, Bohr (1998, 101) points out that "all unambiguous interpretation of the quantum mechanical formalism involves the fixation of the external conditions, defining the initial state of the atomic system and the character of the possible predictions as regards subsequent observable properties of that system." On the other hand, he stresses that "the entire formalism is to be considered as a tool for deriving predictions of determinate or statistical character, as regards information obtainable under experimental conditions described in classical terms" (144). In other words, the criteria for the proper use of state vectors are fixed, on the one hand, by the conditions which must obtain in order for a particular state vector to qualify as an adequate description of a given situation and, on the other hand, by the statistical expectations that legitimately derive from endorsing such a description.

The preceding characterization of state vectors is reminiscent of Dummett's (1991) "two-aspect" model for describing the use of concepts.[8] Blending the semantic insights of pragmatism and verificationism, and generalizing the proof-theoretic approach to the definition of logical connectives, Dummett suggests that both the *circumstances* under which a concept is correctly applied and the appropriate *consequences* of its use contribute to determining its content. As Brandom (2000, 62) notes, "This model can be connected to inferentialism via the principle that the content to which one is committed by using [a] concept or expression may be represented by the inference one implicitly endorses by such use, the inference, namely, from the circumstances of appropriate employment to the appropriate consequences of such employment." More generally, inferentialism identifies the significance of a speech act with the way in which it alters the commitments it is appropriate to attribute to those who partake in the relevant "inferential game."

The inferentialist approach to semantics defines itself by opposition to the "dominant tradition that reads inferential correctnesses off from representational correctnesses, which are assumed to be antecedently intelligible" (Brandom 2000, 46–7). It aims to reverse

> the *representationalist* order of [semantic] explanation, which starts with an independent notion of relations of reference or denotation obtaining between mental or linguistic items and objects and sets of objects in the largely non mental, nonlinguistic environment, and determines from these, in the familiar fashion, first truth conditions for the sentential representings built out of the subsentential ones, and then, from these, a notion of goodness of inferences understood in terms of set-theoretic inclusions among the associated sets of truth conditions. (49; emphasis in the original)

The relation between the notion of *formal* and *material* inference provides a good example of such a reversal. *Material inferences* are the kind of inferences whose correctnesses *determine* the conceptual content of the premises and conclusions (Sellars 1953). Traditionally, material inferences are treated as a derivative category, based on the idea that "wherever an inference is endorsed, it is because of belief in a conditional" (Brandom 2000, 53). Thus, for example, the fact that a competent speaker would endorse the inference from "it is raining" to "the streets will be wet" is explained by postulating that competent speakers believe that the conditional "if it is raining, the streets will be wet" is true in all possible worlds, or, equivalently, that the streets are wet in all worlds in which it is raining. With this "suppressed" premise, the inference becomes an instance of a formally valid scheme of deduction, that is, a scheme that can be explained in terms of model-theoretic consequence.

By contrast, Sellars proposes that we understand the inferential commitment from a premise to a conclusion *as part of the propositional content* of the premise and the conclusion, and that we derive the notion of formal validity from that of material correctness. This inversion results in a strongly holistic account of propositional content:

> Saying or thinking *that* things are thus-and-so is undertaking a distinctive kind of *inferentially* articulated commitment: putting it forward as a fit premise for further inferences, that is, *authorizing* its use as such a premise, and undertaking *responsibility* to entitle oneself to that commitment, to vindicate one's authority, under suitable circumstances, paradigmatically by exhibiting it as the conclusion of an inference from other such commitments to which one is or can become entitled. Grasping the *concept* that is applied in such a making explicit is mastering its *inferential* use: knowing (in the practical sense of being able to distinguish, a kind of knowing *how*) what else one would be committing oneself to by applying the concept, what would entitle one to do so, and what would preclude such entitlement. (Brandom 2000, 11; emphases in the original)

Within the inferentialist semantic framework, therefore, the lack of a suitable model-theoretic account no longer precludes the explanation of our capability to take part in the inferential game associated with quantum physics. In particular, what arguably constitutes the root of the measurement problem, namely, the impossibility of "analyzing" the link between the predictions associated with a given experimental "preparation" and those associated with the occurrence of certain results, no longer demands a justification in terms of the values taken by some observable: such a link constitutes, as it were, a primitive semantic unit. Incidentally, this hints at the need for a semantic construal of such notions as "indivisibility" and "wholeness," which Bohr typically used in connection with "phenomenon" and "quantum of action" and conceived of mostly in ontic or epistemic terms.

Bohr's puzzling attitude with respect to concepts, and more specifically his seemingly inconsistent blend of conservatism and revisionism, might find a natural interpretation within the inferentialist perspective. The conceptual framework required to elucidate the inferential practices associated with quantum physics must include, besides the familiar extensional relations codified by predicate logic, also genuinely *probabilistic* relations. One may conjecture that such probabilistic relations should conform to the geometric model of incompatible observables *as a matter of conceptual necessity*. This idea can, in principle, be accommodated within the inferentialist account of conceptual content, provided that mere assertibility conditions are replaced by *degrees* of assertibility, and commitments are accordingly attributed a measure. This results in a formal characterization of the (indeterminist) conceptual framework corresponding to complementarity, which puts us in a position to assert that "the viewpoint of complementarity represents a wider frame for the description, allowing us to embrace regularities beyond the scope of ordinary physical explanation" (Bohr 1998, 162; see also Bohr 1963, 1; 1958a, 65). What quantum mechanics has taught us, according to this reconstruction, is not that ordinary concepts are inadequate, but that a particular story about how such concepts work (along with the formal structures it deploys to make their inferential content explicit) is unsuited to capture our actual inferential practices.

Finally, we must clarify the sense in which the new conceptual framework enables an *objective* description of the phenomena—indeed "the only possible objective description" (Bohr 1962). Let us grant that reversing the representational paradigm of semantic explanation allows one to account for inferential games that go beyond classical physics. This, however, is achieved at the price of precluding any familiar "correspondence" approach to the definition of what counts as an *objective* claim in the framework of those practices. One is thus faced with a task which is notoriously hard to undertake using nonrepresentational semantic resources alone, namely, that of explaining how the criteria by which the "correctness" of claims is assessed may include a dimension that is entirely independent of the practitioners' epistemic states. After all, the notion of *commitment*, which is central to the inferentialist account of propositional content, is one that seems inextricably linked to subjective attitudes. In particular, given that it is not possible to rely on a consistent representation of possible states of affairs, it is not clear in which sense the statistical expectations to which quantum physicist are committed in a given experimental situation should be regarded as objective.[9]

The same point can be reformulated placing emphasis upon an aspect that was central to Bohr's reflection, namely, the role of communicability as a precondition of scientific knowledge. As we have seen, the inferentialist account of conceptual content is radically holistic. "In the context of a constellation of inferential practices, endorsing or committing oneself to one proposition (claimable) is implicitly endorsing or committing oneself to others which follow from it. Mastery of these inferential connections is the implicit background against which alone explicit claiming is intelligible" (Brandom 2000, 18; see also Brandom 1994, 176–8). Borrowing Bohr's (1985, 263) words: "Every thought, every word, is only suited to underline a connection, which can never be fully described, but always reflected deeper."[10] The problem is then whether it is possible to make sense of *communication* in this framework: "If the conceptual content expressed by each sentence or word is understood as essentially consisting in its inferential relations ... then one must grasp many such contents in order to grasp any. Such holistic conceptual role approaches to semantics potentially face problems concerning ... the possibility of *communication* between individuals who endorse different claims and inferences" (Brandom 2000, 29).

Brandom argues that "such concerns are rendered much less urgent, however, if one thinks of concepts as *norms* determining the *correctness* of various moves." Different people may well share the same conceptual norms "in spite of being disposed to make different claims and inferential moves" (Brandom 2000, 29).[11] Indeed, one might be tempted to reverse the accusation of subjectivism: insofar as it ultimately rests on "indefinables," whose explanatory power has to be justified by intuitive processes (the paradigmatic example of which is Russell's *acquaintance*) (Coffa 1991, 260–3), it is the representationalist approach that owes us a rational account of objectivity.

Insofar as the meaning of statements about incompatible observables boils down to a network of partial commitments to the results of possible measurements, and insofar as such partial commitments can be taken to obey a set of *rules* of inference (ultimately determined by Born's rule), it is appropriate to say that the rules *constitute* the meaning of the statements. This is where Bohr's struggle to make sense of the indeterminacy relations comes closest to the semantic analyses carried out, in the same years, by Wittgenstein and Carnap, from which the notions of philosophical grammar and logical syntax emerged. Although, as far as I know, Wittgenstein never dealt with complementarity in this context, he did explicitly include claims concerning the infinite divisibility and continuity of space (Waismann 1979, 230), as well as the law of causality (Wittgenstein 1979, 16), among his examples of "grammatical sentences" disguised in the "form of empirical propositions" (Coffa 1991, 265). This is relevant, for such claims constitute the backbone of the "conceptual framework" which complementarity was meant to generalize (Bohr 1934, 54).

The semantic approach to the a priori, which resulted from Wittgenstein's revolutionary idea that "the ultimate explanatory level in semantics is not given by reference to ... objects or meanings, but by reference to the meaning-giving activity of human beings, an activity embodied in their endorsement of rules" (Coffa 1991, 167), has been compared by Coffa to Kant's "Copernican turn" in epistemology. Acknowledging that this philosophical breakthrough provides an appropriate background for analyzing Bohr's reflection on quantum mechanics[12] is not, of course, claiming that the inferentialist reading of complementarity provides an acceptable *solution* of the quantum puzzles. For the normative conception of meaning is notoriously as controversial as the interpretation of quantum theory itself (Wright 2008; Hale and Wright 2010). Yet I believe that, by bringing the metasemantic issues raised by quantum probabilities to the fore, such a reading helps identify the questions that need to be addressed in order to implement the philosophical agenda implicit in Bohr's approach.

Notes

1 In a recent book, Faye (2016) argues that Bohr might have been quite sympathetic to an evolutionary explanation of the privileged role of classical concepts. All depends, however, on what one expects the evolutionary argument to establish when the issue of consistency is at stake.

2 A manuscript of 1958 (in all likelihood dictated by Bohr to Petersen) reports the following remark: "Everett's view about a virtual world state described by the wave function does not take into account the conditions for the definition of our experiences" (Petersen 1958). I thank Jan Faye for bringing this document to my attention.

3 Assuming that M could be defined, one would be faced with still another problem. Owing to the condition of mutual exclusiveness, which must hold for contexts designed to measure incompatible observables, the state vectors associated with the various configurations of the apparatus denote mutually exclusive situations. They can accordingly be identified with the eigenvectors of a single observable, namely, M. We now note that, qua quantum observable, M belongs to a family of incompatible observables. The problem with recalcitrant truth-value assignments discussed earlier in this section, which contextualization was intended to fix, reappears therefore on the level of the metaobservable, whose eigenvectors characterize the possible configurations of the original apparatus. This just means that the probabilities for the values of M must themselves be construed as conditionals. Also, the requirement

of mutual exclusiveness for contexts designed to measure incompatible observables must apply, mutatis mutandis, to the family of incompatible observables to which M belongs. Because the argument can be reiterated ad infinitum, this leads to a regress. Moreover, for reasons analogous to those which prevent the straightforward definition of M, it is not clear how such constraints might be formulated recursively. I discuss these issues extensively in a forthcoming paper.

4 These views are expressed in a letter to B. S. DeWitt (May 31, 1957) and in a manuscript entitled *Objective vs Subjective Probability* (1955), both quoted in Osnaghi, Freitas, and Freire (2009, 105–106). Along the same lines, Weinberg (2005) argues that Bohr's approach "is surely wrong; physicists and their apparatus must be governed by the same quantum mechanical rules that govern everything else in the universe." Omnès (1992, 340–1) makes similar remarks.

5 See Jammer (1974, 488–90) and Murdoch (1987, 114–18). In his textbook of 1951, Bohm pointed out that "a measurement process is irreversible in the sense that, after it has occurred, re-establishment of definite phase relations between the eigenfunctions of the measured variable is overwhelmingly unlikely" and he noticed that this irreversibility "greatly resembles that which appears in thermodynamic processes, where a decrease of entropy is also an overwhelmingly unlikely possibility" (608).

6 In his reply to the letter from Schrödinger quoted above, Bohr explained that his "emphasis on the unavoidability of the classical description of the experiment refer[red] in the end to nothing more that the apparently obvious fact that the description of every measuring apparatus basically must contain the arrangement of the apparatus in space and its function in time, *if we are to be able to say anything at all about the phenomena*" (quoted in Moore 1989, 213; my emphasis). He concluded that the argument was "thus above all that the measuring instruments, if they are to serve as such, cannot be included in the actual range of applicability of quantum mechanics" (ibid.).

7 A number of studies (Chevalley 1994; Faye 1991; Brock 2003) have pointed out the roots of Bohr's methodology in nineteenth-century scientific philosophy. To borrow a classic metaphor from literary criticism, no doubt Bohr's model of rationality had more to do with the Romantic archetype of the *lamp* than with the paradigm of the *mirror* inherited from the epistemological tradition of the Enlightenment (see Brandom 2000, 7–10).

8 Some analogies between Bohr's approach and Dummett's are discussed by Murdoch (1987, 222–5) and Faye (1991, 221).

9 Bohr's instrumentalism was sometimes taken to imply "subjectivism" (see Howard 2004).

10 The quoted passage is taken from an unpublished note of 1928, which appears (translated from Danish) in Brock (2003, 269). John Honner (1987) has dealt both with the holistic features of Bohr's approach and with the transcendental structure of his overall argument for the completeness of quantum mechanics. His analysis and conclusions, however, differ substantially from mine. For example, Honner contrasts Bohr's allegedly "moderate" holism to Richard Rorty's "thoroughgoing holism," claiming that what Bohr is ultimately concerned with is the relationship between "picturing" reality and reality itself (18–22). Jan Faye's analysis is more relevant to the present study. According to Faye, Bohr "was keenly opposed to all the ingredients" of the representational theory of knowledge. Faye (1991, 136) argues that Bohr "explicitly or implicitly rejected ... a correspondence theory of truth, a picture theory of knowledge [and] strong objectivism," and he instead adhered to "a coherence theory

of truth, a non-picturing theory of knowledge [and] weak objectivism." See also Bitbol (2001).

11 One of the main (and most challenging) goals of Brandom's inferentialist theory of propositional content is to provide a rigorous account of objectivity along these lines. This can be achieved, according to Brandom, by taking the "normative fine structure of rationality" into account, that is, by introducing the notion of "entitlement to a commitment" and by postulating that there exist commitments to which we are *not* entitled if we endorse a certain claim. This idea is developed at length in Brandom (1994).

12 Edward MacKinnon (1985, 115) argued that Bohr "anticipated some of the key features [of philosophy] later developed in Wittgenstein's Philosophical Investigations," with its major thesis that "the meaning of a word is its use in the language." See also Stapp (1972, 1114) and Bub (1977).

References

Bell, J. S. 1990. "Against 'measurement.'" *Physics World* 3 (8): 33–40.

Bitbol, M. 1998. "Some steps towards a transcendental deduction of quantum mechanics." *Philosophia Naturalis* 35: 253–80.

Bitbol, M. 2001. "Non-representationalist theories of knowledge and quantum mechanics." *Nordic Journal of Philosophy* 2: 37–61.

Bohm, D. 1951. *Quantum Theory*. New York: Prentice-Hall.

Bohr, N. 1934. *Atomic Theory and the Description of Nature*. Cambridge: Cambridge University Press.

Bohr, N. 1958a. *Atomic Physics and Human Knowledge*. New York: Wiley.

Bohr, N. 1958b. "The unity of knowledge." John Franklin Carlson Lecture, Iowa State University. Archive for the History of Quantum Physics, American Philosophical Society, Philadelphia.

Bohr, N. 1962. Interview with Thomas S. Kuhn, Aage Petersen, and Léon Rosenfeld, November 1, 1962, Session V, Niels Bohr Library & Archive, American Institute of Physics, New York.

Bohr, N. 1963. *Essays 1958–1962 on Atomic Physics and Human Knowledge*. New York: Wiley.

Bohr, N. 1985. *Naturbeskrivelse og menneskelig erkendelse*. Copenhagen: Rhodos.

Bohr, N. 1998. *Causality and Complementarity: Supplementary Papers*. Vol. 4 of *The Philosophical Writings of Niels Bohr*, edited by J. Faye and H. J. Folse. Woodbridge, CT: Ox Bow Press.

Brandom, R. 1994. *Making It Explicit*. Cambridge, MA: Harvard University Press.

Brandom, R. 2000. *Articulating Reasons: An Introduction to Inferentialism*. Cambridge, MA: Harvard University Press.

Brock, S. 2003. *Niels Bohr's Philosophy of Quantum Physics*. Berlin: Logos.

Bub, J. 1977. "Reply to Professor Causey." In *The Structure of Scientific Theories*, edited by F. Suppe, 402–408. Urbana: University of Illinois Press.

Bub, J. 1979. "The measurement problem of quantum mechanics." In *Problems in the Foundations of Physics: Proceedings of the International School of Physics Enrico Fermi. Course LXXII*, edited by G. Toraldo di Francia, 71–124. Dordrecht: Reidel.

Chevalley, C. 1994. "Niels Bohr's words and the Atlantis of Kantianism." In *Niels Bohr and Contemporary Philosophy*, edited by in J. Faye and H. J. Folse. Dordrecht: Kluwer Academic Publishers.

Chevalley, C. 1995. "On objectivity as intersubjective agreement." In *Physik, Philosophie und die Einheit der Wissenschaften*, edited by L. Krüger and B. Falkenburg, 333–46. Heidelberg: Spektrum Akademischer Verlag.

Coffa, J. A. 1991. *The Semantic Tradition from Kant to Carnap. To the Vienna Station*, edited by Linda Wessels. Cambridge: Cambridge University Press.

Daneri, A., A. Loinger, and G. M. Prosperi. 1962. "Quantum theory of measurement and ergodicity conditions." *Nuclear Physics* 33: 297–319. Reprinted in *Quantum Theory and Measurement*, edited by J. A. Wheeler and W. H. Zurek, 657–79. Princeton: Princeton University Press, 1983.

Dummett, M. 1991. *The Logical Basis of Metaphysics*. Cambridge, MA: Harvard University Press.

Everett III, H. 1957. "'Relative state' formulation of quantum mechanics." *Reviews of Modern Physics* 29: 454–62. Reprinted in *Quantum Theory and Measurement*, edited by J. A. Wheeler and W. H. Zurek, 315–23. Princeton: Princeton University Press, 1983.

Everett III, H. 1973. "The theory of the universal wave function." In *The Many-Worlds Interpretation of Quantum Mechanics*, edited by B. S. DeWitt and N. Graham, 3–140. Princeton: Princeton University Press.

Faye, J. 1991. *Niels Bohr, His Heritage and Legacy: An Anti-realist View of Quantum Mechanics*. Dordrecht: Kluwer Academic Publisher.

Faye, J. 2016. *Experience and Beyond: The Outline of a Darwinian Metaphysics*. London: Palgrave-Macmillan.

Folse, H. J. 1978. "Kantian aspects of complementarity." *Kant-Studien* 69: 58–66.

Greenberger, D. M., M. A. Horne, A. Shimony, and A. Zeilinger. 1990. "Bell's theorem without inequalities." *American Journal of Physics* 58 (12): 1131–43.

Hale, B., and C. Wright. 2010. "Assertibilist truth and objective content: Still inexplicit?" In *Reading Brandom: On Making It Explicit*, edited by B. Weiss and J. Wanderer, 276–94. London: Routledge.

Honner, J. 1982. "The transcendental philosophy of Niels Bohr." *Studies in History and Philosophy of Sciences* 13: 1–30.

Honner, J. 1987. *The Description of Nature: Niels Bohr and the Philosophy of Quantum Physics*. Oxford: Oxford University Press.

Howard, D. 2004. "Who invented the Copenhagen interpretation? A study in mythology." *Philosophy of Science* 71: 669–82.

Jammer, M. 1974. *The Philosophy of Quantum Mechanics*. New York: Wiley.

Kaiser, D. 1992. "More roots of complementarity: Kantian aspects and influences." *Studies in History and Philosophy of Science* 23 (2): 213–39.

Katsumori, M. 2011. *Niels Bohr's Complementarity: Its Structure, History, and Intersections with Hermeneutics and Deconstruction*. Dordrecht: Springer.

MacKinnon, E. M. 1985. "Bohr on the foundations of quantum theory." In *Niels Bohr, A Centenary Volume*, edited by A. P. French and P. J. Kennedy, 101–20. Cambridge, MA: Harvard University Press.

Moore, W. J. 1989. *Schrödinger: Life and Thought*. Cambridge: Cambridge University Press.

Murdoch, D. 1987. *Niels Bohr's Philosophy of Physics*. Cambridge: Cambridge University Press.

Omnès, R. 1992. "Consistent interpretations of quantum mechanics." *Reviews of Modern Physics* 64: 339–82.

Osnaghi, S. 2008. "Van Fraassen, Everett, and the critique of the Copenhagen view of measurement." *Principia* 12 (2): 155–75.

Osnaghi, S., F. Freitas, and O. Freire Jr. 2009. "The origin of the Everettian heresy." *Studies in History and Philosophy of Modern Physics* 40: 97–123.

Petersen, Aa. 1958. "Philosophical lesson." Manuscript, January 4, 1958. Niels Bohr Archive, Copenhagen.

Petersen, Aa. 1968. *Quantum Physics and the Philosophical Tradition.* Cambridge: MIT Press.

Petersen, Aa. 1985. "The philosophy of Niels Bohr." In *Niels Bohr, A Centenary Volume,* edited by A. P. French and P. J. Kennedy, 299–310. Cambridge, MA: Harvard University Press.

Rosenfeld, L. 1965. "The measuring process in quantum mechanics." *Progress of Theoretical Physics Supplements,* 222–31.

Rovelli, C. 1996. "Relational quantum mechanics." *International Journal of Theoretical Physics* 35: 1637–78.

Schroeder-Heister, P. 2014. "Proof-theoretic semantics." In *The Stanford Encyclopedia of Philosophy* (Summer 2014 Edition), edited by Edward N. Zalta. http://plato.stanford.edu/archives/sum2014/entries/proof theoretic-semantics/.

Sellars, W. 1953. "Inference and meaning." *Mind* 62: 313–38.

Stapp, H. P. 1972. "The Copenhagen interpretation." *American Journal of Physics* 40: 1098–116.

Teller, P. 1981. "The projection postulate and Bohr's interpretation of quantum mechanics." In *PSA 1980: Proceedings of the 1980 Biennial meeting of the Philosophy of Science Association—II,* edited by P. D. Asquith and R. N. Giere, 201–23. East Lansing: Michigan State University.

van Fraassen, B. C. 1972. "A formal approach to the philosophy of science." In *Paradigms and Paradoxes: The Philosophical Challenge of the Quantum Domain,* edited by R. G. Colodny, 303–66. Pittsburgh: University of Pittsburgh Press.

van Fraassen, B. C., and C. A. Hooker. 1976. "A semantic analysis of Niels Bohr's philosophy of quantum theory." In *Foundations of Probability Theory, Statistical Inference, and Statistical Theories of Science—III,* edited by W. L. Harper and C. A. Hooker, 221–41. Dordrecht: Reidel.

van Fraassen, B. C. 1991a. "The problem of measurement in quantum mechanics." In *Quantum Theory of Measurement and Related Philosophical Problems. Proceedings of the Symposium on the Foundations of Modern Physics 1990,* edited by P. J. Lahti and P. Mittelstaedt, 497–503. Singapore: World Scientific Publishing.

van Fraassen, B. C. 1991b. *Quantum Mechanics: An Empiricist View.* Oxford: Oxford University Press.

Waismann, F. 1979. *Ludwig Wittgenstein and the Vienna Circle,* edited by Brian McGuinness, translated by J. Schulte and B. McGuinness. New York: Harper & Row.

Weinberg, S. 2005. "Einstein's mistakes." *Physics Today* 58 (11): 31–5.

Wheeler, J. A. 1957. "Assessment of Everett's 'relative state' formulation of quantum theory." *Reviews of Modern Physics* 29 (3): 463–65. Reprinted in *Quantum Theory and Measurement,* edited by J. A. Wheeler and W. H. Zurek, 324–5. Princeton: Princeton University Press, 1983.

Wigner, E. P. 1961. "Remarks on the mind-body question." In *The Scientist Speculates,* edited by I. J. Good, 284–302. London: Heinemann. Reprinted in *Quantum Theory and Measurement,* edited by J. A. Wheeler and W. H. Zurek, 168–81. Princeton: Princeton University Press, 1983.

Wittgenstein, L. 1979. *Wittgenstein's Lectures: Cambridge, 1932-1935,* edited by A. Ambrose. Totowa: Rowman & Littlefield.

Wright, C. 2008. "Rule-following without reasons: Wittgenstein's quietism and the constitutive question." In *Wittgenstein and Reason*, edited by J. Preston, 123–44. Malden: Wiley-Blackwell.

Zurek, W. H. (2003). "Decoherence, Einselection, and the quantum origins of the classical." *Reviews of Modern Physics* 75: 715–75.

8

Fragmentation, Multiplicity, and Technology in Quantum Physics

Bohr's Thought from the Twentieth to the Twenty-First Century

Arkady Plotnitsky

1. Introduction

I would like to begin with the discovery of the Higgs boson, arguably the greatest event of fundamental physics of the twenty-first century thus far, and, thus far, a culminating event in the history of quantum physics. This discovery has been discussed at all levels and in all media, with photographs of the "events" testifying to the discovery, of various components of the Large Hadron Collider (LHC), and the relevant parts of the mathematical formalism of quantum field theory (e.g., "The Higgs Boson," Wikipedia 2016). This event, I shall argue, confirms the understanding of quantum theory, from quantum mechanics (QM) to quantum electrodynamics (QED) to quantum field theory (QFT), in "the spirit of Copenhagen," initiated by Bohr's thinking (Heisenberg 1930, iv).[1] As such, it reflects the continuing significance and impact of this thinking, which was shaped more by, and has shaped more, the development of QFT than is customarily acknowledged. The development of QED and QFT, beginning with Dirac's pioneering work, served to Bohr as a confirmation of his understanding of quantum phenomena and quantum theory, and as giving new dimensions to this understanding. In particular, QFT reaffirms what I shall call the technological understanding of both developed by Bohr, in the broad sense of "technological." Thus, the discovery of the Higgs boson is the result of the joint workings of three technologies:

1. The experimental technology of the LHC
2. The mathematical technology of QFT (sometimes coupled to the technology of philosophical thought)
3. Digital computer technology

Any quantum event, I shall argue, is made possible through the joint workings of the first two *types* of technology, with the third becoming increasingly more prominent. I am of course not suggesting (hence my emphasis on "types") that all quantum events are reducible to those definable by any particular experimental technology or by QFT, but only that the first type of technology should be capable of detecting quantum events and the second of predicting them. Mathematical technology or the technology of philosophical thought are not conventional concepts, and their use here will become clear as I proceed. For now, I use these terms in the general sense of technology as a set of tools that enable us to do something—"to get from here to there." Thus, the mathematics of QM enabled us to predict the hydrogen spectrum, and QFT enabled us to predict the Higgs boson. Although the role of digital technology is one of the defining aspects of contemporary physics, I can only mention it in passing. There is yet another technology involved: that of science as a cultural project, which, however, will be only mentioned in passing as well.

Bohr's philosophy for the twenty-first century is, I argue, defined by its contribution to our understanding the role of technology (experimental, mathematical, and philosophical) in quantum physics, at this point, most significant the high-energy quantum physics. However, in order to understand this contribution, I shall, first, reassess its contribution to the twentieth century. This contribution is, I would argue, defined most significantly by a new concept of fragmentation, as that of "parts" without a "whole," corresponding to Bohr's most famous concept, complementarity, and the non-realist philosophy that accompanies it. I understand this philosophy as that of "reality without realism." It places the reality of quantum objects and processes beyond representation or even thought, and establishes our relations to this reality as probabilistically or statistically predictive, rather than representational, as in classical physics or relativity. This philosophical position is brought to its radical limit in Bohr's later thinking, via his concepts of phenomenon and atomicity, which supplemented his concept of complementarity. QED and QFT added a dimension of radical multiplicity to this situation, by making it no longer possible to control the identity of "parts," even in the absence of wholeness, which was still possible in the low energy quantum regimes, governed by QM. High-energy quantum regimes do not allow one to maintain the identity of a given object, such as an elementary particle, say, an electron, which can transform into a different particle or set of particles (a positron, an electron-positron pair, a photon, etc.) in the course of a given experiment. I would argue that this conjunction of the irreducibly unthinking and the irreducible multiple defines the twenty-first-century thinking in fundamental physics and philosophy, and beyond.

Bohr does not always expressly define the concepts he uses, even his own concepts, which, specifically complementarity, phenomenon, and atomicity, I shall discuss in Section 2. Here I outline the key general concepts used in this chapter, most especially reality and realism, and causality. These concepts could be defined in different ways, and they have been debated from the pre-Socratics on, in modern times, especially following I. Kant's philosophy, and with a new vigor in the wake of quantum mechanics.[2]

By "reality" I refer to that which exists or is assumed to exist, without making any claims concerning the character of this existence. I understand existence as the capacity to have effects upon the world with which we interact. In the case of physics, it is nature or matter, which is usually assumed to exist independently of our interactions with it, and to have existed when we did not exist and to continue to exist when we will no longer exist.

I define "realism" as a specific set of claims concerning what exists and, especially, *how* it exists. In this definition, any form of realism is more than only a claim concerning the existence, or *reality*, of something, but rather a claim concerning the *character* of this existence. Realist theories are sometimes also called ontological theories. The term "ontological" carries additional connotations. They are, however, not important for the present discussion, and these terms will be used interchangeably here. What defines realism most generally is the assumption that a *structure of reality*, rather than only reality itself, exists independently of our interactions with it, or at least that the concept of structure would apply to reality. In other words, realism is defined by the assumption that the ultimate constitution of the domain considered possesses attributes and the relationships among them, which may be either (a) known in one degree or another and, hence, represented, at least ideally, by a theory or model; or (b) unknown or even unknowable.

Non-realist interpretations of quantum phenomena and quantum mechanics, at least that of Bohr and others in the spirit of Copenhagen, not only do not make any of these realist assumptions, but also, in Bohr's ([1958] 1987b, 62) language, "*in principle* [exclude]" them. Quantum objects exist, are *real*, but the nature of this existence places them beyond representation, at least by quantum theory, or even conception, although Bohr might not have been willing to go that far. The reality of quantum objects is a reality without realism. By the same token, only predictions, of probabilistic or statistical nature, concerning quantum phenomena are possible, even in dealing with *elemental individual quantum processes*, such as those associated with elementary particles. This character of quantum predictions is an experimental fact, because the repetition of identically prepared experiments, in general, leads to different outcomes (73). Unlike in classical physics, this difference cannot be diminished beyond a certain limit (defined by Planck's constant, h) by improving the conditions of measurement, as reflected in the uncertainty relations, which are correlative to the probabilistic or statistical nature of quantum predictions.

There is still a form of realism associated with non-realist interpretations, at least that of Bohr. It is defined by the interpretation of the physics of measuring instruments in which the outcomes of quantum experiments are registered as the effects of the interaction between them and quantum objects. These instruments, or rather *their observable parts*, are assumed to be describable by classical physics. The interaction between quantum objects and measuring instruments is quantum and, hence, is not amenable to a realist treatment. In each experiment, this interaction leaves, as an effect, a mark or a set of marks in measuring instruments, which could be very complex, as they were in the case of the Higgs boson. The numerical data associated with such marks can be predicted in the probabilistic or statistical terms by QM or, in high-energy regimes, by QFT. The origin of this "trace" is beyond the reach of experiment or theory. These data, by contrast, can be made part of a permanent record that can be unambiguously represented, communicated, and so forth, and in this sense, is objective (73–4).

The lack of causality, *as it is classically understood*, is an automatic consequence, especially if one places the reality of quantum objects and processes beyond conception, because causality would imply a partial conception of this reality. However, even if one adopts a weaker assumption, which only precludes a representation of this reality, causality is difficult to maintain. E. Schrödinger (1935, 154) expressed this difficulty as follows: "if a classical state does not exist at any moment, it can hardly change causally." I need, however, to define causality, by which I mean classical causality (the only form of causality considered here, or by Bohr). Causality is an ontological category, characterizing reality. It relates to the behavior of physical systems whose evolution is defined by the fact that the state of a given system (as idealized by a given theory or model) is determined at all moments of time by their state at a particular moment of time, indeed at any given moment of time. I define "determinism" as an epistemological category, which denotes our ability to predict the state of a system, at least as defined by an idealized model, exactly, rather than probabilistically, at any moment of time once we know its state at a given moment of time. This concept of causality is in accord with the principle of causality, commonly used from Kant on, which states that if an event takes place, it has a determinable cause or set of causes of which this event is an effect. It is also commonly (although not universally) assumed that the cause must be antecedent to, or at least simultaneous with, the effect. Quantum phenomena, at least in non-realist interpretations, violate the principle of causality, because no determinable event could be established as the cause of a given event, and only statistical correlations between events could be ascertained. Bohr captured this situation in his reply to the famous paper of Einstein, Podolsky, and Rosen (EPR) (Einstein, Podolsky, and Rosen [1935] 1983). He argues for "the necessity of a final renunciation of the classical ideal of causality and a radical revision of our attitude towards the problem of physical reality" (Bohr 1935, 697). Thus, while causality is renounced, the existence, *reality*, of quantum objects is assumed, without our being able to represent this reality.

By "randomness" I refer to a manifestation of the unpredictable. A random event may or may not result from some underlying causal dynamics. The first eventuality defines what may be called "classical randomness," an appearance of randomness underlain by a hidden causal process. This view has been a dominant form of realist thinking throughout the history of Western thought. Thus, in classical statistical physics, randomness and the resulting recourse to probability or statistics are due to our insufficient information concerning systems that are at bottom causal but whose mechanical complexity prevents us from accessing their causal behavior and making deterministic predictions concerning them. The situation in quantum physics is different, given the difficulties of sustaining arguments for the causality of the independent behavior of quantum objects. If an interpretation is non-realist, the absence of causality is, again, automatic, and the recourse to probability or statistics is unavoidable, even in the case of elemental quantum processes. According to Bohr ([1958] 1987b, 34):

It is most important to realize that the recourse to probability laws under [quantum] circumstances is essentially different in aim from the familiar application of statistical considerations as practical means of accounting for the properties of mechanical systems of great structural complexity. In fact, in quantum physics we are presented not with intricacies of this kind, but with the inability of the classical frame of concepts to comprise the peculiar feature of indivisibility, or "individuality," characterizing the elementary processes.

I close this introductory section with the principle of locality, central to the Bohr-Einstein debate. The principle states that no instantaneous transmission of physical influences between spatially separated physical systems ("action at a distance") is allowed, or equivalently that physical systems can only be physically influenced by their immediate environment. Nonlocality is usually seen as undesirable. Under certain circumstances, such as those of the EPR-type experiments, quantum mechanics can make *predictions* concerning the state of spatially separated systems, while allowing one nevertheless to maintain locality. Bohr (1935), by using the concept of complementarity, argued that quantum mechanics is both a complete and a local theory, which Einstein saw as two alternatives for quantum mechanics,[3] By "complete," Bohr meant that quantum mechanics was as complete as nature allows a theory of quantum phenomena to be. It is *incomplete* insofar as it does not represent and (exactly) predict the behavior of individual quantum systems, which Einstein saw as a requirement for a fundamental theory. The question that defined the Bohr-Einstein debate and is still with us now is whether nature would allow for such a theory in the case of quantum phenomena. Einstein thought it should. Bohr thought that it *might not*, which is not the same as that it never will.

2. Fragmentation, complementarity, and the unrepresentable

In Bohr's work, the term "complementarity" designates a concept, a principle, and an interpretation of both quantum phenomena and quantum mechanics. I shall focus on Bohr's ultimate, fully non-realist, interpretation, in place by the later 1930s. A few revisions were added later on, but they were minor. However, several earlier revisions were significant. In particular, while Bohr's concept of complementarity was introduced, in the Como lecture, in 1927, it was modified by Bohr, along with his interpretation itself, already by 1928–29 under the impact of his exchanges with Einstein. Bohr thinking was characterized by a progressive movement away from realism, finally abandoning it. In mid-to-late 1930s, two new (correlative) concepts, "phenomenon" and "atomicity," were introduced by Bohr, in conjunction with the concept of "reality without realism," admittedly, not Bohr's terms ([1937] 1998; [1938] 1998).

Although the concept of complementarity was never given a single sharp definition by Bohr, such as the one offered below, this definition may be surmised from several of his statements. I would argue that by the time of his reply to the EPR argument and even, or just about, by 1929, Bohr gave the concept this type of meaning. Bohr ([1958] 1987b, 40) comes closest to such a formulation in his 1949 "Discussion with Einstein": "Evidence obtained under different experimental conditions cannot be comprehended within a single picture, but must be regarded as *complementary* in the sense that only the totality of the phenomena exhausts the possible information about the objects." Complementarity is defined by:

(a) a mutual exclusivity of certain phenomena, entities, or conceptions; and yet
(b) *the possibility of considering each one of them separately at any given moment of time; and*
(c) *the necessity of using all of them at different moments of time for a comprehensive account of the totality of phenomena that one must consider in quantum theory.*

Part (b) is not stated in the above formulation by Bohr. It can, however, be established on the basis of Bohr's other elaborations there. Parts (b) and (c) are just as important as part (a), and by missing them one would miss much of the import of Bohr's concept.

Complementarity is a specific *concept*, and it should be treated by respecting its specificity. According to Bohr ([1937] 1998, 87), it "does not belong to our daily concepts" and "serves to remind us of the epistemological situation here encountered, which at least in physics is of an entirely novel character." In its ultimate form, the concept is correlative to this situation, which, as that of reality without realism, prevents us from ascertaining the "whole" composed from or assumed to exist behind the complementary "parts." At any given moment of time only one of these parts and not the other could be ascertained, as an effect of quantum-level reality manifested in a measuring device. This ascertainable part is the only "whole" at this moment.

As a *quantum-theoretical concept*, complementarity also acquires probabilistic or statistical aspects and, it follows, mathematical aspects (one needs mathematics to make quantum predictions). That we have a free choice as concerns what kind of experiment we want to perform, as represented in (c), is in accordance with the very idea of experiment, a choice that, as Bohr (1935, 699) notes, also defines classical physics. Contrary to classical physics, however, implementing our decision allows us to make only certain types of predictions and excludes the possibility of other, *complementary*, types of predictions. We actively shape, define the course of, events. In this sense, complementarity may be seen, as it was by Bohr, as a generalization of causality, in the absence of classical causality.

The definition of complementarity given above is more general and allows for applications of the concept beyond quantum physics. Bohr and others inspired by him, such as W. Pauli, K. G. Jung, and M. Delbrück, proposed using the concept in philosophy, biology, and psychology, where the concept acquired new attention recently, as part of the general surge of interest in the use of quantum-mechanical-like modeling.[4] I shall not discuss these extensions here. I would like, however, to comment on the genealogy of complementarity, which does extend beyond physics, including to psychology, Bohr's early scientific interest (Bohr 1962, session 5; Folse 1985, 175; Plotnitsky 2012, 172–9).[5] This may help us to understand Bohr's thinking leading him to complementarity and the key aspects of his use of the concept in quantum theory.

W. James's use of "complementary" (as an adjective) in referring to a split of "the total possible consciousness" into parts in his seminal 1890 work *The Principles of Psychology* (2007) is often cited in this connection. James said: "in certain persons, at least, the total possible consciousness may be split into parts which coexist but mutually ignore each other, and share the objects of knowledge between them. More remarkably still, they are complementary" (206). Whether Bohr was influenced by James in this regard, or otherwise, has been debated (the evidence is sparse). My own sense is that possible influences of James's thinking on that of Bohr or independent parallel traits in their thinking are related less to complementarity than to certain epistemological aspects of James's pragmatism. This influence would be conjoined with those of Hume, Kant, and Nietzsche. There are also junctures of Bohr's thinking, where the connections to and possibly influences of Bergson, Husserl, Heidegger, and Whitehead are likely. A friend of Bohr's father, Georg Brandes, whom Bohr admired, was one of the early champions of Nietzsche and taught the first ever university course on Nietzsche at the University of Copenhagen. In addition, according to Bohr, even in his early thinking, along the lines of complementarity, concerning psychological epistemology, Riemann's ideas about the functions of complex variables might have been more significant than James's ideas (Bohr 1972–96, Vol. 1, 513; 1962; Plotnitsky 2012, 172–9). In any event, Bohr's concept of complementarity is very different from that of James. That Bohr makes complementarity into a noun (James only uses "complementary" as an adjective) is significant, because doing so helped Bohr to define it as a concept. Most crucial, however, is the specific architecture of Bohr's concept, which arises from physical considerations. Probabilistic or statistical, or correlatively mathematical, aspects of this architecture give it its quantum-theoretical specificity, equally important for Bohr's concepts of phenomenon and atomicity. It was quantum physics that led Bohr to these concepts, although Bohr's thinking concerning quantum physics also helped him to define them as philosophical concepts.

In this case, one deals with *extending* these concepts beyond physics. In tracing their genealogy in Bohr's thought, the traffic proceeds in the other direction, although such an investigation, too, might want to avoid "vague analogies" and explore the "conceptual means shared by different fields," as urged by Bohr ([1958] 1987b, 2). Consider, for example, the case of cubism, which presents both a risk of the first and a possibility of the second. Cubism is sometimes invoked as an influence on Bohr, who owed a 1911 cubist painting, "*La Femme au Cheval*" (A woman on a horse), by Jan Metzinger, which Bohr bought at an auction in 1932 (this date is, as will be seen, significant) and apparently liked to explain to his visitors (Anderson 1967, 321–3; Pais 1991, 335). In commenting on this influence, Wikipedia draws on A. I. Miller (2005):

> The artist has broken down the picture plane into facets, presenting multiple aspects of the subject simultaneously. This concept first pronounced by Metzinger in 1910—since considered a founding principle of Cubism—would soon find its way, via complementarity, into the foundations of the Copenhagen interpretation of quantum mechanics; the fact that a complete description of one and the same subject may require diverse points of view which defy a unique description. ("*La Femme au Cheval*," Wikipedia)

The article then cites Miller (2005, 44):

> Cubism directly helped Niels Bohr to discover the principle of complementarity in quantum theory, which says that something can be a particle and a wave at the same time, but it will always be measured to be either one or the other. In analytic cubism, artists tried to represent a scene from all possible viewpoints on

one canvas.... An observer picks out one particular viewpoint. How you view the painting, that's the way it is. Bohr read the book by Jean Metzinger and Albert Gleizes on cubist theory, *Du "Cubisme"* [1912]. It inspired him to postulate that the totality of an electron is both a particle and a wave, but when you observe it you pick out one particular viewpoint.

Is there much more than a vague analogy to these connections, or is much more possible than suggested here in considering the connections between Bohr's thinking and cubism? It depends on how one develops these connections. Miller's way of expressing the situation would not be a good starting point, and in fact it is in conflict with Bohr's thinking. Bohr would never "postulate that the totality of an electron is both a particle and a wave, but when you observe it you pick out one particular viewpoint." This statement is at odds with Bohr's concept as defined by him, because complementary parts never reflect a totality, or a whole, so that one could, arbitrarily, at will, pick a particular point to observe one part of the other. Bohr (1935, 699) expressly says as much: "we are not dealing with an incomplete description characterized by the arbitrary picking up of different elements of physical reality at the cost of sacrificing other such elements." A single electron is *never both* a particle and a wave, for one thing because there would not be complementarity then. For Bohr it is, ultimately, neither.

While there are affinities between cubism in Bohr's thinking and complementarity, the claim that "cubism *directly helped* Bohr to discover the principle of complementarity in quantum theory" is a stretch. Apart from the fact that Bohr had other sources and inspirations, complementarity has a specific set of features, especially as a quantum-theoretical concept, which would be difficult to find in cubism. The ultimate reason for Bohr's invention of complementarity is, I argue, a specific set of interpretive problems found in quantum phenomena and quantum mechanics, and the overall architecture of the concept arises from these problems, even its philosophical architecture that could be extended beyond quantum theory. However, unlike, it is worth noting, Einstein (Pais 1991, 335), Bohr had interests in, and intellectual affinities with, cubism, or modernist art in general, and Metzinger's painting is too tempting not to consider in this connection, even though it would be difficult to make definitive claims here. It is not clear, for example, at what point Bohr read Metzinger and Gleizes's book. My surmise is that Bohr developed his interest in cubism after quantum mechanics and the introduction of complementarity (as noted above, he bought the painting in 1932). He is likely to have seen some cubist paintings earlier and they might have impacted his thinking. It is also likely that he understood cubist paintings in his own, quantum-mechanical way, and found things there that the painters themselves were unlikely to contemplate. But then, it is the essence of great art (or of anything great) to allow others to move beyond how one thought in creating it.

That said, do the relationships between cubism and complementarity or quantum theory, nevertheless, reflect the fact that, to return to Bohr's formulation, "we are not dealing here with more or less vague analogies, but with an investigation of the conditions for the proper use of our conceptual means shared by different fields"? Could they be related in a way "helpful in clarifying the conditions logical relations which, in different contexts, are met with in wider fields" (Bohr [1958] 1987b, 2)? They might be, in particular in considering the problem of fragmentation, which extends across art, philosophy, and mathematics and science, and which is clearly at stake in complementarity. Cubism is an *image* of fragmentation. Complementarity is a *concept* of fragmentation, a fragmentation without a whole that this fragmentation fragments.

Once we deal with fragments we could only make a probabilistic (if not necessarily numerical) assumption of the whole of which these fragments are parts, and then try to justify why our estimate is correct or is more or less likely. That these fragments are the fragments of a whole is already an assumption that may or may not be true. This fact is crucial to complementarity, in which case it is not true. Cubist paintings still appear to rely on our Euclidean intuition of objects and relations between them in space. Or do they? Perhaps what they are instead aiming to tell is that the whole or whatever it is that is behind the parts is different than our Euclidean thinking tells us it is. Our Euclidean intuition and images of objects and their relations are the constructions of our brain from much more fragmented pictures that our eyes "see," to the degree one could speak of seeing apart from this construction. Given that this was discovered before cubism, contemporary with the rise of modern neuroscience, cubists were aware of this. Bohr must have known this when he undertook his early psychological investigations. Or perhaps, cubist paintings tell us, closer to "this seeing without seeing," that these fragments is all there is—This is what a woman on a horse is!—and we should not trust our Euclidean intuition, Euclidean *realism*, which tells us what is the nature of *reality* behind this image.

Did cubists take the next step toward reality without realism, taken by Bohr with complementarity? It is doubtful, in my view, but a secondary matter here. What is crucial is that, in the case of complementary quantum phenomena, "fragments" or "parts" do not add up to a whole, are not "fragments" or "parts" of a "whole." This *breakdown* is the essence of complementarity, and it makes the application of such terms of "whole" and "parts" provisional and ultimately inapplicable. Each complementary part is the only wholeness there is at this point.

In Bohr's ultimate interpretation, the classical properties of complementary parts are manifested, as *effects*, in measuring instruments impacted by quantum objects. Accordingly, even a partial representation of quantum reality is no longer available. Each such configuration—an effect cum the specific experimental arrangement in which this effect is observed—defines a phenomenon in Bohr's ([1958] 1987b, 64) sense:

> I advocated the application of the word phenomenon exclusively to refer to the *observations* obtained under specified circumstances, including an account of the whole experimental arrangement. In such terminology, the observational problem is free of any special intricacy since, in actual experiments, all observations are expressed by unambiguous statements referring, for instance, to the registration of the point at which an electron arrives at a photographic plate. Moreover, speaking in such a way is just suited to emphasize that the appropriate physical interpretation of the symbolic quantum-mechanical formalism amounts only to predictions, of determinate or statistical character, pertaining to individual phenomena

appearing under conditions defined by classical physical concepts [describing the observable parts of measuring instruments]. (Emphasis added)

Referring to *observed* physical situations explains Bohr's choice of the term "phenomenon." This choice is in accord with Kant's philosophy or, historically closer to Bohr, Husserl's phenomenology of consciousness, which also involves a rigorous specification of the circumstances in which phenomena occur. In this case, these circumstances are different: they are phenomenological and do not involve the specification of measuring instruments. This is an important difference, because phenomenally identical entities (such a spot on a silver-bromide screen) must be considered as different depending on the setup of the experiment (such as the double-slit experiment) that defines them. Consciousness is crucial because our conscious experience is necessary to have phenomena in Bohr's sense as registered phenomena and to guarantee that we can unambiguously communicate our knowledge concerning them. Otherwise, no physics would be possible. However, quantum phenomena place insurmountable limits upon how far our consciousness or thinking can reach. Because the observed parts of the measuring arrangements are described classically, phenomena and, physically, measuring instruments, as objects (not quantum objects!), could be considered as identical to each other, as they are in classical physics. By contrast, the *emergence* of these phenomena, due to the interaction between measuring instruments and *quantum objects*, and quantum objects and processes themselves are no longer available to a representation.

Technically, objects and phenomena are also different in classical physics, as was realized by Kant. There, however, this difference could be disregarded. The difference between quantum phenomena and quantum objects in Bohr is, thus, beyond Kant's difference between phenomena (representation) and things-in-themselves (objects), which ultimately allows for realism (Kant 1997). It is also beyond Husserl's phenomenology, or even Heidegger's ontology, which emphasized the difference between (phenomenally experienced) entities, "things," and processes (Being-becoming) that bring them about (Heidegger [1927] 2010). In Heidegger, this process is still in principle available to thought, as things-in-themselves are in Kant. The difference between Heidegger and Kant is Heidegger's emphasis, which is closer to Hegel than to Kant, on the process or becoming and emergence, similar to the view of "process and reality" advanced by Whitehead ([1929] 1979), from whom Bohr might have borrowed his concept of atomicity. In Bohr, the *reality* of quantum objects and *processes*, which leads to the *emergence* of phenomena, is beyond phenomenal representation, if not conception. Terms like object and process, or quantum, are provisional and ultimately inapplicable any more than any other terms. This makes Bohr's conception of reality and his epistemology more radical than that of these thinkers just mentioned and brings it closer to Nietzsche and his twentieth-century descendants, such as J. Derrida, G. Bataille, and J.-F. Lyotard (Plotnitsky 1994).

This situation also led Bohr to his associated concept of "atomicity," in the original Greek sense of an entity that is not divisible any further, which, however, now applies at the level of phenomena, rather than refers, on Democretean lines, to indivisible physical entities, "atoms," of nature. Bohr's rethinking of the quantum-mechanical situation in terms of phenomena enabled him to transfer to the level of observable configurations manifested in measuring instruments all the key features of quantum physics—discreteness, discontinuity, individuality, and atomicity (indivisibility)—previously associated with quantum objects themselves. Both concepts emerged at the same time and are more or less equivalent. Both are defined in terms of *individual* effects of quantum objects upon the classical world. Thus, "atomicity" refers to physically complex and hence *physically* subdivisible entities, and no longer to elemental physical entities. These "atoms" are individual phenomena in Bohr's sense, rather than indivisible quantum objects, to which one can no longer ascribe atomic properties any more than any other properties.[6]

It follows that when one deals with complementary phenomena one never deals with complementary properties of quantum objects or their independent behavior, because no attribution of such properties, single or joint, is possible. No complementary phenomena can ever be associated with a single quantum object. One always needs two quantum objects in order to enact, in two separate complementary experiments, say, those associated, respectively, with the position or the momentum measurement, with each corresponding quantity physically pertaining strictly to the measuring instruments. The uncertainty relations, too, now apply to the measuring instruments and not quantum objects. Furthermore, in Bohr's interpretation, the uncertainty relations mean that one cannot define but rather only measure both variables simultaneously. The recourse to probability or statistics, to which the uncertainty relations are correlative, is the cost of our active role in *defining* events, rather than merely tracking them, as in classical physics or relativity.

3. The unrepresentable and the multiple

Heisenberg's ([1925] 1969) initial approach to quantum mechanics and Bohr's interpretation of the theory were defined by the following main principles:

1. The principle of quantum discreteness, according to which all quantum phenomena, defined as what is observed in measuring instruments, are individual and discrete, which is not the same as the (Democretian) atomic discreteness of quantum objects;
2. The principle of the probabilistic or statistical nature of quantum predictions, even in the case of quantum processes and events associated with elemental individual constituents of nature; and
3. The correspondence principle, which, as initially defined by Bohr, required that the predictions of quantum theory must coincide with those of classical mechanics at the classical limit, but was given a mathematical form by Heisenberg, which required that both the equations and variables used converts into those of classical mechanics in the classical limit.

Bohr's interpretation, first, proposed in 1927 in the Como lecture, added:

4. The complementarity principle, which stems from the concept of complementarity, considered in the preceding section.

Bohr's (Bohr [1925] 1987a, 48) initial comments on matrix mechanics show a clear grasp of Heisenberg's approach:

> In contrast to ordinary mechanics, the new quantum mechanics does not deal with a space-time description of the motion of atomic particles. It operates with manifolds of quantities which replace the harmonic oscillating components of the motion and symbolize the possibilities of transitions between stationary states in conformity with the correspondence principle. These quantities satisfy certain relations which take the place of the mechanical equations of motion and the quantization rules.

Part of the mathematical architecture of Heisenberg's scheme was provided by the equations of classical mechanics by the mathematical correspondence principle. Classical variables, however, had to be replaced by new variables, which were complex-valued matrix variables (essentially, operators in a Hilbert space over complex numbers), never used in physics previously.

Heisenberg's thinking not only introduced a new type of mathematical model in physics but, as a result, also redefined the nature of theoretical physics. Indeed, it redefined the nature of experimental physics as well. *The practice of experimental physics* no longer consists, as in classical physics or relativity, in tracking the independent behavior of the systems considered. Instead it consists in creating configurations of experimental technology, which reflects the fact that what happens is defined by what experiments we perform. These configurations embody the effects of the interactions between quantum objects and measuring instruments, through which effects quantum objects are defined and, when possible, distinguished from one another, while themselves remaining beyond the reach of quantum theory.

The practice of theoretical physics, accordingly, no longer consists in offering an idealized mathematical representation of quantum objects and their behavior, but in developing a mathematical technology able to predict the outcomes of quantum events and correlations between some of these events. These predictions are unavoidably probabilistic or statistical.

The situation takes a more radical form in high-energy quantum physics. While, in this interpretation, retaining the non-realist epistemology of QM, QFT (including QED) supplements the irreducibly unrepresentable nature of quantum processes with their irreducible multiple nature, due to the following circumstances:

1. more complex configurations of observed phenomena, defined by the effects of the interaction between quantum objects and measuring instruments, configurations reflecting the loss of the identity of elemental quantum objects even within a single experiment;
2. a more complex structure of the theoretical predictions and, thus, of the relationships between the mathematical formalism of the theory and the effects in question;
3. a more complex structure of the mathematical formalism, responding to the situation described in (1) and (2).

While the discovery of the Higgs boson may be the most spectacular recent example of this situation, many other remarkable examples are found throughout the history of high-energy physics, beginning with the discovery of antimatter, a consequence of Dirac's equation for the (free) relativistic electron, and the inaugural event of this history.

Dirac's discovery was based on the fundamental principles of both quantum theory, as stated above, and relativity. Dirac's (1928, 611–12) theory enabled him to answer, for the relativistic electron, the question: "What is the probability of any dynamical variable at any specified time having a value laying between any specified limits, when the system is represented by a given wave function ψ_n?" This is the main question of quantum theory, as understood, beginning with Bohr, in the spirit of Copenhagen. Bohr quickly came to realize the significance of Dirac's equation and QED, and then QFT. Bohr ([1958] 1987b, 63) came to see

> Dirac's ingenious quantum theory of the electron as a most striking illustration of the power and fertility of the general quantum-mechanical way of description. In the phenomena of the creation and annihilation of electron pairs we have in fact to do with new fundamental features of atomicity, which are ultimately connected with non-classical aspects of quantum statistics expressed in the exclusion principle, and which demanded a still more radical [than in quantum mechanics] renunciation of explanation in terms of a pictorial representation.

By "new fundamental features of atomicity" Bohr clearly refers (the statement dates from 1949) to his concept of atomicity, perhaps influenced as much by the developments of QED as by the development of quantum mechanics, such as the EPR experiment. QED and QFT played major roles in the development of Bohr's thinking. Bohr made a major contribution to the field by a classic treatment of quantum-field measurement (Bohr and Rosenfeld [1933] 1983) and even an updated version of the 1933 article, which takes into account the intervening developments, such as the renormalization of QED (Bohr and Rosenfeld [1950] 1983).[7] Dirac's equation is

$$\left(\beta mc^2 + \sum_{k=1}^{3} \alpha_k p_k c \right) \psi(x,t) = i\hbar \frac{\partial \psi(x,t)}{\partial t}$$

The new mathematical elements here are the 4×4 matrices α_k and β and the four-component wave function ψ. The Dirac matrices are all Hermitian,

$$\alpha_i^2 = \beta^2 = I_4$$

(I_4 is the identity matrix), and they mutually anticommute:

$$\alpha_i \beta + \beta \alpha_i = 0,$$
$$\alpha_i \alpha_j + \alpha_j \alpha_i = 0.$$

The above equation unfolds into four coupled linear first-order partial differential equations for the four quantities that make up the wave function. The matrices form a basis of the corresponding Clifford algebra. p is the momentum operator in Schrödinger's sense. The wave function $\psi\ (t,\ \boldsymbol{x})$ takes value in a Hilbert space $X = \mathrm{C}^4$ (Dirac's spinors are elements of X). For each t, $\psi\ (t,\ \boldsymbol{x})$ is an element of $H = \mathrm{L}^2\ (\mathrm{R}^3;\ X) = \mathrm{L}^2\ (\mathrm{R}^3) \otimes X = \mathrm{L}^2\ (\mathrm{R}^3) \otimes \mathrm{C}^4$. Spin is contained by the theory automatically.

QFT mathematically responded and, in Dirac's case, led to the discovery of the following situation. Suppose that one arranges for an emission of an electron, at a given high-energy, from a source and then performs a measurement at a certain distance from that source. Placing a photographic plate at this point would do. The probability of the outcome would be properly predicted by QED. But what is the outcome? It is not what our classical or even our quantum-mechanical intuition would expect. One might find, in the corresponding region, not only an electron or nothing, as in QM regimes (in classical physics one would find the same object), but also other particles: a positron, a photon, an electron–positron pair, and so on. QED correctly predicts which among such events can occur, and with what probability. In order to do so, one needs more complex Hilbert-space machinery than in QM. Thus, in Dirac's theory, the wave function ψ is a four-component Hilbert-space vector, as opposed to a one-component Hilbert-space vector, as in QM. This Hilbert space is, as noted, $H = \mathrm{L}^2\ (\mathrm{R}^3;\ X) = \mathrm{L}^2\ (\mathrm{R}^3) \otimes X = \mathrm{L}^2\ (\mathrm{R}^3) \otimes \mathrm{C}^4$.

Once we move to still higher energies or different domains governed by QFT, the panoply of possible outcomes becomes even greater. The Hilbert spaces would be given a still more complex structure. It follows that in QFT an investigation of a particular type of particles involves not only other particles of the same type but also other types of particles. The identity of particles within each type is strictly maintained in QFT, as it is in QM. One could never be certain that one encounters the same electron in the experiment just described even in the quantum-mechanical situation, although the probability that it would be a different electron is low in the QM regime in comparison to that in the regime of QED. In QED or QFT one encounters a continuous emergence and disappearance of particles from point to point, the process governed by the concept of virtual particle formation. The operators used to predict the probability of such events are called the creation and annihilation operators. In this way, the high-energy physics, captured by QFT, adds the irreducibly multiple to the irreducibly unrepresentable of QM.

The discovery of this situation was a momentous event in the history of quantum physics, and, as noted above, it was recognized by Bohr as such. Heisenberg (1989, 31–3) spoke of Dirac's theory as "perhaps the biggest change of all the big changes in physics of our century. It was a discovery of utmost importance because it changed our whole picture of matter, … It was one of the most spectacular consequences of Dirac's discovery that the old concept of the elementary particle collapsed completely."

Could this situation be brought in accord with Bohr's concepts of phenomenon and atomicity? Yes, it can, and it may need to be. Already low-energy QM regimes allow for the concept of elementary particles consistent with these concepts, beginning with the fact that elementary particles within the same type (electrons, photons, and so forth) cannot be distinguished from each other, while these types themselves are rigorously distinguishable. Both features are consistently defined by the corresponding sets of effects manifested in measuring instruments, and thus are consistent with the view that the nature of elementary particles and their behavior is beyond representation, precluding one from associating any independent properties with them. An elementary particle of a given type, say, an electron, is defined by a given set of possible phenomena (the same for all electrons) observable in measuring instruments in the experiments associated with the particles of this type. Speaking of the "same" electron detected in the course of a given experiment, again, loses its meaning in high-energy regimes. The *elementary* character of an elementary particle is defined by the fact that there is no experiment that allows one to associate the corresponding effects on their interactions with measuring instruments with more elementary individual quantum objects. Of course, once such an experiment is performed or becomes possible, the status of a given object as an elementary particle could be experimentally challenged or disproved, as it happened in hadron physics, when baryons and mesons were shown to be composed of quarks and gluons. Elementary particles are defined not in terms of their representation but in technological terms of particular types of effects of the interactions between them and measuring instruments.

QFT made a remarkable progress since the time of Heisenberg's assessment in 1970s, as is manifested in the electroweak unification and the quark-gluon model of nuclear forces, both underlined by the Higgs field, developments that commenced around the time of these remarks. QFT also led to string and brane theories. However, the essential epistemological aspects of QFT, as here considered, have remained in place.[8]

This does not mean that QFT does not need further advancement or does not have problems, such as and in particular those that led to the necessity of renormalization. While exceptionally successful in their experimental predictions, QFTs still are mathematically unsatisfactory, if legitimate at all, although it is possible to see renormalization in more benign ways, via the so-called renormalization group and effective quantum field theories, introduced in the 1970s. It was realized by the early 1930s that the computations provided by QED were reliable only as a first order of perturbation theory and led to the appearance of infinities when one attempted to use the formalism for calculations that would provide closer approximations matching the experimental data. These difficulties were eventually handled through the renormalization procedure. Roughly speaking, the procedure might be seen as manipulating infinite integrals that are divergent and, hence, mathematically illegitimate. At a certain stage of calculation, these integrals are replaced by finite integrals through artificial cut-offs that have no proper mathematical justification within the formalism and are performed by putting in, by hand, experimentally obtained numbers that make these integrals finite, which removes the infinities from the final results of calculations, which are experimentally confirmed to a very high degree. In the case of QED, renormalization was performed in the later 1940s by S.-I. Tomonaga, J. Schwinger, and R. Feynman (which brought them a joint Nobel Prize in 1965), with important contributions by others, especially F. Dyson. The Yang-Mills theory, which grounds the standard model, was eventually shown to be renormalizable by M. Veltman and G. 't Hooft in the 1970s (bringing them their Nobel Prize). This allowed for the development of the standard model of all forces of nature, except for gravity, which has no quantum form thus far.

The technical aspects of renormalization and of the subsequent developments mentioned above are beyond my scope, and I refer to Kuhlman (2014), Cao (1999), and Teller (1995), and, for a historical account, Schweber (1994, 595–605). Instead I would like to ask a somewhat different question, which brings Bohr back into the picture: "Why does quantum field theory, say, quantum electrodynamics, contain such infinities, in contrast to quantum mechanics?" Bohr's argumentation, arising from the analysis given in Bohr and Rosenfeld ([1933] 1983), suggests a possible reason. The mathematical formalism of QFT, Bohr and Rosenfeld argued, is essentially linked to the same idealization of measurement that is used in the low energy of the QM regimes, because it disregards "the atomic structure of the field sources and the measuring instruments" (520). As Bohr noted in 1937, with his work with Rosenfeld in mind:

> On closer consideration, the present formulation of quantum mechanics in spite of its great fruitfulness would yet seem to be no more than a first step in the necessary generalization of the classical mode of description, justified only by the possibility of disregarding in its domain of application the atomic structure of the measuring instruments themselves in the interpretation of the results of experiment. For a correlation of still deeper laws of nature involving not only the mutual interaction of the so-called elementary constituents of matter but also the stability of their existence, this last assumption can no longer be maintained, as we must be prepared for a more comprehensive generalization of the complementary mode of description which will demand a still more radical renunciation of the usual claims of so-called visualization. (88)

In classical physics, then, we can disregard measuring instruments altogether, at least ideally and in principle, which is, however, how mathematical models work in physics. In low-energy quantum physics, while we cannot disregard the measuring instruments, we can disregard the atomic constitution of measuring instruments without any mathematical problems in predicting the outcomes of experiments by QM. However, these predictions are now probabilistic or statistical. In high-energy quantum physics, governed by QED and QFT, which is equally probabilistic or statistical, we still disregard the atomic structure of the measuring instruments, using the same idealization that we use in quantum mechanics, but now at a mathematical cost.

Could this idealization, which poses no problems in QM, be responsible for the appearance of the infinities in QFT? Dyson thought so, under the impact of Bohr and Rosenfeld's analysis:

> We interpret the contrast between the divergent Hamiltonian formalism and the finite S-matrix as a contrast between two pictures of the world, seen by two observers having a different choice of measuring equipment at their disposal. The first picture is of a collection of quantized fields with localizable interactions, and is seen by a fictitious observer whose apparatus has no atomic structure and whose measurements are limited in accuracy only by the existence of the fundamental constants c and h. This ["ideal"] observer is able to make with complete freedom on a sub-microscopic scale the kind of observations which Bohr and Rosenfeld

employ ... in their classical discussion of the measurability of field-quantities. The second picture is of collections of observable quantities (in the terminology of Heisenberg) and is the picture seen by a real observer, whose apparatus consists of atoms and elementary particles and whose measurements are limited in accuracy not only by c and h, but also by other constants such as α [the fine-structure constant] and *m* [the mass of the electron]. (Dyson 1949, 1755; cited in Schweber 1994, 547–8)

In other words, in our interactions, via measuring instruments, with quantum objects we are, thus far, "ideal observers," subject only to the uncertainty relations.[9] The observable strata of these instruments are described classically, just as they are in QM measurements, by disregarding their atomic constitution, which would compel us to take into account α and *m*. The necessity of doing so does not mean that we need to describe this atomic constitution and the quantum interaction between quantum objects and the measuring instruments, which may not be possible. However, even in the absence of any representation of quantum objects and processes (including those occurring in the measuring instruments interacting with quantum objects), taking into account the atomic structure of the measuring instruments would require a different type of theory. Would such a theory, assuming it is possible, be able to avoid the infinities? Perhaps! The idealized conceptualization of measurability in question is only one of many idealizations of QED and QFT, some of which are shared with QM. Even if no representational claim is made concerning quantum objects, these theories inherit their Hamiltonian or Lagrangian formalism from classical mechanics, which depends on the idealization of classical objects as dimensionless points endowed with mass. Renormalization may be due to the insufficiency of these idealizations in high-energy quantum regimes.

But then, how important is avoiding the infinities or getting rid of renormalization? Dyson did not think that his considerations, just cited, implied that QED should be replaced by an essentially different theory. In fact he did not think that there is, in view of his argument, anything especially wrong with the infinities found in quantum electrodynamics or renormalization. Bohr and Rosenfeld's ([1950] 1983) updated version of their analysis, which addressed the renormalizability of QED, adopts a similar view. As Schweber (1994, 548) explains:

If [Dyson's] notion of measurability is accepted, the correlation between expressions which are unobservable to a real observer and expressions which are infinite "is a physically intelligible and acceptable feature of the theory. The paradox is the fact that it is necessary ... to start from the infinite expressions in order to deduce finite ones." What may be therefore looked for in "a finite theory" is not necessarily a modification of the present theory which will make all infinite quantities finite "but rather a turning around of the theory so that all the finite quantities shall become primary and the infinite quantities secondary." (Dyson 1949, 1755)

Dyson concluded that "there is no longer [once renormalization is shown to work], as there seemed to be in the past, a compelling necessity for a future theory to abandon some essential features of the present electrodynamics. The present [quantum] electrodynamics is certainly incomplete, but is no longer certainly incorrect" (ibid.; cited in Schweber 1994, 548).

Nevertheless, Dyson's conjecture that renormalization might arise because we are ideal observers and use (or because we use) idealized measuring instruments is worth pondering a bit further. The Hamiltonian character of the quantum formalism was initially brought into QM from classical mechanics, via Bohr's correspondence principle in its mathematical form, by directly transferring the equations of classical mechanics, while changing their variables to those of QM and, by the same token, making the formalism to lose its representational capacity and become probabilistically or statistically predictive. The formalism was then "transferred" by Dirac into QED, via the adjusted mathematical correspondence principle, which requires that Dirac's equation convert into Schrödinger's equation at the QM limit. Hence, unlike in QM, the formalism was changed formally as well. Nevertheless, it was still a descendent of the Hamiltonian formalism of classical mechanics. Neither classical mechanics nor QM needs to take into account the atomic structure of measuring instruments, classical mechanics naturally (because it can disregard the role of measuring instruments entirely), and QM without any real justification, but luckily still allowing us to make correct predictions of the outcomes of quantum experiments. With QED we run out of luck as concerns the finite nature of the formalism, although not out of luck as concerns its predictive capacity, because renormalization works.

Thus, while the correspondence principle, Bohr's great discovery, which was given a mathematical form by Heisenberg, was crucial for the development of QM and then QED, it may have also been responsible for the divergent nature of QED and QFT. The Hamiltonian framework may also be responsible for us not being able to turn it around the way Dyson suggested—by making primary the finite rather than the infinite quantities in the theory. We continue to remain lucky, because renormalization works, at least insofar as gravity is disregarded.

Our future fundamental theories might prove to be finite, thus making the necessity of renormalization a result of the limited reach of our theories at present. While, however, a finite theory may be preferable, renormalization may not be a big price to pay for QFT's extraordinary capacity to predict the manifold of phenomena that it has confronted throughout its history. But then, a finite theory may not be possible, in which case renormalization will continue to be our main hope. For now, QFT continues to provide an extraordinarily effective mathematical technology to handle ever-new types of phenomena observed in ever-new configurations of the high-energy measuring technology. The Linear Hadron Collider (LHC) is the latest machine (in either sense) for creating such configurations, one of which (a very complex one) confirmed the existence of the Higgs boson. This confirmation brought the 2013 Nobel Prize to F. Englert and P. Higgs, who were among those involved in development of the mathematical theory of the Higgs field. Some of the earlier discoveries, such as those of the electroweak (W+, W−, and Z) bosons, the top quark, and the tau-lepton are hardly less significant, and the experimental technologies accompanying them were equally spectacular. Of course, as all technologies, the mathematical technology of QED and the experimental technology of LHC might, and even one day must, become obsolete and be replaced by others. Thus far, however, both have worked marvelously, from Dirac's discovery of antimatter to the discovery of the Higgs boson and most recently a pentaquark. New discoveries are bound to follow: they may confirm the power of QFT or defeat it. A defeat might be more exciting, because it would require a new theory, perhaps an entirely new kind of theory.

4. "The question concerning technology": From Bohr to Heidegger and from Heidegger to Bohr

I began my discussion of quantum mechanics with Bohr's ([1925] 1987a) comments on the non-realist nature of matrix mechanics in the last section, "The Development of a Rational Quantum Mechanics," of "Atomic Theory and Mechanics." Now, in closing, I would like to consider Bohr's concluding postscript there, which might be unexpected, given his subsequent emphasis on the role of measuring instruments. The measuring instruments came to replace "the mathematical instruments," invoked in Bohr's passage, in playing the most "essential part" in his interpretation of quantum phenomena and quantum mechanics from the 1927 Como lecture on. In 1925, however, in the immediate wake of Heisenberg's discovery, Bohr wrote:

> It will interest mathematical circles that the mathematical instruments created by the higher algebra play an essential part in the rational formulation of the new quantum mechanics. Thus, the general proofs of the conservation theorems in Heisenberg's theory carried out by Born and Jordan are based on the use of the theory of matrices, which go back to Cayley and were developed especially by Hermite. It is to be hoped that a new era of mutual stimulation of mechanics and mathematics has commenced. To the physicists it will at first seem deplorable that in atomic problems we have apparently met with such a limitation of our usual means of visualization. This regret will, however, have to give way to thankfulness that mathematics in this field, too, presents us with the tools to prepare the way for further progress. (51)

Bohr was too optimistic in his last hope. There has, it is true, been "thankfulness that mathematics in this field, too, presents us with the tools to prepare the way for further progress." However, the discontent with "the limitation" in question has never subsided. Einstein, again, led the way. Schrödinger was quick to join, with a majority of physicists and philosophers to follow.

Be it as it may on this score, central here is Bohr's view of the essential role of the mathematics in quantum mechanics or physics in general. It is also significant that Bohr speaks of "a new era of mutual stimulation of *mechanics* and mathematics," rather than physics and mathematics, although Heisenberg's discovery redefined the relationships between them as well. Most fundamentally at stake are elemental *individual* quantum processes and events. The mathematical science of these processes in classical physics is mechanics. It is, correspondingly, classical mechanics that is now replaced by quantum mechanics, as a theory that only predicts, in probabilistic or statistical terms, such events as effects of the interactions between quantum objects and measuring instruments without representing these processes.

Heisenberg (1930, 11) argued that

> it is not surprising that our language [or concepts] should be incapable of describing the processes occurring within atoms, for … it was invented to describe the experiences of daily life, and these consist only of processes involving exceedingly large numbers of atoms…. Fortunately, mathematics is not subject to this limitation, and it has been possible to invent a mathematical scheme—the quantum theory [quantum mechanics]—which seems entirely adequate for the treatment of atomic processes.

That Heisenberg *found* a mathematical scheme that could predict the data in question was as fortunate as that mathematics is free of this limitation, for this is also the case in classical physics and in relativity, beginning at least with Lagrange's and Hamilton's analytical mechanics. It is true that matrix algebra was introduced in mathematics before Heisenberg, who was, famously, unaware of it and had to reinvent it, although unbounded infinite matrices that he used were not previously studied in mathematics. But, even if Heisenberg had been familiar with it, his scheme would still have needed to be invented as a mathematical model dealing with quantum phenomena. Mathematics now becomes in a certain sense primary, even though quantum mechanics cannot be reduced to mathematics and, as against classical physics or relativity, contains an irreducible nonmathematical remainder, because no mathematics can apply to quantum objects and processes. It only predicts what is observed in measuring instruments.

Bohr's elaboration shows his profound understanding of this situation. Although it may appear to announce a program that is different from the one Bohr came to pursue later, by taking this view one underestimates the complexity of Bohr's thinking. Bohr's view of quantum mechanics has been defined by recognizing the essential roles of both measuring and mathematical instruments, of both technologies, in their relationships. The very appeal to "instruments" is hardly casual. Apart from the fact that such choices of expression are rarely casual in Bohr, the point is consistent with Bohr's general view of mathematics (e.g., Bohr [1958] 1987b, 68). It suggests that, even if considered apart from physics, mathematics is a technology, a technology of thought, rather than something absolute or ideal, along Platonist lines. Bohr's statement establishes the connections between the mathematical and experimental technologies of quantum physics, and makes technology in its broader sense a foundation of quantum physics.

Technology is a means of doing something and doing it more successfully. For example, the experimental technology of quantum physics enables us to understand how nature works at the ultimate level of its constitution or obtaining and using information through quantum phenomena, which quantum objects enable us to have and to manipulate, as in quantum cryptography and computing. Quantum objects themselves are not technology; they are something technology helps us to discover, understand, work with, and so forth. However, they can become part of technology, beginning with the quantum parts of measuring instruments through which the latter interact with quantum objects, or again, as part of devices we use elsewhere, such as lasers, MRI machines, and so forth. It also follows that the mathematical technology of physics or of mathematics itself is not only or even primarily a way of *representing* reality but a way of experimenting with it, and creating new realities, physical and mathematical, or philosophical.

One might also see philosophy or even thinking in general as technology. This subject, however, may lead us into "metaphysical depths," which would require an engagement with philosophical works that would be difficult to undertake here. H. Weyl (1994, 7) once said: "We cannot set out here in search of a definitive elucidation of what is to be a state of affairs, a judgment, an object, or a property. This task leads into metaphysical depths. And concerning it one must consult men, such as Fichte, whose names may not be mentioned among mathematicians without eliciting an indulgent smile." Nevertheless, I would like to finish by moving from Bohr to Heidegger, by citing Heidegger's conclusion in "The Question of Concerning Technology," and then return to Bohr. According to Heidegger (2004, 34–5):

> There was a time when it was not technology alone that bore the name *techne*. Once that revealing that brings forth truth into the splendor of radiant appearing also was called *techne*.
>
> Once there was a time when the bringing-forth of the true into the beautiful was called *techne*. And the *poiesis* of the fine arts also was called *techne*.
>
> In Greece, at the outset of the destining of the West, the arts soared to the supreme height of the revealing granted them.... And art was simply called *techne*. It was a single, manifold revealing. It was ..., *promos*, i.e., yielding to the holding-sway and the safekeeping of truth.
>
> The arts were not derived from the artistic. Art works were not enjoyed aesthetically. Art was not a sector of cultural activity.
>
> What, then, was art—perhaps only for that brief but magnificent time? Why did art bear the modest name *techne*? Because it was a revealing that brought forth and hither, and therefore belonged within *poiesis*. It was finally that revealing which holds complete sway in all the fine arts, in poetry, and in everything poetical that obtained *poiesis* as its proper name ...
>
> Whether art may be granted this highest possibility of its essence in the midst of the extreme danger [of modern technology], no one can tell. Yet we can be astounded. Before what? Before this other possibility: that the frenziedness of technology may entrench itself everywhere to such an extent that someday, throughout everything technological, the essence of technology may come to presence in the coming-to-pass of truth.
>
> Because the essence of technology is nothing technological, essential reflection upon technology and decisive confrontation with it must happen in a realm that is, on the one hand, akin to the essence of technology and, on the other, fundamentally different from it.
>
> Such a realm is art. But certainly only if reflection on art, for its part, does not shut its eyes to the constellation of truth after which we are *questioning*.

Thus questioning, we bear witness to the crisis that in our sheer preoccupation with technology we do not yet experience the coming to presence of technology, that in our sheer aesthetic-mindedness we no longer guard and preserve the coming to presence of art. Yet the more questioningly we ponder the essence of technology, the more mysterious the essence of art becomes. (Emphases in the original)

I would argue that experimental and mathematical technologies of quantum physics are *techné* in the sense that Heidegger wants to give this term. They certainly were in the hands of Heisenberg and other founding figures, such as Schrödinger and Dirac, and many of their followers, or their predecessors, such as Einstein and Bohr. This is equally true about many experimenters, involved in the discoveries of quantum physics, that of the Higgs boson, among them. By the same token, contrary to Heidegger's view, implied here, fundamental physics, experimental and theoretical, is "an experience of the coming to presence of technology" that "guards and preserves the coming to presence of art," the art of physics. This returns us from Heidegger to Bohr, because the spirit of Copenhagen is the spirit of guarding and preserving this art.

Notes

1 This rubric is preferable to that of "the Copenhagen interpretation," because there is no such single interpretation, even in the case of Bohr, who changed his view a few times. For the discussion of the development of Bohr's views by the present author, see Plotnitsky (2012).

2 More recently they have been reconsidered yet again in the context of science as a cultural project, which, as noted above, could be seen in terms of "technology of culture." I am referring to such authors as T. Kuhn, I. Lakatos, and P. Feyerabend, and their followers in the constructivist studies of science. The subject, however, requires a separate treatment.

3 For further discussion and references, see Plotnitsky (2009, 237–312). The question of the locality of quantum mechanics or quantum phenomena is a matter of much controversy, especially after the Bell and Kochen-Specker theorems and related findings. Among the standard treatments are Cushing and McMullin (1989) and Ellis and Amati (2000).

4 See Pauli (1994, 149–64), Bohr (1987a, 3–12; 1987b, 23–29), Plotnitsky (2014), and further references therein.

5 Among the works addressing Bohr's philosophical background are Favrholdt (1992), Faye (1991), and Folse (1987).

6 Bohr's concept of atomicity has rarely been examined. Among exceptions are H. Folse (1985; 1987; 2002) and H. Stapp (2007). The philosophical genealogy of this concept is not easy to trace, although one can find a few parallels, in particular Whitehead's ([1929] 1979) conception of atomicity ("drops of experience") in *Process and Reality*. It is unclear whether Whitehead's ideas had an actual impact on Bohr's thought.

7 For a detailed discussion, see Plotnitsky (2015).

8 For an introduction the current state of QFT, see Kuhlman (2014) and references therein.

9 The latter were the main subject of Bohr and Rosenfeld's paper, written in response to L. Landau and R. Peierls's argument, contested by Bohr and Rosenfeld ([1933] 1983), concerning a possible inapplicability of the uncertainty relations in QFT.

References

Anderson, M. 1967. "An impression." In *Niels Bohr: His Life and Work as Seen by His Friends and Colleagues*, edited by S. Rozental. New York: Interscience Publishers.

Bohr, N. [1925] 1987a. "Atomic theory and mechanics." In *Atomic Theory and the Description of Nature*. Cambridge: Cambridge University Press. Reprinted as *The Philosophical Writings of Niels Bohr*, Vol. 1, 25–51.Woodbridge, CT: Ox Bow Press.

Bohr, N. [1934] 1987b) *Atomic Theory and the Description of Nature*. Cambridge: Cambridge University Press. Reprinted as *The Philosophical Writings of Niels Bohr*, Vol. 1. Woodbridge, CT: Ox Bow Press.

Bohr, N. 1935. "Can quantum-mechanical description of physical reality be considered complete?" *Physical Review* 48: 696–702.

Bohr, N. [1937] 1998. "Causality and complementarity." In *The Philosophical Writings of Niels Bohr, Volume 4: Causality and Complementarity, Supplementary Papers*, edited by J. Faye and H. J. Folse, 83–91. Woodbridge, CT: Ox Bow Press.

Bohr, N. [1938] 1998. "The causality problem in atomic physics." In *The Philosophical Writings of Niels Bohr, Volume 4: Causality and Complementarity, Supplementary Papers*, edited by J. Faye and H. J. Folse, 83–91. Woodbridge, CT: Ox Bow Press.

Bohr, N. [1958] 1987c). *Essays 1932–1957: Atomic Physics and Human Knowledge*. New York: Wiley. Reprinted as *The Philosophical Writings of Niels Bohr*, Vol. 2. Woodbridge, CT: Ox Bow Press.

Bohr, N. 1962. "Interview with Thomas Kuhn, Aage Petersen, and Eric Rüdinger, 17 November 1962." *Niels Bohr Archive*, Copenhagen and American Institute of Physics, College Park, MD.

Bohr, N. 1972–96. *Niels Bohr: Collected Works*, 10 vols. Amsterdam: Elsevier.

Bohr, N., and L. Rosenfeld. [1933] 1983. "On the question of the measurability of electromagnetic field quantities." In *Quantum Theory and Measurement*, edited by J. A. Wheeler and W. H. Zurek, 479–522. Princeton, NJ: Princeton University Press.

Bohr, N., and L. Rosenfeld. [1950] 1983. "Field and charge measurements in quantum electrodynamics." In *Quantum Theory and Measurement*, edited by J. A. Wheeler and W. H. Zurek, 523–34. Princeton, NJ: Princeton University Press.

Cao T. Y. (eds). 1999. *Conceptual Foundations of Quantum Field Theories*. Cambridge, UK: Cambridge University Press.

Cushing, J., and E. McMullin (eds). 1989. *Philosophical Consequences of Quantum Theory: Reflections on Bell's Theorem*. Notre Dame, IN: Notre Dame University Press.

Dirac, P. A. M. 1928. "The quantum theory of the electron." *Proceedings of the Royal Society of London A* 117: 610–24.

Dyson, F. 1949. "The S-matrix in quantum electrodynamics." *Physical Review* 75: 1736–55.

Einstein, A., B. Podolsky, and N. Rosen. [1935] 1983. "Can quantum-mechanical description of physical reality be considered complete?" In *Quantum Theory and Measurement*, edited by J. A. Wheeler and W. H. Zurek, 138–41. Princeton, NJ: Princeton University Press.

Ellis, J., and D. Amati (eds). 2000. *Quantum Reflections*. Cambridge, UK: Cambridge University Press.

Favrholdt, D. 1992. *Niels Bohr's Philosophical Background*. Copenhagen: Det Kongelige Danske Videnskabernes Selskab.

Faye, J. 1991. *Niels Bohr: His Heritage and Legacy. An Anti-Realist View of Quantum Mechanics*. Dordrecht: Kluwer Academic Publishers.

"*La Femme au Cheval.*" Wikipedia 2016.

Folse, H. J. 1985. *The Philosophy of Niels Bohr: The Framework of Complementarity*. Amsterdam: North Holland.

Folse, H. J. 1987. "Niels Bohr's concept of reality." In *Symposium on the Foundations of Modern Physics 1987: The Copenhagen Interpretation 60 Years after the Como Lecture*, edited by P. Lahti and P. Mittelstaedt, 161–80. Singapore: World Scientific.

Folse, H. J. 2002. "Bohr's conception of the quantum-mechanical state of a system and its role in the framework of complementarity." In *Quantum Theory: Reconsiderations of Foundations*, edited by A. Khrennikov, 83–98. Växjö Sweden: Växjö University Press, 1987.

Heidegger, [1927] M. 2010. *Being and Time*, translated by Joan Stambaugh. Albany, NY: SUNY Press.

Heidegger, M. 2004. *The Question Concerning Technology, and Other Essay*. New York: Harper.

Heisenberg W. [1925] 1969. "Quantum-theoretical re-interpretation of kinematical and mechanical relations." In *Sources of Quantum Mechanics*, edited by B. L. van der Warden, 261–77. New York: Dover.

Heisenberg, W. 1930. *The Physical Principles of the Quantum Theory*, translated by C. Eckhart and F. C. Hoyt. New York: Dover.

Heisenberg W. 1989. *Encounters with Einstein, and Other Essays on People, Places, and Particles*. Princeton, NJ: Princeton University Press.

"The Higgs Boson." Wikipedia 2016.

James, W. [1890] 2007. *The Principles of Psychology*, Vol. 1. New York: Cosimo Classics.

Kant, I. 1997. *Critique of Pure Reason*, translated by P. Guyer and A. W. Wood. Cambridge, UK: Cambridge University Press.

Kuhlman, M. 2014. "*Quantum Field Theory.*" *Stanford Encyclopedia of Philosophy*, Edited by E. Zalta. http://plato.stanford.edu/entries/quantum-field-theory/.

Miller, A. 2005. "One culture." *New Scientist*, October 25 (web).

Pais, A. 1991. *Niels Bohr's Times: In Physics, Philosophy, and Polity*. Oxford: Clarendon.

Pauli, W. 1994. *Writings on Physics and Philosophy*. Berlin: Springer.

Plotnitsky, A. 1994. *Complementarity: Anti-Epistemology after Bohr and Derrida*. Durham, NC: Duke University Press.

Plotnitsky, A. 2009. *Epistemology and Probability: Bohr, Heisenberg, Schrödinger, and the Nature of Quantum-Theoretical Thinking, Fundamental Theories in Physics*. Berlin and New York: Springer.

Plotnitsky, A. 2012. *Niels Bohr and Complementarity: An Introduction*. New York: Springer.

Plotnitsky, A. 2014. "Are quantum-mechanical-like models possible, or necessary, outside quantum physics?" *Physica Scripta* T163, 014011: 1–20 (web).

Plotnitsky, A. 2015. "A matter of principle: The principles of quantum theory, Dirac's equation, and quantum information." *Foundations of Physics* 45:1222–68.

Schrödinger, E. 1935. "The present situation in quantum mechanics." In *Quantum Theory and Measurement*, edited by J. A. Wheeler and W. H. Zurek. 152–67. Princeton, NJ: Princeton University Press, 1983.

Schweber, S. S. 1994. *QED and the Men Who Made It: Dyson, Feynman, Schwinger, and Tomonaga*. Princeton, NJ: Princeton University Press.

Stapp, H. P. 2007. *Mindful Universe: Quantum Mechanics and Participating Observer*. Heidelberg: Springer.

Teller, P. 1995. *An Interpretive Introduction to Quantum Field Theory*. Princeton, NJ: Princeton University Press.

Weyl, H. 1994. *The Continuum: A Critical Examination of the Foundation of Analysis*, translated by S. Pollard and T. Bole. New York: Dover.

Whitehead, A. N. [1929] 1979. *Process and Reality*. New York: Free Press.

Part Two

Bohr's Interpretation of Quantum Mechanics in Twenty-First-Century Physics

Complementarity and Quantum Tunneling

Slobodan Perović

Niels Bohr's complementarity principle is a tenuous synthesis of seemingly discrepant theoretical approaches (the wave mechanical approach, and that of Heisenberg and early Bohr) based on a comprehensive analysis of relevant experimental results. Yet the role of complementarity and the experimentalist-minded approach behind it were not confined to a provisional best-available synthesis of well-established experimental results alone. They were also pivotal in discovering and explaining the phenomenon of quantum tunneling in its various forms. The core principles of Bohr's method and the ensuing complementarity account of quantum phenomena remain highly relevant guidelines in the current controversial debate and in experimental work on quantum tunneling times.

1. Niels Bohr's method and the complementarity framework for understanding quantum phenomena

Although complementarity was developed as a multifaceted account, including the complementarity of space-time coordination and the laws of dynamical conservation, the principle of superposition, and the laws of dynamical conservation, its most general aspect is the complementarity of the wave mechanical approach to quantum phenomena and the approach based on matrix mechanics formalism. At the time of its inception, Bohr's complementarity principle was a result of his experimentalist-minded methodology that sought to provide the most encompassing general account of microphysical phenomena as they were revealed in relevant experiments (Perović 2013). Any metaphysical aims, if they existed, were secondary to this primary goal of Bohr's method. The method is in fact akin to a scientific strand of experimentalism already anticipated and elaborated on by Bacon (ibid.). Understanding complementarity outside of the context in which Bohr devised it can be detrimental to understanding it as a theoretical framework. In fact, the resulting theoretical framework is best understood in the context of the experimentalist-minded inductive methodology that produced it.

In constructing his complementarity principle, Bohr emphasized the two-stage nature of the experimental process. In the first stage, experimental results are gathered and described within the framework of everyday experiences and its constraints, for example, clicks of the Geiger counter or ionization tracks are recorded. This is why experimental reports, which include descriptions and accounts of observations and relevant aspects of the apparatus, inevitably stick to the language of local physical interactions and discrete physical properties. And that is why classical physical concepts are useful in the shaping of the reports, in order to make them more precise. The second interpretive or theoretical stage aims at making sense of diverse records of the results. Sometimes, as in the case of the formation of quantum mechanics, this will involve seemingly mutually exclusive concepts—for example, tracks in a cloud chamber can be interpreted either as an ionization path left by a particle that whizzed by, or as a footprint of the ionization of an atom. And the results of the second stage may also collide with the structure of our immediate experience and perception. Thus quantum concepts collide with habituated intuitions (e.g., those concerning the locality of physical interactions) concerning basic physical objects and their interactions. In pursuing this second stage, those physicists developing classical physics never had to push their concepts as far from intuitions based on immediate experience as those who were concerned with microphysical phenomena. They never had to conceptualize nonlocal interactions and entangled systems, nor was an adherence to both continuity and discreteness of elementary physical blocks forced upon them by experimental results.

The complementarity account of quantum phenomena was a result of this two-stage method. It was an early, satisfying, general, albeit provisional account of microphysical phenomena, aimed at reconciling diverse experimental results and an only partially successful theoretical framework: in the words of Heisenberg, Bohr was committed to "the requirement of doing justice at the same time to the different experimental facts which find expression in the corpuscular theory on the one hand and the wave theory on the other" (Heisenberg in Bohr 1985, Vol. 6, 20–1). The result did not resonate with the metaphysically motivated customary intuitions of either wave mechanical or corpuscular interpretations, since, to paraphrase Francis Bacon, such experimentalist-minded accounts rarely do.

Several authors (Perović 2013; Chevalley 1994; Kaiser 1992; Hooker 1972) have pointed to the two-stage nature of Bohr's method, although they disagree on the origin and its underlying philosophical grounds. Such methodological reassessments and their conclusions have put Bohr's complementarity on a much more firm footing than many other attempts to interpret its nature and aims. Thus complementarity is not a metaphysical account forced onto the experimental phenomena, but rather a provisional conceptual framework for making sense of numerous and diverse experimental results (scattering experiments, experiments with atomic collisions, light refraction experiments, etc.) that tries to avoid hasty conclusions based on one's metaphysically preferred concepts. Thus, for instance, the complementarity approach avoids commitment to the principle of spatiotemporal continuity, to which the wave-mechanical interpretation adheres based on isolated experimental results with light interference; but also refrains from taking the discovery of the conservation of momentum in particle interactions as a proof of inherently discrete corpuscular units at the quantum level.

Once we understand complementarity from this methodological perspective, rather than as an obscure metaphysical view, it is clear that much of the sometimes harsh criticism it receives (Beller 1999; Bub and Demopoulus 1974) is unjustifiably directed at the original formulation and its initially intended role. This deflationary perspective sidelines metaphysical pretentions of complementarity's primary goals, if indeed there were any—at least initially. It places it in a category of methodologically and experimentally driven useful tentative accounts of experimental work, and that of a dependable working hypothesis. Whether harsh criticisms are justifiably directed at its later forms and development is another question. But there are two questions of more immediate concern, raised by this particular view of complementarity and Bohr's general approach as outlined:

1. Along with its primary goal of satisfyingly encompassing existing experimental results was the complementarity approach, both as a method and as a theoretical standpoint, also a guide for the study of unexplored phenomena at the time, and how exactly did it perform such a function?
2. Does and can complementarity play the same useful role today that it presumably played at the time of its inception, with respect to not fully understood phenomena subject to ongoing experimental research?

2. Complementarity and the discovery of quantum tunneling

The discovery of quantum tunneling and wave mechanics

In this section we will explore the two questions posed above by focusing on the case of quantum tunneling. Physicists were beginning to discover quantum tunneling at the same time that Bohr was establishing his complementarity principle. And Bohr played a direct role in encouraging early research into the phenomenon. Yet quantum tunneling is still the focus of theoretical debate and experimental research—especially with regard to the speed of tunneling. All this makes it a particularly good test case for probing the questions posed above.

Quantum tunneling was initially conceptualized as a penetration of, or leaking through, a barrier or a hill. A particle with energy lower than the energy it takes to overcome an obstacle, a potential hill can occasionally "tunnel" through it. It is the effect of a particle tunneling through an obstacle forbidden in classical mechanics. A typical textbook presentation focuses on an electron and a rectangular potential barrier, which classical-mechanically should reflect the electron back because the energy of the electron is insufficient to go over it. Yet the electron can tunnel through the barrier quantum-mechanically. This typical presentation was established by Nordheim (1927), who made the initial steps in the discovery of quantum tunneling—to which we will return below.

The case of so-called frustrated total internal reflection, unaccounted for by geometric optics, is responsible for the phenomenon of evanescent light waves at the overlap of two transparent media. When light rays reflect within a denser medium (glass) and the medium is interrupted by a barrier of a less dense medium (e.g., air), they will not all reflect back and stay within the portion of the denser medium preceding the barrier. Wave optics suggests that a small portion of spherical waves will break into the air. Louis de Broglie realized early on that by analogy with light waves and the wave optics of refraction, matter-waves in a similar situation will penetrate into classically inaccessible regions of energy, with the energy of the resulting wave decreasing.

Yet it was F. Hund (1927a, b, c) who gave the first elaborate account of quantum tunneling at the atomic level and understood its significance with respect to the nature of chemical bonds. While Hund analyzed tunneling between bound atomic states, Nordheim's (1927) subsequent analysis focused on continuous states and the tunneling of free electrons. Later on, Gamow (1928) and Gurney and Condon (1929) independently discovered tunneling effects in solid-state physics and its intrinsic connection with α-decay.

Before we look at the details of these discoveries let us briefly turn to Merzbacher's (2002) view of them. He states that "[a] quantitative analysis of the physical implications of this tunneling effect had to await Erwin Schrödinger's wave mechanics and Max Born's probability interpretation of the quantum wave function" (44), hinting at the indispensability of the wave-mechanical approach to tunneling. He also insists that

> transmission of particles through a potential barrier of finite height and width is less easily visualized in the Heisenberg–Bohr formulation of quantum mechanics, which speaks of particles going over the top of the barrier with transient violation of conservation of energy. In both formulations, the language that permeates most descriptions of quantum transmission through a potential barrier has the anachronistic ring of Newtonian mechanics, with its underlying assumption that a particle always moves in a continuous orbit. (Ibid.)

Thus, the wave mechanical approach to tunneling was not only indispensable but superior to that of Bohr and Heisenberg, which introduces orbits of electrons and quantum jumps of elementary particles (electrons) as transitions from one stationary atomic state to another.

The wave mechanical approach pioneered by Schrödinger (1926) introduced a central formalism of quantum mechanics, namely, the wave equation, as well as a treatment of microphysical systems based on the assumption of their continuous nature. Heisenberg's (1925) approach, in contrast, introduced the less elegant matrix mechanical formalism and relied on Bohr's (1913) model of the atom, which postulates stationary atomic states—states in which electrons are bound with the nucleus. The model's apparatus of atomic dynamics (quantum rules) was also in general accord with the quantization of microphysical systems (hence the label "Bohr–Heisenberg approach"—"Bohr" referring here to Bohr's work prior to complementarity).

If Merzbacher is correct, then it seems that Hund's initial discovery of tunneling— and the subsequent developments even more so—represented a departure from Bohr's model and its associated concepts, signifying an embracement of the wave-mechanical framework of quantum systems. The question we wish to examine is whether perhaps both theoretical approaches played a role in the discovery: was Schrödinger's account truly superior in the case of tunneling, as Merzbacher suggests, or was its use limited in the way Bohr's complementarity suggested at the time? Did the general framework of complementarity perhaps play the same role, if any, in reconciling these two approaches in this particular case, as it did in other cases? And, in general, how informed were the ongoing theoretical considerations of not only Hund, but also Nordheim and later Gamow, Gurney, and Condon in terms of the framework of complementarity?

Hund's work: Bound states and tunneling

In his first paper in the series of three, Hund (1927a) looks at the so-called luminous electron and classically impenetrable barriers, namely, pairs of atoms as double potential wells. He analyzes oscillations of the electron between two atomic bound states. In his third, breakthrough paper (1927c), he analyzes the molecules and the dynamics of the atoms that compose them. A necessary assumption of such analysis was the separability of the motion of the electron from the vibration and rotation of atoms. He looked at the case of the reflection-symmetric potentials of classically impenetrable barriers: two stationary states, even and odd, that is, grounded and excited, of two atoms in a molecule; in one case two of the same atoms, and in the other case two different atoms (e.g., an atom with an additional electron). The two even states are symmetric, and the two odd states are antisymmetric potentials.

Now, the superposition of these particular stationary states turns the system into a non-stationary state oscillating between them. The distance between the atoms in the molecule is defined as a width and length of the barrier that determines the exact beat due to tunneling. This results in a transition in the chirality state of the molecule; that is, a beat due to tunneling is manifested as a transition in molecular configuration (chirality), namely, the transitions between the optically active right or left handedness of the molecule. How excessive tunneling will be depends on the nature of the molecule. Atoms in ammonia molecules tunnel considerably, while those in organic molecules do not.

It seems that this treatment of the bound states, and the very choice of exploring the as-yet-unexplored phenomenon of tunneling at the level of atomic states, rather than, say, free electrons in a potential well, is in the spirit of Bohr's general line of argument for complementarity as outlined in the paper published in *Nature* in 1925. So Hund starts off with Bohr's model of the atom and the appropriate description of its dynamics, only to extend it using the wave mechanical account of the behavior of the potential well as a nonstationary oscillation of the state of superposition between two bound states. Wave mechanics was obviously a very welcome new approach, but Merzbacher somewhat exaggerates its importance by treating it as a preferred newcomer among a selection of formal-theoretical treatments of quantum states. Actually, complementarity spelled out its role as well as its limitations very precisely, as it did for Heisenberg–Bohr's approach. The wave mechanical treatment was a very welcome addition to the study of microphysical states, but Bohr warned that attempts to overgeneralize it and to pronounce it superior rather than complementary to other approaches such as that of Heisenberg–Bohr were unjustified. Nothing in Hund's exploration suggests that he failed to understand Bohr's argument.

Bohr's role

It is certainly possible and indeed likely that a general framework such as that outlined by Bohr and explorations such as Hund's guided by it were simply up in the air,
and that Bohr simply articulated this rather well. Yet with respect to the first steps in
discovering tunneling, how direct could Bohr's influence have been? Did Bohr understand the physical significance of tunneling, at least in general terms?

Bohr explicitly announced his complementarity account in his Como lecture of
1927. Yet his article in *Nature* (Bohr 1925) was in effect a precursor of the complementarity approach, since he offered a detailed case for the unavoidability of both particle
and wave approaches. Now, in this paper he already offers a general outline of the
phenomenon of quantum tunneling in the context in which Hund was interested at
the time. Even though he emphasizes the classical aspect of the electron in dynamics,
led by collision and scattering experiments where the energy is conserved in individual processes, he introduces a general caveat (590). He leaves the issue open, as he is
aware that the particle-like description has clear limits. He then specifically mentions
the ongoing work of Hund on molecular spectra, the work he personally encouraged
Hund (1927a) to undertake. Quantum tunneling was very much a matter of theoretical
rather than experimental consideration at the time. It is thus not surprising that Bohr
dedicates limited time to it in his *Nature* paper, given that he bases his argument on the
subtleties of well-established experimental cases, but he understands its significance
within the overall picture of quantum phenomena.

Moreover, Bohr's shaping of complementarity was a two-way process, from the first,
experimental stage, to the second stage of theory-formation and back. Theoretical concepts had to be employed to properly account for the experimental records, but only
very cautiously. Bohr employed the wave-mechanical and Heisenberg–Bohr accounts
as such. Even though they were in their infancy at the time, considerations of tunneling
were still part of highly theoretical considerations, and Bohr was clearly aware of this.
Judging by the abovementioned published comments, the emerging tunneling effect
was one of the caveats constraining the approach that tendentiously insisted on the
particle-like aspects of microphysical interactions. This caveat played the same role,
although not as fully, because physicists were only just starting to understand it at the
time, since the well-known experiments with light interference played a role, along with
some others such as scattering and collision experiments, in constraining the meaning
and use of the wave-mechanical approach. In any case, Bohr's encouragement of Hund
reflects his conviction with regard to the rising importance of the phenomenon, which
he tentatively and cautiously inserts as a piece of the larger puzzle he is assembling.

3. Later work on tunneling and the role of complementarity

Nordheim's work: Quantum tunneling, continuous quantum systems, and wave mechanics

Nordheim and Gamow's work represents later developments that came when the
complementarity argument was already well known and established. Does their

work develop in accord with the complementarity framework as much as Hund's, if at all?

Nordheim explored tunneling in continuous quantum systems, that is, when an unbound electron is acted upon by an external electric field acting as a barrier. The subject of Nordheim's (1927) first paper was the thermionic emission of electrons and their reflection off metals. It was essentially an application of wave-mechanics to Sommerfeld's electron theory of metals. The assumption was that metals behave as an ideal Fermi gas.

Nordheim modeled a surface barrier of metal that keeps in the electron—he used, now famously, a rectangular model for the potential—and calculated the wave function of the electron across steep potential rises and drops. The result of his calculations was that near the top of the barrier reflections or transmissions are probable, while classically only reflections could occur, since the electron does not have enough potential energy to overcome the barrier. He concluded, however, that the emission through the barrier is negligibly small. Yet in the second paper, coauthored with Fowler (1928), it becomes apparent that the emission of electrons from a metal surface in a strong electric field is connected to tunneling. Thus "emission begin[s] to be sensible for fields of rather more than 10exp7 Volts/cm" (180). They also realize that the emission "will depend essentially on the exact form of the potential energy curve" (181), that is, on its height and width.

As the main general point regarding the physics of the case, the authors state that "it seems fair to conclude that the phenomenon of electron emission in intense fields is yet another phenomenon which can be accounted for in a satisfactory quantitative manner" (Fowler and Nordheim 1928, 180). And as far as the formal treatment goes, "in order to study the emission through the potential energy step … we have only to solve the wave equation" (175), which accounts for the energy of the electron in the external field with a particular form of the potential.

A treatment in the vein of wave mechanics was more appropriate technically than that of Bohr–Heisenberg and the framework of Bohr's model. This is not surprising given that the problem is set up precisely as a continuous interaction of an electron in an electric field, making use of Bohr–Heisenberg's quasi-particle framework, based on Bohr's model of the atom, redundant. In other words, the setup of the problem within the context of unbound states with continuum energy eigenvalues could not benefit from Bohr's model of the atom and the Bohr–Heisenberg approach. Instead, one accounts for the case simply by solving appropriate wave equations. In contrast, Hund's treatment of tunneling in molecular bonds crucially relies on Bohr's model and its concepts, precisely because of the nature of the case.

This apparent discrepancy was not an exception, since the wave-mechanical and Heisenberg–Bohr treatments were each variously adequate for different isolated experimentally tackled phenomena (Perović 2013). This was a key general insight that led to the complementarity principle. The principle mapped the theoretical work precisely in terms of competing frameworks, each accounting better than the other for particular isolated phenomena. Their limitations are reached and become transparent when one has to draw conclusions and explore quantum phenomena in a more comprehensive way. In fact, both approaches have to be invoked when a comprehensive account of the phenomenon is required, when trying to explain α-decay or determine duration of tunneling.

α-decay, quantum tunneling, and complementarity

High energy α-particles are emitted by radioactive elements. Experimentalists noticed that such elements regularly emit α-particles of much lower energy than the energy required for a particle to break out from the forces (which we now know are strong forces) that keep them together in the nucleus. Independently, Gamow (1928) and Gurney and Condon (1929) explained this phenomenon as a direct consequence of quantum tunneling, synthesizing Nordheim's account of a free particle in a potential barrier and Hund's approach to tunneling in bound states.

Rutherford pioneered research on α radiation by shooting the particles into atoms. He soon realized that the particles were nothing but positively charged helium atoms of high velocities. In his scattering experiments, he let a beam of α-particles go through the diaphragm and a metal foil, and detected their angles of impact with a scintillation detector observed with a microscope. Some particles reached the detector at a considerable angle and some even turned back. This indicated that the nucleus, also positively charged, was tightly packed rather than spread out in the atom; otherwise the particles could not be repelled by it. As an electron circling around such a nucleus would quickly lose its energy according to Coulomb's law, it became obvious that Bohr's model for quantizing the mechanical energy of an electron was more accurate than Thomson's classical model.

Now, it was not clear how a tightly packed nucleus kept together by obviously very strong nuclear forces could absorb a particle *of the same charge* with much lower energy, which results in the nucleus transformation. It was also mysterious how the same much less energetic (compared to the nuclear forces) individual α-particles could fly off the nucleus despite the strong forces keeping them together. "Wave mechanics," Gamow (1966, 92) stated years after his celebrated paper that accounted for these two phenomena, "could explain [these] phenomena well beyond the reach of Old Quantum theory."

Gamow's (1928) idea was to describe, within the wave-mechanical framework, the penetration of α-particles into the nucleus as an overcoming of a potential barrier: "In wave mechanics a particle always has a finite probability, different from zero, of going from one region to another region of the same energy, even though the two regions are separated by an arbitrarily large but finite potential barrier" (5).[1] The analogy between wave mechanics and wave optics, as opposed to Newtonian mechanics and geometrical optics, offered a model for the phenomenon. As waves penetrate glass but also to some extent reach the outside air according to the wave mechanics, as opposed to staying within the denser medium in the sense of geometrical optics, the matter-waves slightly penetrate the potential well and reach into the nucleus. These must be very energetic particles, and the barrier has to be between certain width sizes; and even when these conditions are satisfied the likelihood of such penetration is very small. But with a large number of collisions, the number of α-particles that overcome the potential barrier and reach the nucleus is significant for radioactive elements. Similarly, some particles that will leave the nucleus as the matter-wave will reach beyond the potential well.

Now, Gamow realizes that there is *no sharp stationary state* due to the leakage of the matter-wave. But he also realizes that this leakage is very small, so the state of the system is most accurately described as a *quasi-stationary state*. This *nearly bound state* had to be described so that it gave accurate decay rates for the radioactive elements. This meant treating the matter-wave as conserving probability, which Gamow (1928, 210) did by introducing appropriate damping of vibrations by a complex energy expression (i.e., a small imaginary contribution to the energy equation).

Thus, due to the context of the case, the wave-mechanical approach and that of Heisenberg–Bohr are brought closely together, unlike in Hund and Nordheim's treatments. Gamow starts off with a wave-mechanical treatment but ends up, motivated purely by physical reasons (small leakage, and relative stability of the state), introducing a boundary condition for a quasi-stationary state.

The nature of Gamow's work *shows that there is a threshold of comprehensiveness for accounting for phenomena, beyond which one must use the two complementary approaches.* Gamow treated such an inherently comprehensive phenomenon rather than its isolated aspects, in contrast to Hund and Nordheim. In fact, he tried and succeeded in putting together the isolated aspects treated by Hund and Nordheim in order to explain spontaneous α-decay and α-ray penetration into the nucleus.

Gamow was searching for a more general theory encompassing various phenomena, and thus adopted the same stance that led Bohr to suggest the complementarity of wave and particle approaches while trying to bring together a number of various experimental results. Gamow's method was really very similar to Bohr's in this respect. Bohr's early enthusiasm with Gamow's (1966, Section IV) ideas is not at all surprising; nor is Gamow's awe at Bohr's approach. In fact, in the case of Gamow's work, and perhaps even more so in the case of Gurney and Condon's solution, a general account synthesizes Nordheim's continuous treatment of a barrier and Hund's bound-states treatment. Both such comprehensive approaches to the phenomenon, which aimed at uniting such seemingly disparate phenomena only accounted for in isolation until then, relied on the complementarity, as Bohr characterized it, of the stationary state model and the wave-mechanical treatment. As much as the physicists were impressed by the potential of the wave-mechanical treatment, they continued to fall back on the old Heisenberg–Bohr model as well.

This falling back is even more obvious in Gurney and Condon's treatment. Merzbacher (2002, 49) states that they "less consistently applied quantum-mechanical problem." This is one way of putting it, but perhaps a more accurate characterization is that they made more obvious the complementary features of the phenomenon, as their approach was much more of a formal analysis compared to Gamow's, which was an intuitive application of the analysis of tunneling. They viewed α-decay and the penetration of the nucleus as natural consequences of quantum mechanics, concluding that the "α particle slips away almost unnoticed" (Gurney and Condon 1928, 439). Such a formulation of the conclusion of their seminal paper reveals the elegance of the quantum mechanical solution, but also the central place that the notion of the stationary state, albeit slightly modified with respect to the notion of tunneling offered in Bohr's initial model, plays in such an explanation. In their semiclassical approach, α-particles are presupposed to perform periodic and aperiodic motions in classically accessible regions—that is, within the nucleus and outside the barrier. The entire domain can perhaps in principle be treated wave mechanically, but the system is best accounted for as a complementary feature of a stationary state and its energy: "One can think of the particle as executing its classical motion in range I [before the barrier], but as having at each approach to the barrier the probability of escaping to range II [after the barrier] given by expression ... above" (133). And they find an inherent connection between their approach and that of Oppenheimer, precisely in this complementary feature: "His [Oppenheimer's] formula for the mean time required for dissociation of the atom by a steady electric field splits naturally into a factor which is the classical frequency of motion in the Bohr orbit multiplied by an exponential probability factor of the type of expression ... used in this paper" (ibid.).

4. The methodological principles behind complementarity that contributed to the discovery and explanation of quantum tunneling

Perhaps one does not have to work within the complementarity framework when theoretically synthesizing experimental phenomena the way Gamow and Gurney and Condon did. One could perhaps more consistently develop the wave mechanical approach or work within the confines of Bohmian mechanics. But the fact is that an approach in the spirit of complementarity turned out to be fruitful in this case.

Perhaps these developments not only took place within the complementarity framework and were done in its spirit, rather than that of Schrödinger's intended general framework, but grew directly out of Bohr's argument. This would hardly be surprising given the magnitude of Bohr's influence in general at the time and on the physicists who did the work in particular, and given the largely positive reactions to his *Nature* paper and complementarily as it was spelled out within it. In fact, in the paper by Condon and Morse (1931, 59), where the time of tunneling is explicitly mentioned for the first time, the authors understand the role of their own work within the two-stage methodological spirit of the complementarity framework:

> There are two parts to the study of the theory. One is the weaving of the new canvas of purely mathematical relations on which the picture of nature is to be painted. The other is the painting of the picture. Any given set of experimental operations leading to numerical results of an observation of the system, i.e. pointer-readings, will be called an observable. (45)

This set will be inserted into the theoretical picture in an appropriate way.

If Bohr's complementarity indeed exerted such a direct influence, and based on our previous analysis of the different stages of discovering tunneling, can we extract *the key principles of Bohr's method that led to the complementarity framework*, which constituted such a strong methodological force at different stages of the discovery?

First, a particularly important point was the principle (let us label it P1) that *synthesizing theoretical accounts that seem opposed in light of particular metaphysical presuppositions can be beneficial in terms of explaining the known and empirically examined phenomena and predicting new ones.* This was something that Bohr demonstrated with respect to the use of the wave mechanical and the Heisenberg–Bohr approach, where the advantages and shortcomings of both in accounting for particular phenomena were identified. The two accounts were deemed complementary rather than mutually exclusive—they could be deemed exclusive only if one stuck to the metaphysical presuppositions that seemed to underlie them. Thus, the wave-mechanical approach was treated as complementary to Heisenberg–Bohr's particle-like treatment, despite Schrödinger's insistence that the continuity (of physical processes and objects) principle that underlies it, which Heisenberg–Bohr's discontinuous quantum jumps violate, is the baseline of any approach to quantum phenomena. Similarly, in the exploration of the tunneling with respect to continuous phenomena, wave mechanics sufficed. But it was not treated as an exclusive and superior account in general; in fact, the treatment of tunneling in bound atomic states was supplemented by the framework of Heisenberg–Bohr's atomic model and its posits. This amalgamated approach led to new insights and predictions at the level of molecular bonds, and probably even more so with respect to the case of α-decay.

Second, *experimental limits were placed on the scope of theoretical frameworks, that is, on the understanding of physical properties* (P2). Thus, even though the quantum states could be understood as waves, this understanding was limited by insights from relevant experimental results (e.g., scattering and collision experiments, and the experiments with spectral lines) as much as it was warranted by other experiments (e.g., light interference). This warranted acknowledging the wave aspect of quantum states but not pronouncing that quantum states were waves, and everything else that may follow from such a strong proposition. And accordingly, accounting for the phenomena with the help of the wave aspect of quantum states could be complemented by the particle aspect of quantum states displayed by these limiting experiments. Thus, following this general framework, the use of wave mechanics in accounting for the tunneling of a free electron through the potential well did not lead to the generalization of the wave-mechanical approach. Although innovative and useful, it was never meant to be a general or even a superior account, but only a step toward synthesis with bound states accounts of tunneling. Again, this is particularly obvious in the treatment of α-decay.

5. The present: The role of the complementarity framework in the controversy over the velocity of quantum tunneling (i.e., the tunneling time)

Let us now turn to the question of whether complementarity offers similar useful methodological guidelines and theoretical insight for current research and the ongoing debate on the speed of tunneling.

The problem of the time or velocity of tunneling was recognized early on. Condon and Morse (1931, 59) stated that "there is no way of telling ... what the mean duration in this region [i.e., within the barrier] is for those [particles] that penetrate." They also provide an equation for the mean time that the packet of particles penetrating the barrier spends within the barrier.

Since then, the very concept of the time tunneling takes has proved a nontrivial problem. Chiao (1998, 2) remarks, regarding his experiments with photon tunneling that aimed to measure the transition time, that "we learned the important lesson that a clear definition of the experimental method by which the tunneling time is measured is necessary before the ... question can even be well formulated." And "in fact, different operational procedures will lead to conflicting experimental outcomes, so that the time or duration of a process in quantum physics, such as tunneling, is no longer unique, in contrast to the situation in classical physics" (ibid.).

Here Chiao refers to four different approaches to the time of tunneling that yield different results. The first approach follows the wave packet crossing the barrier. Typically, the incoming and the outgoing peak are compared, or the wave packet's center of mass ("centroid"). The phase or group delay, a delay in the arrival of the reflected and transmitted packet, is predicted and experimentally recorded. The second approach looks at an "internal clock" by following certain degrees of freedom within the barrier-particle system that indicates the time the particle spends within the barrier. The third follows semi-classical trajectories of the particle through the barrier, using Feynman diagrams, Bohm's mechanics, or Wigner distribution. Finally, the fourth approach looks at the relation between the probability density within the barrier and the density of the incoming flux; this is the so-called dwell time (Winful 2006). The dwell time of a particle under the barrier, however, cannot be treated as traversal time, nor are the flux delays in emptying the barrier in either direction transit times. However, the dwell time can be shown to be equal to the group delay (Winful 2006, Section 2.5).

Now, each approach emphasizes distinct physical assumptions within which it defines the tunneling time. Assessing the plausibility and the exact meaning of these physical presuppositions is no trivial matter; nor is discerning the mutual relationship between them, squaring the various definitions of tunneling times that they imply, or understanding the discrepancies between the results they produce. Moreover, the experiments with individual quantum particles are difficult to perform and interpret. In contrast, there are many more experiments with electromagnetic, optical, and acoustic waves.[2] The tunneling effects experiments with light are feasible as the required times are in the domain of pintoseconds. Following electrons through the barrier requires the femtosecond scale, which is presently much harder to realize experimentally.

Thus, these diverse experiments, combined with diversity of approaches and their mutual disagreements, raise the question of what exactly is being measured in each. On the one hand, the different approaches, other than the one focusing on the inner-clock measurement, seem to converge on the so-called Hartman effect: the independence of the mean tunneling time from the width of the barrier implies that for arbitrarily large barriers the tunneling velocity can become infinitely large. Many physicists believe that this implies a superluminal group delay for thin enough and distant enough barriers: in such conditions group velocity is greater than the equal time, that is, the time the packet would need to traverse the same distance in a vacuum. For a group of photons this implies faster-than-light traversal, although, the argument goes, this does not necessarily imply that individual particles actually traverse faster than light (Büttiker and Washburn 2003). Other physicists, however, refute such traversal times as a relevant time-scale for the tunneling effect, or deny that phase delays have anything to do with traversals (Winful 2006). Thus, different approaches are empirically equivalent with respect to the limiting, and at the same time a controversially interpreted factor exhibited in the Hartman effect; that is, they disagree on the physical meaning of the tunneling time.

It seems that we face a similar situation to the one Bohr confronted when he developed complementarity (and for the same physical reasons). The tunneling controversy is similar to the situation where wave mechanics and the Heisenberg–Bohr approach captured different experimental results, harbored disparate concepts, and yet remained equivalent with respect to Bohr's model of the atom (Perović 2008). In both situations, as P1 suggests, overgeneralizations may be as misleading. Thus, for instance, insisting on the Bohmian approach that solely semiclassical trajectories are adequate, based on the general appeal of Bohmian mechanics, may be useful only if we carefully connect its results and concepts to the results of other approaches. Given the context of the problem, rather than being guided by a particular theoretical approach favoring only particular experimental results while sidelining others, one should first and foremost gather and comprehensively assess the experimental results as the initial step. This is exactly what Winful (2008, 39) suggests in the spirit of the experimentalist-minded methodology condoned by Bohr that we outlined in Section 1.

This is still uncharted territory experimentally, which is an additional reason to be wary of hasty generalizations that could stifle insights from other approaches. Thus Bohr's insistence on limiting theoretical accounts by relevant experimental content from the very beginning of experimentation (P2), that is, on careful gathering of experimental data that includes information on experimental setups, becomes especially relevant. Biases are already present at the level of gathering observational particulars and always threaten to sway the experimental work in a particular narrow and biased direction, while instead diverse experimental inputs should offer a comprehensive theoretical outlook. This is why the experimentalist community should construct various experimental setups carefully in order to avoid falling for one such bias, while theoreticians should take these setups into account when they are interpreting the data (Perović 2013). This is what actually happens in a good scientific community when it faces controversy. In the case of tunneling, then, experiments that measure the amount of rotation of electron spins caused by tunneling through the barrier could accompany common experiments on phase delay times: "Experiments of this kind would provide a key to the question of whether superluminal tunneling is an important development, or just a misnomer" (Büttiker and Washburn 2003, 272).

Other than these methodological guidelines in the spirit of complementarity, the conclusions Bohr drew with respect to microphysical systems seem to be quite apparent in the current work on tunneling. Thus, for instance, Olkhovsky, Recami, and Jakiel (2004) develop a formalism that follows semiclassical trajectories of particles using Feynman diagrams in order to calculate tunneling and collisions times. Based on results with atomic and nuclear collisions, they state a few conclusive results, among them acknowledging "the coincidence of the quasi-classical limit of our QM definitions of time durations with analogous well-known expressions of classical mechanics" (168). Moreover, they conclude that the Feynman approach of studying the time-evolution of collisions following semi-classical trajectories and Schrödinger's purely wave-mechanical approach both "lead to the same results" (ibid.). They both confirm a reinforced Hartman effect, where velocity does not depend on either the widths of the barriers nor on the distance between them.

The authors also emphasize something we pointed out earlier with respect to the difference between the processes analyzed by Hund and Nordheim: "Let us add that for discrete energy spectra the time analysis of the processes (particularly in the case of wavepackets composed of states bound by two well potentials, with a barrier between the wells) is rather different from the time analysis of processes corresponding to continuous energy spectra" (ibid.). Moreover, it is precisely this realization concerning the disparate approaches, and their limits, first articulated in the complementarity playbook, that should guide further synthetic investigations of the tunneling effect: "One can expect that the time analysis of more complicated processes, in the quasi-discrete (resonance) energy regions, with two (or more) well-potentials, such as the *photon or phonon-induced tunnelings* from one well to the other, could be performed by a suitable combination, and generalization, of the methods elaborated for continuous and for discrete spectra" (ibid.).

Again, the complementarity framework and its current methodological derivatives is not a solution to the important problem of the quantum tunneling times and the dilemmas it raises, but it is a useful methodological guideline that has arisen from careful and comprehensive reflections on insights into the physics of the microphysical world, which is invaluable in the context of novel experimental research. It is hard to see how it can be sidelined, and exactly why it should be, unless a completely new theory that can account for scores of relevant phenomena in a new reliable way replaces quantum mechanics.

Notes

This work was supported by the project "Logical and Epistemological Foundations of Science and Metaphysics" (Ref. number 179067) financed by the Ministry of Education, Science, and Technological Development of the Republic of Serbia.

1 Translation from German at http://web.ihep.su/dbserv/compas/src/gamow28/eng.pdf.
2 See Winful (2006) for a comprehensive review of relevant experiments.

References

Beller, M. 2001. *Quantum Dialogue: The Making of a Revolution*. Chicago: University of Chicago Press.
Bohr, N. 1913. "On the constitution of atoms and molecules." *The London, Edinburgh, and Dublin Philosophical Magazine and Journal of Science* 26 (151): 1–25.

Bohr, N. 1925. "Atomic theory and mechanics." *Nature* 116 (2927): 845–52.

Bohr, N. 1985. *Niels Bohr: Collected Works, Vol. 6: Foundations of Quantum Physics I.* Amsterdam: North Holland Publishing Company.

Bub, J., and W. Demopoulos. 1974. *The Interpretation of Quantum Mechanics.* Dordrecht: Reidel.

Büttiker, M., and S. Washburn. 2003. "Optics: Ado about nothing much?" *Nature* 422 (6929): 271–2.

Chevalley, C. 1994. "Niels Bohr's words and the Atlantis of Kantianism." In *Niels Bohr and Contemporary Philosophy*, edited by J. Faye and H. J. Folse, 33–55. Dordrecht: Kluwer Academic Publishers

Chiao, R. Y. 1998. "Tunneling times and superluminality: A tutorial." *arXiv preprint quant-ph/9811019.*

Condon, E. U., and P. M. Morse. 1931. "Quantum mechanics of collision processes I: Scattering of particles in a definite force field." *Reviews of Modern Physics* 3 (1): 43.

Fowler, R. H., and L. Nordheim. 1928. "Electron emission in intense electric fields." *Proceedings of the Royal Society of London A: Mathematical, Physical and Engineering Sciences* 119 (781) (May): 173–81.

Gamow, G. 1928. "Zur Quantentheorie des Atomkernes." *Zeitschrift für Physik* 51 (3–4): 204–12.

Gamow, G. 1966. *Thirty Years that Shook Physics: The Story of Quantum Theory.* Chelmsford, MA: Courier Corporation.

Gurney, R. W., and E. U. Condon. 1928. "Wave mechanics and radioactive disintegration." *Nature* 122 (3073): 439.

Gurney, R. W., and E. U. Condon. 1929. "Quantum mechanics and radioactive disintegration." *Physical Review* 33 (2): 27.

Heisenberg, W. 1925. "Quantum-theoretical re-interpretation of kinematic and mechanical relations." *Zeitschrift für Physik* 33: 879–93.

Hooker, C. A. 1972. "The nature of quantum mechanical reality." In *Paradigms and Paradoxes*, edited by R. G. Colodny, 206–207. Pittsburgh: University of Pittsburgh Press.

Hund, F. 1927a. "Zur Deutung der Molekelspektren I." *Zeitschrift für Physik* 40 (10): 742–64.

Hund, F. 1927b. "Zur Deutung der Molekelspektren II." *Zeitschrift für Physik* 42 (2–3): 93–120.

Hund, F. 1927c. "Zur Deutung der Molekelspektren III." *Zeitschrift für Physik* 43 (11–12): 805–26.

Kaiser, D. 1992. "More roots of complementarity: Kantian aspects and influences." *Studies in History and Philosophy of Science Part A* 23 (2): 213–39.

Merzbacher, E. 2002. "The early history of quantum tunneling." *Physics Today* 55 (8): 44–50.

Nordheim, L. 1927. "Zur Theorie der Thermischen Emission und der Reflexion von Elektronen an Metallen." *Zeitschrift für Physik* 46 (11–12): 833–55.

Olkhovsky, V. S., E. Recami, and J. Jakiel. 2004. "Unified time analysis of photon and particle tunnelling." *Physics Reports* 398 (3): 133–78.

Perović, S. 2008. "Why were matrix mechanics and wave mechanics considered equivalent?" *Studies in History and Philosophy of Science Part B: Studies in History and Philosophy of Modern Physics* 39 (2): 444–61.

Perović, S. 2013. "Emergence of complementarity and the Baconian roots of Niels Bohr's method." *Studies in History and Philosophy of Science Part B: Studies in History and Philosophy of Modern Physics* 44 (3): 162–73.

Schrödinger, E. 1926. "An undulatory theory of the mechanics of atoms and molecules." *Physical Review* 28 (6): 1049.

Winful, H. G. 2006. "Tunneling time, the Hartman effect, and superluminality: A proposed resolution of an old paradox." *Physics Reports* 436 (1): 1–69.

10

Bohr and the Problem of the Quantum-to-Classical Transition

Maximillian Schlosshauer and Kristian Camilleri

1. Introduction

Niels Bohr famously insisted on the indispensability of what he called "classical concepts," as well as on the necessity of giving a classical description of measuring instruments. What is known as Bohr's "doctrine of classical concepts" refers to the need to use classical concepts (such as position and momentum) in the description of microscopic quantum objects (like electrons) in mutually exclusive experimental arrangements. In perhaps his most frequently quoted account of the doctrine, Bohr declared: "It is decisive to recognize that, however far the phenomena transcend the scope of classical physical explanation, the account of all evidence must be expressed in classical terms," and to this extent "the account of the experimental arrangement and of the results of the observations must be expressed in unambiguous language with suitable application of the terminology of classical physics" (Bohr 1949, 209).

As we have argued in an earlier publication—from which the present chapter has been adapted (Camilleri and Schlosshauer 2015)—and in another contribution by one of us (K. C.) in this volume, Bohr's doctrine of classical concepts should be understood as an epistemological thesis based on a functional understanding of experiment. Put simply, it is an attempt by Bohr to elaborate an *epistemology of experiment*. Indeed, it is in this context that one should place Bohr's contention that the experimental apparatus must be described in terms of the concepts of classical physics. Bohr (1939) acknowledged that the "quantum mechanical formalism" can be applied to "any intermediate auxiliary agency employed in the measuring process." However, for the purposes of *performing a measurement*, the experimental apparatus and its interaction with the object of investigation "must be described on classical lines" (23–4). Bohr's central insight was that if a measuring instrument is to serve its purpose of furnishing us with knowledge of an object, it must be described classically. That is to say, any functional description of the experimental apparatus, in which it is treated as a means to an end and not merely as a dynamical system, must make use of the concepts of classical physics.

This view of Bohr's (1948, 317) was significantly motivated by his early recognition that the quantum-mechanical inseparability of measured system and measurement apparatus brought about by entanglement constituted a difficulty as far as the concept of observation is concerned, since the impossibility of "separating the behaviour of the objects from their interaction with the measuring instruments" in quantum mechanics "implies an ambiguity in assigning conventional attributes to atomic objects." While Bohr would be happy to allow that it is always possible to represent the experimental apparatus from a purely structural point of view, without any reference to its function, as a quantum-mechanical system and thus give a quantum-mechanical treatment of the observational interaction, he also recognized that this would render the very distinction between object and instrument ambiguous. However, as Bohr repeatedly emphasized, such a distinction is a necessary condition for empirical inquiry. Therefore, for Bohr measuring instruments must admit of a classical description, otherwise they could not perform their epistemic function as measuring instruments (Camilleri and Schlosshauer 2015).

To this extent, any "quantum mechanical treatment" of a measuring instrument will, by virtue of its *function* as a measuring instrument, "be essentially equivalent with a classical description," as Bohr (1939, 23) explained at the Warsaw conference in 1938. However, this raises the question of why, from a purely dynamical point of view, the quantum-mechanical treatment is "essentially equivalent" to a classical description. Bohr's *epistemological* explanation for why we must use a classical description thus begs the question of what *dynamical* features of a macroscopic system entitle us to neglect the "quantum effects." Bohr (1935, 701) here appears to simply assume that there exists a macroscopic "region where the quantum-mechanical description of the process concerned is effectively equivalent with the classical description." Thus we are led to ask: How is it that classical physics can be employed, at least to a very good approximation, under certain dynamical conditions (typically those corresponding to measuring scenarios)? Bohr never provided a satisfactory answer to this question.

As these introductory remarks should have made clear, it is crucial to distinguish Bohr's doctrine of classical *concepts*, as it applies to the mutually exclusive experimental conditions in the description of quantum objects, from the use of classical dynamical *theories* in the description of macroscopic systems. In this chapter, we will focus on the latter issue—and the associated problem of the quantum-to-classical transition—and recount how a number of Bohr's followers attempted to provide a dynamical reformulation of Bohr's epistemological doctrine (Section 2). These attempts took the form of attempts to explain why macroscopic systems (such as measuring apparatuses) behave "classically" (or at least approximately classically), and they assumed crucial importance in the 1950s and 1960s and generated much debate and discussion. What is perhaps remarkable about this situation is that both critics and defenders of Bohr's viewpoint saw the need to arrive at a deeper understanding of the quantum-to-classical transition. Notably, Heisenberg took the crucial step of emphasizing the importance of the openness of quantum systems in his brief remarks on the problem of the quantum-to-classical transition in the 1950s (Section 3). Yet it was only with the deeper understanding of the role of entanglement with the environment, brought about by decoherence theory (Zeh 1970; Zurek 1981, 1982, 2003; Joos et al. 2003;

Schlosshauer 2007), that the dynamical problem of the quantum-to-classical transition could be properly addressed. In Section 4, we comment on the relationship between Bohr's views and decoherence.

2. Irreversibility and the dynamical problem of the quantum-to-classical transition

As mentioned, Bohr appears to have simply assumed that any macroscopic system serving as a measuring instrument must be describable by means of a classical approximation—otherwise we could not rely on such a system to perform experiments. For him, it is simply the case that without such a presupposition, experimental knowledge would be rendered impossible. The task then fell to Bohr's followers to provide an adequate dynamical account of why this is so. Throughout the 1950s and 1960s, a number of Bohr's disciples, including Rosenfeld, Petersen, and Groenwald, became increasingly preoccupied with the dynamical problem of the quantum-to-classical transition amid the new wave of criticisms of quantum mechanics. While most of these attempts ended in failure, they did not signify a departure from Bohr's general epistemological viewpoint, and they did yield some surprising insights. Commenting on attempts in this direction in 1965, Rosenfeld (1979, 536) argued that "it is understandable that in order to exhibit more directly the link between the physical concepts and their mathematical representation, a more formal rendering of Bohr's argument should be attempted." In spite of their philosophical differences, many of Bohr's followers, such as Weizsäcker and Rosenfeld, pursued this kind of approach to the physics of the quantum-to-classical transition as entirely in keeping with the spirit in which Bohr had intended his doctrine of classical concepts.

As Weizsäcker (1971, 25) put it at a colloquium in 1968, "The crucial point in the Copenhagen interpretation" is captured, "but not very luckily expressed, in Bohr's famous statement that all experiments are to be described in classical terms." As a devotee of Bohr, this was a view that Weizsäcker endorsed wholeheartedly, but which he now wished to justify. "My proposed answer is that Bohr was essentially right" in arguing that the instrument must be classically describable, "but that he did not know why" (28). The paradox at the heart of the Copenhagen interpretation for Weizsäcker is therefore to be stated: "Having thus accepted the falsity of classical physics, taken literally, we must ask how it can be explained as an essentially good approximation" when describing objects at the macrolevel. He spells this out:

> This amounts to asking *what physical condition must be imposed on a quantum-theoretical system in order that it should show the features which we describe as "classical."* My hypothesis is that this is precisely the condition that it should be suitable as a measuring instrument. If we ask what that presupposes, a minimum condition seems to be that irreversible processes should take place in the system. For every measurement must produce a trace of what has happened; an event that goes completely unregistered is not a measurement. Irreversibility implies a description of the system in which some of the information that we may think of as being present

in the system is not actually used. Hence the system is certainly not in a "pure state"; we will describe it as a "mixture." I am unable to prove mathematically that the condition of irreversibility would suffice to define a classical approximation, but I feel confident it is a necessary condition. (29; emphasis in original)

Weizsäcker's remarks here point to a program of dynamical explanation that had been pursued since the 1950s. After the Second World War, a number of physicists had devoted themselves to investigating the thermodynamic conditions of irreversibility that would need to hold in order for a measurement to be registered macroscopically as "classical." On a number of occasions Bohr (1958b, 310) referred explicitly to "the essential irreversibility inherent in the very concept of observation," but he never suggested, at least in his published writings, that this might hold the key to a thermodynamic explanation of the "classicality" of quantum systems at the macroscopic level. "The amplification of atomic effects," as he noted, "emphasizes the irreversibility characteristic of the very concept of observation" (Bohr 1958a, 170). But such allusions to "irreversibility" were not developed further. Indeed, as Everett pointed out in 1956, "there is nowhere to be found any consistent explanation of this 'irreversibility' attributed to the measuring process" (Everett's notes on Stern's letters 1956, as quoted in Osnaghi, Freitas, and Freire 2009, 106).

A more rigorous "thermodynamic approach" was developed independently by Günther Ludwig in the second half of the 1950s (Jammer 1974, 488). Ludwig attempted to explain measurement by means of a thermodynamic analysis of the irreversible amplification process triggered by a microscopic event (490). As Osnaghi, Freitas, and Freire (2009, 103) explain, many physicists saw Ludwig's approach as opening up "the possibility of providing a rigorous foundation for Bohr's approach, thereby dispelling the misunderstandings surrounding the alleged subjectivism of the Copenhagen view." A number of attempts to develop further Ludwig's basic program of a thermodynamic approach were pursued during the 1960s, the most elaborate of which was the work carried out by Daneri, Loinger, and Prosperi (1962). Their aim was to find the exact ergodicity conditions for the validity of the ergodic theorem in quantum statistical mechanics. As Jeffery Bub (1971, 66) explained, the question the authors wished to answer was, "How does the *theory* guarantee that quantum theoretical macrostates will not exhibit interference effects?" The answer Daneri, Loinger, and Prosperi gave was that the physical structure of a large body implies ergodicity conditions, which in turn prevent the system from exhibiting quantum superpositions at the macroscopic scale (see also Ludwig 1953; 1955).

In Rosenfeld's (1979, 537) view, this work represented a major step forward in clearing up the misunderstandings and the "extravagant speculations" on the measurement problem, which had arisen through "the deficiencies in von Neumann's axiomatic treatment." Importantly, Rosenfeld declared that the Daneri–Loinger–Prosperi approach was "in complete harmony with Bohr's ideas" (539). Jeffery Bub (1968), however, saw it as "basically opposed to Bohr's ideas" insofar as it treated classical mechanics as an approximation to a quantum theory of macroscopic systems (see also Bub 1971, 65). In a similar vein, Max Jammer (1974, 493) called into question whether the approach of Daneri, Loinger, and Prosperi was "really congenial, or at least not incompatible, with the basic tenets of the Copenhagen interpretation." Yet as we have seen, the *epistemological* primacy of classical physics, on which the functional description of experiment rested, was for Bohr perfectly compatible with, and indeed depended on, the view that classical mechanics was an approximation of quantum mechanics. Those physicists closest to Bohr appreciated this point. As Weizsäcker (1987, 283) put it, Bohr's emphasis on "the classical description of an instrument just meant that only so far as it was an approximation would the instrument be of use *as* an instrument." There was never any suggestion for Bohr that measuring instruments could not in principle be described by quantum mechanics, or that classical dynamics was somehow more fundamental in an ontological sense than quantum mechanics.

Yet the subtleties of Bohr's point of view were typically lost on other physicists, many of whom saw little value in Bohr's forays into epistemology. This episode is indicative of the state of confusion surrounding Bohr's doctrine of classical concepts, which persisted well into the 1960s and early 1970s. In a qualified sense, Rosenfeld was right—there was no essential conflict between Bohr's *epistemological* view of the doctrine of classical concepts and the attempt to find a *dynamical* solution to the problem of the quantum-to-classical transition.

Nevertheless, an adequate dynamical solution turned out to be far more complicated than Bohr or many of his followers had originally thought. Rosenfeld's endorsement of the ergodic solution as having provided the solution to the problem was based on his view that "the reduction rule" is "essentially a thermodynamic effect, and, accordingly, only valid to the thermodynamic approximation" (Rosenfeld to Belinfante, July 24, 1972, Rosenfeld Papers, quoted in Osnaghi, Freitas, and Freire 2009, footnote 243). In hindsight, such pronouncements appear to have been overly optimistic. Eugene Wigner (1995, 65), who retained a key interest in foundational questions of quantum mechanics throughout the 1970s, rightly pointed out that the ergodicity conditions of Daneri–Loinger–Prosperi, which purported to explain "the transition to a classical description of the apparatus," rested on "an arbitrary step," which only served to "postulate the miracle which disturbs us." In a similar vein, Bub (1971, 70) argued that the authors had simply reintroduced "von Neumann's infinite regress all over again." The problem of the quantum-to-classical transition remained unsolved. On this point, Wigner and Bub were right.

3. The relation between object, instrument, and the external world

Looking back over the history of the foundations of quantum mechanics, it is easy to see that the crucial obstacle to an understanding of the quantum-to-classical transition was the erroneous assumption that we can treat quantum systems as isolated from the environment. There do, however, appear to be anticipations of the relevance of the environment in the 1950s and 1960s. In his contribution to the Bohr *Festschrift* edited by Pauli in 1955, Heisenberg reminded his readers that a quantum-mechanical treatment of the "interaction of the system with the measuring apparatus" does not of itself "as a rule lead to a definite result (e.g., the blackening of a photographic plate at a given point)" (22). Inclusion of further systems, such as a secondary apparatus or a human observer, will not terminate the resulting von Neumann chain (1955) if these systems are treated as interacting quantum systems.

Heisenberg observed that this situation had prompted a number of physicists to attempt to develop a modified dynamics of wave-packet collapse, or to suggest that the conscious observer plays an essential role in the process of measurement. Yet, as Heisenberg (1955, 22) explained, such attempts failed to grasp the real nature of the problem. In the von Neumann scheme, "the apparatus and the system are regarded as cut off from *the rest of the world* and treated as a whole according to quantum mechanics" (emphasis added). Yet the measuring apparatus is in reality never isolated from its environment: "If the measuring device would be isolated from the rest of the world, it would be neither a measuring device nor could it be described in the terms of classical physics at all" (Heisenberg [1958] 1989, 24). Heisenberg (1955, 27) emphasized that "the connection with the external world is one of the necessary conditions for the measuring apparatus to perform its function." He spelled out this position in more detail in *Physics and Philosophy*. Here it is worth quoting Heisenberg (1989, 121–2) at some length:

> Again the obvious starting point for the physical interpretation of the formalism seems to be the fact that mathematical scheme of quantum mechanics approaches that of classical mechanics in dimensions which are large compared to the size of atoms. But even this statement must be made with some reservations. Even in large dimensions there are many solutions of the quantum-mechanical equations to which no analogous solutions can be found in classical physics. In these solutions the phenomenon of the "interference of probabilities" would show up ... [which] does not exist in classical physics. Therefore, even in the limit of large dimensions the correlation between the mathematical symbols, the measurements, and the ordinary concepts is by no means trivial. In order to get at such an unambiguous correlation one must take another feature of the problem into account. It must be observed that the system which is treated by the methods of quantum mechanics is in fact a part of a much bigger system (eventually the whole world); it is interacting with this bigger system; and one must add that the microscopic properties of the bigger system are (at least to a large extent) unknown. This statement is undoubtedly a correct description of the actual situation ... The interaction with the bigger system with its undefined microscopic properties then introduces a new statistical element into the description ... of the system under consideration. In the limiting case of the large dimensions this statistical element destroys the effects of the "interference of probabilities" in such a manner that the quantum-mechanical scheme really approaches the classical one in the limit.

In this intriguing passage, Heisenberg seems to have recognized that one cannot simply appeal to the macroscopic dimensions or the mass of the measuring apparatus to explain why the apparatus is never found in a quantum superposition. Such a view marks a departure from Bohr's suggestion that the heaviness of the apparatus renders it effectively classical. Heisenberg enlarges the system–apparatus composite to include couplings to further degrees of freedom in the environment (the "external world"). This is a most interesting point. Of course, for practical purposes, Heisenberg admits that we often treat the quantum system and the measuring apparatus as isolated from the rest of the world. But, in Heisenberg's (1955, 23) view, a proper account of the quantum-to-classical transition must in the end rest on "the underlying assumption," which Heisenberg took to be implicit in the Copenhagen interpretation, "that the interference terms are in the actual experiment removed by the partly undefined interactions of the measuring apparatus, with the system and with the rest of the world (in the formalism, the interaction produces a 'mixture')."

4. Decoherence theory and "emergent classicality"

Heisenberg's remarks might be read, somewhat charitably, as anticipating certain results of the decoherence program. After all, it is the hallmark of decoherence that it proceeds from the recognition that it is practically impossible to isolate the quantum system and the apparatus from the surrounding environment, and moreover that it is precisely this feature that results in the emergence of classicality. But while Heisenberg (1955, 23) emphasizes the importance of environmental interactions, he gives no detailed account of precisely how "the interference terms are ... removed by the partly undefined interactions of the measuring apparatus, with the system and with the rest of the world." In particular, nowhere does he explicate the specific role of entanglement between the system and the environment as the crucial point in the dynamical emergence of classicality in the system.

While physicists such as Bohr, Heisenberg, and Schrödinger had recognized the nonseparability of quantum systems—that is, entanglement—as a characteristic feature of quantum mechanics as early as the 1920s and 1930s, the feeling prevailed that entanglement was something unusual and a peculiarly microscopic phenomenon that would have to be carefully created in the laboratory (such as in an EPR-type experiment). Entanglement was regarded as an essential quantum feature that would necessarily have to be irreconcilable with classicality. These long-held beliefs likely contributed to the comparably late "discovery" of the theory of decoherence, which has its roots in Zeh's work from the 1970s but did not receive broader attention until Zurek's contributions appeared in the 1980s (Camilleri 2009).

It is indeed a particular irony that entanglement would turn out to be not something that had to be tamed to ensure classicality but would instead assume an important role in our understanding of aspects of the quantum-to-classical transition. Decoherence proceeds from the realization that realistic quantum systems are never completely isolated from their environment (Zeh 1970; Zurek 1981, 1982 and 2003; Joos et al. 2003; Bacciagaluppi 2012; Schlosshauer 2004 and 2007). When a quantum system interacts with its environment, it will in general become entangled with a large number of environmental degrees of freedom. This entanglement influences what we can locally observe upon measuring the system. Thus, decoherence can be viewed as a dynamical filter on the space of quantum states: It provides a quantitative, dynamical explanation, wholly within standard mechanics, of the difficulty of creating and observing many "nonclassical" superposition states and interference phenomena, especially for mesoscopic and macroscopic systems whose many degrees of freedom typically lead to strong coupling to the environment. Thus, decoherence provides a partial solution to the dynamical problem of the quantum-to-classical transition, insofar as it describes effective restrictions on the superposition principle for subsystems interacting with other systems. Decoherence thus addresses what Bub (2012) has called a *consistency problem*: the agreement between the probability distributions obtained from quantum mechanics and classical mechanics in the relevant cases. It is in this narrow qualified sense that we may say that "classicality," so to speak, "emerges" out of the quantum formalism.

Recently, a number of physicists have suggested that Bohr's views on the primacy of classical concepts, and by extension his doctrine of an (ostensibly) fundamental quantum–classical divide, amount to little more than superfluous semantic or philosophical baggage, much of which has been discredited by recent developments. Dieter Zeh, for example, has contrasted the dynamical approach of decoherence with the "irrationalism" of the Copenhagen school (Joos et al. 2003, 27). Erich Joos (2006, 54), who attributes the origins of decoherence to a dissatisfaction with the "orthodoxy of the Copenhagen school" and "the desire to achieve a better understanding of the quantum–classical relation," has argued that "the message of decoherence" is that "we do not need to take classical notions as the starting point for physics," given that "these emerge through the dynamical process of decoherence from the quantum substrate" (77).

However, a purely dynamical account such as decoherence must be viewed in the context of the series of similar attempts made by Bohr's followers as discussed in Section 2. Its relationship to Bohr's views regarding the object–instrument divide and the primacy of classical concepts should therefore be judged in the same way as those earlier attempts: as something that was believed by many of Bohr's followers to be in natural harmony with Bohr's views. The dynamical problem of the quantum-to-classical transition, as many of Bohr's followers recognized, in fact addressed a problem quite different from the chiefly epistemological problem that had preoccupied Bohr. Of course, if classical concepts are understood in a purely pragmatic sense—as something we simply *do* use when we perform a measurement—decoherence may supply a justification for their use. Furthermore, decoherence explains why macroscopic objects, such as measuring apparatus, permit a classical description, in the sense of the applicability of classical dynamical theories. Thus, decoherence can be seen as providing a dynamical (and finally satisfactory) explanation for Bohr's (1935, 701) assumption that there exists a macroscopic "region where the quantum-mechanical description of the process concerned is effectively equivalent with the classical description." As Zurek (1993, 311) put it:

> The role of decoherence is to establish a boundary between quantum and classical. The boundary is in principle moveable, but in practice largely immobilized by the irreversibility of the process of decoherence ... The equivalence between "macroscopic" and "classical" is then validated by the decoherence considerations, but only as a consequence of the practical impossibility of keeping objects which are macroscopic perfectly isolated.

It should be kept in mind, however, that while decoherence theory provides a dynamical account of the applicability of classical and quantum-mechanical descriptions and their relationship to the distinction between macroscopic and microscopic, Bohr's (1935, 701) discrimination "between those parts of the physical system … which are to be treated as measuring instruments and those which constitute the objects under investigation" was determined by functional (epistemological) considerations and not by dynamical (ontological) considerations. In Bohr's view, once the aims of the experiment had been decided upon and the experimental apparatus has been set up accordingly, the "cut" between system and apparatus was effectively fixed— it could not be moved around arbitrarily. Indeed this was precisely the point that Bohr insisted upon in an exchange of correspondence with Heisenberg in September 1935 (Camilleri and Schlosshauer 2015, 77–9). To this extent, Bohr argued that the cut corresponds to something "objective" in the sense that the object–instrument distinction was essentially fixed by the *functional-epistemological* considerations dictated by the choice of the particular experimental arrangement.

5. Concluding remarks

Bohr's epistemological thesis concerning a functional description of experiment should be distinguished from the efforts, which gathered momentum in the 1960s, to supply a dynamical explanation of the emergence of classicality from quantum theory. What such approaches have aimed for—and, in the case of decoherence, with great success—is to ensure, through consideration of the dynamics, a consistency between the quantum and classical descriptions (expressed in terms of the relevant probabilistic predictions), particularly for mesoscopic and macroscopic systems.

While such approaches can provide, post facto, a physical, pragmatic justification for the applicability of classical concepts, they do not touch on Bohr's main concern of an epistemology of experiment that underlies his doctrine of classical concepts. Indeed, Bohr's doctrine was premised on the assumption that a quantum-mechanical treatment of the measurement apparatus would be "essentially equivalent with a classical description," an assumption for which decoherence can now provide a dynamical explanation. This suggests that Bohr's philosophy and the insights brought about by decoherence theory may not only coexist peacefully, but also that decoherence may be seen as supplying the dynamical justification for the distinction, which Bohr held to be epistemically necessary, between the "classical system" serving as the measuring apparatus and the "quantum object."

References

Bacciagaluppi, G. 2012. "The role of decoherence in quantum mechanics." In *The Stanford Encyclopedia of Philosophy*, edited by E. N. Zalta. http://plato.stanford.edu/archives/win2012/entries/qmdecoherence.

Bohr, N. 1935. "Can quantum-mechanical description of physical reality be considered complete?" *Physical Review* 48: 696–702.

Bohr, N. 1939. "The causality problem in atomic physics." In *New Theories in Physics*, 11–30. Paris: International Institute of Intellectual Cooperation.

Bohr, N. 1948. "On the notions of causality and complementarity." *Dialectica* 2, 312–19.

Bohr, N. 1949. "Discussions with Einstein on epistemological problems in atomic physics." In *Albert Einstein: Philosopher-Scientist*, edited by P. A. Schilpp, 201–41. Evanston, IL: Library of Living Philosophers. Reprinted in Wheeler and Zurek (eds), 1983, 9–49.

Bohr, N. 1958a. "On atoms and human knowledge." *Dædalus* 87: 164–75.

Bohr, N. 1958b. "Quantum physics and philosophy: Causality and complementarity." In *Philosophy at Mid-Century: A Survey*, edited by R. Klibanksy, 308–14. Florence: La Nuova Italia Editrice.

Bub, J. 1968. "The Danieri–Loinger–Prosperi quantum theory of measurement." *Nuovo Cimento* 57B: 503–20.

Bub, J. 1971. "Comment on the Daneri–Loinger–Prosperi quantum theory of measurement." In *Quantum Theory and Beyond*, edited by T. Bastin, 65–70. Cambridge: Cambridge University Press.

Bub, J. 2012. "Bananaworld: Quantum mechanics for primates." E-print arXiv:1211.3062 [quant-ph].

Camilleri, K. 2009. "A history of entanglement: Decoherence and the interpretation problem." *Studies in History and Philosophy of Modern Physics* 40: 290–302.

Camilleri, K., and M. Schlosshauer 2015. "Niels Bohr as philosopher of experiment: Does decoherence theory challenge Bohr's doctrine of classical concepts?" *Studies in History and Philosophy of Modern Physics* 49: 73–83.

Daneri, A., A. Loinger, and G. M. Prosperi 1962. "Quantum theory of measurement and ergodicity conditions." *Nuclear Physics* 33: 297–319.

Heisenberg, W. 1955. "The development of the interpretation of the quantum theory." In *Niels Bohr and the Development of Physics: Essays Dedicated to Niels Bohr on the Occasion of his Seventieth Birthday*, edited by W. Pauli, L. Rosenfeld and V. Weisskopf, 12–29. New York: McGraw Hill.

Heisenberg, W. [1958] 1989. *Physics and Philosophy: The Revolution in Modern Science*. London: Penguin.

Jammer, M. 1974. *The Philosophy of Quantum Mechanics*, first edition. New York: John Wiley & Sons.

Joos, E. 2006. "The emergence of classicality from quantum theory." In *The Re-emergence of Emergence: The Emergentist Hypothesis from Science to Religion*, edited by P. Clayton, and P. Davies, 53–77. Oxford: Oxford University Press.

Joos, E., H. D. Zeh, C. Kiefer, D. Giulini, J. Kupsch, and I. O. Stamatescu. 2003. *Decoherence and the Appearance of a Classical World in Quantum Theory*, second edition. New York: Springer.

Ludwig, G. 1953. "Der Messprozess." *Zeitschrift für Physik* 135: 483–511.

Ludwig, G. 1955. "Zur Deutung der Beobachtung in der Quantenmechanik." *Physikalische Blätter* 11: 489–94.

Osnaghi, S., F. Freitas, and O. Freire Jr. 2009. "The origin of the Everettian heresy." *Studies in History and Philosophy of Modern Physics* 40: 97–123.

Rosenfeld, L. 1979. "The measuring process in quantum mechanics." In *Selected Papers of Leon Rosenfeld*, edited by R. S. Cohen, and J. J. Stachel, 536–46. Dordrecht: D. Reidel.

Schlosshauer, M. 2004. "Decoherence, the measurement problem, and interpretations of quantum mechanics." *Reviews of Modern Physics* 76: 1267–305.

Schlosshauer, M. 2007. *Decoherence and the Quantum-to-Classical Transition*. Berlin: Springer.

von Neumann, J. 1955. *Mathematical Foundations of Quantum Mechanics*. Princeton: Princeton University Press.

Weizsäcker, C. F. v. 1971. "The Copenhagen interpretation." In *Quantum Theory and Beyond*, edited by T. Bastin, 25–31. Cambridge: Cambridge University Press.

Weizsäcker, C. F. v. 1987. "Heisenberg's philosophy." In *Symposium on the Foundations of Modern Physics. The Copenhagen Interpretation 60 Years after the Como Lecture*, edited by P. Lahti and P. Mittelstaedt, 277–93. Singapore: World Scientific.

Wheeler, J. A., and W. H. Zurek (eds). 1983. *Quantum Theory and Measurement*. Princeton: Princeton University Press.

Wigner, E. 1995. *The Collected Works of Eugene Wigner. Part B. Vol. 6: Philosophical Reflections and Syntheses*. Berlin: Springer.

Zeh, H. D., 1970. "On the interpretation of measurement in quantum theory." *Foundations of Physics* 1: 69–76.

Zurek, W. H. 1981. "Pointer basis of quantum apparatus: Into what mixture does the wave packet collapse?" *Physical Review D* 24: 1516–25.

Zurek, W. H. 1982. "Environment-induced superselection rules." *Physical Review D* 26: 1862–80.

Zurek, W. H. 1993. "Preferred states, predictability, classicality, and the environment-induced decoherence." *Progress of Theoretical Physics* 89: 281–312.

Zurek, W. H. 2003. "Decoherence, einselection, and the quantum origins of the classical." *Reviews of Modern Physics* 75: 715–75.

11

On Bohr's Epistemological Contribution to the Quantum-Classical Cut Problems

Manuel Bächtold

1. Introduction: Bohr and the quantum-classical cut

Although microscopic systems are described by means of quantum mechanics, the outcomes obtained when performing measurements on such systems "appear" to us as being "classical" (Giulini et al. 2003). This situation is puzzling. How does the classical appearance of these measurement outcomes emerge (problem A)? And why, after all, couldn't we describe these measurement outcomes by means of quantum mechanics (problem B)? Problem A can be tackled within the domain of physics. As a matter of fact, it has been partly solved by means of decoherence theory (Zurek 1981, 1982, 1991; Blanchard et al. 2000; Joos et al. 2003). Problem B falls under the competency of epistemology, as it deals with the respective role of classical physics and quantum mechanics. Concerning this topic of the quantum-classical cut, the contribution of Bohr and especially his thesis of the necessary use of classical concepts have been very influential during the twentieth century. Many scholars in the past have examined Bohr's writings concerning this thesis (Folse 1985; Honner 1987; Murdoch 1987; Faye 1991; Chevalley 1991; Scheibe 1993). Recently, this thesis has been discussed in relation to decoherence theory. More precisely, Camilleri and Schlosshauer (2015) asked if Bohr's thesis is challenged by decoherence theory. Their inquiry leads them to conclude that this is not the case. In other words, Bohr's view is still defendable.

This chapter aims at showing that Bohr's view is not only defendable but can throw light on the problems associated with the quantum-classical cut. First, I will recall that Bohr's way of dealing with this cut is not ontological but epistemological. I will then put forward four conceptual distinctions and argue that those are essential to understand and discuss Bohr's thesis of the necessary use of classical concepts. Finally, on the basis of this clarification, the relation between Bohr's thesis and decoherence theory will be discussed.

2. Two ways of considering the quantum-classical cut

The ontological approach

The quantum-classical cut is described by some physicists as being a cut between two "worlds" or two "realms" (e.g., Bub 1997, 120; Zurek 1991, 37), namely, the "quantum world" and the "classical world." These worlds are assumed to be characterized by different laws: the laws governing the quantum world being depicted by quantum physics (i.e., by Schrödinger's equation), while those of the classical world by classical physics (i.e., Newton's equations). Problem A becomes then a problem concerning the transition between both sets of laws: how are the quantum laws changing into the classical laws when a measurement is performed?

Decoherence theory is often presented as having yielded keys to understanding the relation between both worlds, or more precisely, as having partially explained the transition between the two kinds of laws of nature. "Decoherence" is described as a process resulting from the interaction of the apparatus and its (external and internal) environment, and by means of which the laws are changing from quantum laws to classical laws. More precisely, decoherence theory shows it is possible for quantum formalism to determine the "preferred" observable, the one which is measured as a matter of fact. However, this theory has only "partially" explained the transition because it still faces two problems: first, when taking account of the decoherence process, we obtain a density operator associated with the system composed of the studied system and the apparatus ($S+A$), at the end of the measurement, which represents an "improper" mixture; second, nothing justifies that the final density operator associated with $S+A$ represents our incomplete knowledge of its physical state, and not directly its physical state, as it does at the beginning of the measurement, what is referred to as the "and/or" problem (van Fraassen 1991; d'Espagnat 1995; Mittelstaedt 1998; Bächtold 2008a).

This way of considering the quantum-classical cut is tacitly relying on scientific realism (Psillos 1999), or more precisely on an ontological interpretation of the physical theories: an interpretation which assumes that the equations of the theory are describing laws of nature, and the entities postulated by the theories (e.g., atoms, electrons …) are representing real entities in the physical world, beyond the observable phenomena.

The epistemological approach

Yet, there is an alternative way of considering the quantum-classical cut, namely, an epistemological one based on an instrumentalist interpretation of the physical theories. To say that the quantum-classical cut is "epistemological" amounts to supporting the view that it has to do primarily with the theories we use to describe and predict the phenomena and not with the reality itself: this cut means that some phenomena can be described and predicted in a satisfactory manner by means of quantum mechanics and other by means of classical physics. Problem A has then to be viewed as a problem concerning our means of knowledge, that is, our theories. It can be stated as follows: how can we account, in the context of a measurement, for the transition between the quantum description and the classical description of the phenomena?

In the frame of such an epistemological approach, decoherence theory appears to have yielded keys to understand the differences between the theoretical tools used to predict the measurement outcome before the measurement and those used to describe the outcome that has been obtained and that can be observed. From this point of view, there is no and/or problem: once the measurement has been performed, one may legitimately make use of a density operator corresponding to mixture in order to represent our incomplete knowledge concerning the outcome that has been obtained and that can be observed, if one has not yet looked at the pointer of the apparatus (Bächtold 2008b). However, it remains puzzling why, according to decoherence theory, the quantum formalism leads to an improper mixture.

Bohr's approach and the problem he addressed

Regarding the quantum-classical cut, Bohr's approach was clearly epistemological, An expression such as "quantum world" is fully foreign to Bohr's writings. Camilleri and Schlosshauer (2015, 79) write that "strictly speaking, as Bohr recognized, the world *is* non-classical" (emphasis in the original). However, it seems that Bohr never made such a claim concerning how the world is or might be. In his philosophical writings on quantum mechanics, he was concerned not with the new properties the world might have in the light of this theory, but with the new situation concerning our means to describe the phenomena, that is, our concepts and our theories: "The extension of physical experience in our days has, however, necessitated a radical revision of the foundation for the unambiguous use of our most elementary concepts, and has changed our attitude to the aim of physical science. Indeed, from our present standpoint, physics is to be regarded not so much as the study of something *a priori* given, but rather as the development of methods for ordering and surveying human experience" (Bohr 1963, 9–10). In the second part of this quotation, Bohr supports an instrumentalist interpretation of the physical theories, which in his view was suggested by the very rise of quantum mechanics. This instrumentalist interpretation is also stated as follows: "the mathematical formalism of quantum mechanics and electrodynamics merely offers rules of calculation for the deduction of expectations about observations" (Bohr 1963, 60). In Bohr's view, the quantum-classical cut was not one between two worlds, but between two ways to account for the phenomena, or in other words, between "two kinds of descriptions: one classical and another quantum" as Kauark-Leite (2012, 53; my translation here) puts it.

What are precisely the questions Bohr dealt with in relation to the quantum-classical cut? He primarily addressed the problem of how to describe the outcomes of measurements performed on microscopic systems (problem B), while making only some few commentaries concerning the transition problem (problem A). As Camilleri and Schlosshauer (2015, 75) have stressed, one has to "distinguish between Bohr's epistemological thesis concerning a functional description of experiment, and the efforts to provide a dynamical explanation of the emergence of classicality from quantum theory. While Bohr offered some remarks on the latter through his oblique references to the 'heaviness' of the apparatus, his main preoccupation was with the former." The transition problem was tackled by several followers of Bohr (Rosenfeld, von Weizsäcker, Ludwig …) but not so much by Bohr himself (Osnaghi, Freitas, and Freire 2009; Camilleri and Schlosshauer 2015). The problem at the core of Bohr's (1963, 10) epistemological approach is the "observational problem," that is, the problem of how to provide an "objective description" of the measurement outcomes (Bohr 1963, 11). This problem has become one because of the features of quantum mechanics, the new theory used to describe or predict the phenomena under study (i.e., the phenomena putting into play microscopic systems): "Notwithstanding the power of quantum mechanics as a means of ordering an immense amount of evidence regarding atomic phenomena, its departure from accustomed demands of causal explanation has naturally given rise to the question whether we are here concerned with an exhaustive description of experience" (Bohr 1963, 3). The rise of quantum mechanics, according to Bohr, led to a "revision of the very problem of observation" (18).

It is in relation to this epistemological problem of observation that Bohr supported his thesis of the necessary use of classical concepts. In other words, this thesis was central in his attempt to clarify the question of how to offer an objective description of the experiments performed on microscopic systems. Let us turn now to this thesis.

3. Bohr's thesis of the necessary use of classical concepts

This thesis was claimed by Bohr (1934, 17) many times in various papers with slightly different words: "only with the help of classical ideas is it possible to ascribe an unambiguous meaning to the results of observation," "all experience must ultimately be expressed in terms of classical concepts" (94), and "the account of all evidence must be expressed in classical terms" (1958, 39).

What precisely are "classical concepts"? What is Bohr's argument to justify the necessary use of these concepts? And in what respect does this thesis bring insight into the problem of observation? To answer these questions, I will argue, it is essential to make the following four conceptual distinctions:

1. Between "classical concepts" and "classical physical description"
2. Between the description of the measurement apparatus and the description of the measurement outcomes
3. Between the system and the apparatus
4. Between the system and the observer

Let us consider these distinctions one by one, and, thereby, present and discuss Bohr's thesis.

"Classical concepts" and "classical physical description"

To avoid misunderstandings concerning Bohr's thesis, it is an important point first to clarify the term "classical." Bohr makes a subtle distinction between a "classical physical description" and a description by means of "classical concepts."

What are "classical concepts" according to Bohr? This is not such an easy question to answer, as Faye (1991, 133–4) writes: "Bohr does not make explicit which classical concepts he has in mind when referring to them." One way to define these concepts amounts to specify how they can be applied. This is Honner's (1987, 63) proposal who provides a "succinct definition of what Bohr mean by 'classical': classical concepts are those which apply univocally in the macroscopic world, but which cannot be applied in precisely the same way in the domain of quantum physics." This is a negative definition.

Let us try to clarify positively the meaning of the expression "classical concepts." Obviously, the term "classical" refers to "classical physics." In turn, "classical physics" refers to Newtonian mechanics, Maxwell's electrodynamics, and Einstein's (special and general) relativity theory (Bohr 1934, 3; 1963, 2). Let us remark that Bohr (1963, 11), in his later writings, restated slightly his thesis and replaced "classical concepts" by "common language" or "common language, suitably refined by the terminology of classical physics" (1963, 24). Is this an important shift in Bohr's thesis? This seems not to be the case. This change of terminology might result merely from Bohr's wish to clarify what he had in mind when speaking of "classical concepts." What matters is not so much that these concepts belong to classical physics, but the very origin and role of these concepts. Classical concepts correspond to an improvement of the everyday concepts which have been elaborated so as to adapt ourselves to the world and to enable us to communicate with each other, or in Bohr's (1963, 88) words, which have been "developed for orientation in our surroundings and for the organization of human communities."

Another way to grasp what Bohr had in mind consists in looking at his examples of "classical concepts." The set of examples is in fact very narrow: space, time, and causality are the three recurrent examples. It has to be stressed that the concepts of space and time are always mentioned together; they seem to correspond, in Bohr's (1934, 54; 1958, 19; 1963, 5) view, to one single concept: that which he called "space-time coordination." This space-time coordination corresponds to the idea of the location of an object in space and time, or in Bohr's (1963, 5) words, in a "limited space-time domain." As for the concept of causality, it is always related to the "laws of conservation of momentum and energy" (Bohr 1963, 5), and sometimes to the "definition of the state of a physical system" (Bohr 1934, 54). The explanation of this relation is not obvious. According to Folse's (1985, 68–9) interpretation, "The conservation principles make possible applying the claim of causality": when two systems are interacting and their total energy and momentum are conserved, it is possible to determine exactly how their states have changed.

Let us turn now to the "classical physical description." What defines such a classical physical description is always stated by Bohr (1934, 11) precisely in the same manner: it consists in applying to the systems under study *simultaneously* the classical concepts mentioned above, namely, space-time coordination, *and* the principles of conservation of momentum and energy: this description makes a "simultaneous use of space-time concepts and the laws of conservation of energy and momentum." Bohr appeals to various alternative expressions in order to refer to such a description: "causal space-time description" (1934, 55), "mechanical mode of description" (1934, 11), "pictorial deterministic description" (1958, 85), and "classical physical description" (1963, 5).

Each of these expressions brings a slightly different light on this description: "causal space-time description" means simultaneous application of the laws of causality and space-time coordination; "classical physical description" and "mechanical mode of description" indicate that this description is characteristic of classical physics, or more precisely, of the mechanistic mode of description associated with the classical physical theories (i.e., Newtonian mechanics, Maxwell's electrodynamics, and Einstein's relativity theory); and "pictorial deterministic description" means that such a description enables us to provide a pictorial representation of the phenomena, or in other words, with a "visualization" of the phenomena (Bohr 1934, 90) (on this point, see Scheibe [1973, 14]).

In which respect do these clarifications concerning the "classical concepts" and the "classical physical description" help us to understand Bohr's thesis? His claim is that a measurement outcome must be described *either* by means of the concepts of space-time *or* by the concept of causality. He does not support the view that it must be described by these concepts of space-time and causality *simultaneously*. Indeed, in the context of microphysics, the "classical physical description" is no longer valid to account for the measurement outcome.

Bohr's argument for this negative point relies on Heisenberg's relations which express mathematically a new limitation put into light by quantum mechanics: the experimental conditions which allow us to apply space-time coordination forbid us to apply the principles of conservation of momentum and energy; and conversely, the experimental conditions which allow us to apply the principles of conservation of momentum and energy forbid us to apply space-time coordination. Indeed, according to Bohr (1958, 72),

> any unambiguous use of the concepts of space and time refers to an experimental arrangement involving a transfer of momentum and energy, uncontrollable in principle, to fixed scales and clocks which are required for the definition of the reference frame. Conversely, the account of phenomena that are characterized by the laws of conservation of momentum and energy involves in principle a renunciation of detailed space-time coordination. These circumstances find quantitative expression in Heisenberg's indeterminacy relations.

In other words, these experimental arrangements are defining "mutually exclusive conditions for the unambiguous use of the very concepts of space and time on the one hand, and of dynamical conservation laws on the other" (Bohr 1958, 73).

By stressing a limitation concerning our means of knowledge, Bohr's epistemology seems to be in line with Kant's philosophy. However, there is a novelty here, as Kauark-Leite (2012, 51) has pointed out: the new limitation discussed by Bohr concerning the applicability of our conceptual means of knowledge is imposed by the very nature of phenomena at the atomic scale. This limitation has not been identified a priori, but through scientific inquiry, that is, a posteriori: "the exploration of the world of atoms was ... to reveal inherent limitations in the mode of description embodied in common language developed for orientation in our surroundings and the account of events in daily life" (Bohr 1963, 59).

The description of the measurement apparatus and the description of the measurement outcomes

According to Camilleri and Schlosshauer (2015, 75), Bohr's thesis consists in supporting the view that classical concepts are necessary to describe the *measurement apparatus*: Bohr's "contention [is] that the experimental apparatus must be described in terms of the concepts of classical physics." Why is that the case? Camilleri and Schlosshauer interpret Bohr's thesis as follows: if the apparatus is not described by means of classical concepts, it cannot yield information concerning the system under study: "Bohr's central insight was that if a measuring instrument is to serve its purpose of furnishing us with knowledge of an object, it must be described classically" (76).

In my view, this reading of Bohr is incomplete and thereby might be misleading. It is incontestable that Bohr thought that the measurement apparatus must be described by means of classical concepts. Indeed, in some of his writings, he emphasizes that the functioning of the apparatus and the pointer of the apparatus, or any equivalent part of the apparatus (e.g., a spot on a photographic plate caused by the system under study), have to be described with classical concepts: "it is an obvious demand that the recording of observations as well as the construction and handling of the apparatus, necessary for the definition of the experimental conditions, be described in plain language" (Bohr 1963, 59).

However, in other parts of his writings, Bohr claimed that classical concepts are also necessary to *interpret* the information carried by the pointer of the apparatus (for instance, the spot produced on a photographic plate), or in other words, to express the knowledge yielded by the apparatus concerning the microscopic system under study: "The unambiguous *interpretation* of any measurement must be essentially framed in terms of classical physical theories, and we may say that in this sense the language of Newton and Maxwell will remain the language of physics for all time" (Bohr 1931, 692; my emphasis); "only with the help of classical ideas is it possible to ascribe an unambiguous *meaning* to the results of observation" (1934, 17; my emphasis); it is necessary to use "classical concepts in the *interpretation* of all proper measurements, even though the classical theories do not suffice in accounting for the new types of regularities with which we are concerned in atomic physics" (1935, 701; my emphasis).

Bohr (1958, 3) made a distinction between the mark on the apparatus produced by its interaction with the microscopic system under study and the interpretation of this mark which corresponds to what physicists are used to calling the "measurement outcome": "it is also essential to remember that all unambiguous information concerning atomic objects is derived from the permanent marks—such as a spot on a photographic plate, caused by the impact of an electron—left on the bodies which define the experimental conditions." In regard to the above quotations, Bohr made two distinct claims: (i) the apparatus and the permanent mark must be described by means of classical concepts; (ii) this mark has to be interpreted by means of classical concepts. Are these two claims independent? On the contrary, they seem to be closely related in Bohr's mind: the apparatus and the permanent mark must be described by means of classical concepts because otherwise this mark cannot be interpreted by means of classical concepts and yield knowledge of the microscopic system under study (on this point, see Bächtold [2008c]).

The system/apparatus distinction

Let us turn now to Bohr's justification of his thesis. Why are classical concepts *necessary* in order to describe the measurement outcomes? "The argument," Bohr (1958, 39) writes, "is simply that by the word 'experiment' we refer to a situation where we can tell others what we have done and what we have learned and that, therefore, the account of the experimental arrangement and of the results of the observations must be expressed in unambiguous language with suitable application of the terminology of classical physics." Classical concepts are required because they enable us *to communicate to others* the information about the microscopic system under study which has been yielded by the measurement. Here is another passage where Bohr (1958, 26) formulates this point: "the aim of every physical experiment—to gain knowledge under reproducible and communicable conditions—leaves us no choice but to use everyday concepts, perhaps refined by the terminology of classical physics, not only in all accounts of the construction and manipulation of the measuring instruments but also in the description of the actual experimental results."

Being communicable seems to be for Bohr a requirement: we are able to extract information on a microscopic system from a measurement if we are able to communicate it to others. In turn, this requirement implies another requirement: to be communicable, this information must be *unambiguous*. This chain of requirements may explain why Bohr (1958, 67) came to identify the notion of the *objectivity* of the description of a system with the one of *unambiguous* communication: "Every scientist … is constantly confronted with the problem of objective description of experience, by which we mean unambiguous communication."

Why is Bohr concerned with this requirement of nonambiguity? Because in the frame of microphysics, with regard to quantum mechanics, the fulfilment of this requirement is no more obvious. For, if we want to gain information on a microscopic system, we have to make it interact with an apparatus. Yet, once they have interacted, the microscopic system and the apparatus form an indivisible whole. This is what Bohr labeled "individuality" (1934, 53; 1958, 62) or later "wholeness" (1963, 4; 1963, 60). Bohr (1934, 54) described this new feature of the physical phenomena at the microscopic scale as follows: "any observation of the atomic phenomena will involve an interaction with the agency of observation not to be neglected. Accordingly, an independent reality in the ordinary physical sense can neither be ascribed to the phenomena nor to the agencies of observation." He came to support a renewed notion of phenomena: "the interaction between atomic objects and measuring instruments forms an integral part of quantum phenomena" (1958, 2).

The problem in microphysics is then: how can we say something concerning the microscopic system if it forms an indivisible whole with the apparatus? We might think that no unambiguous information concerning the microscopic system can be obtained and communicated. How did Bohr tackle this problem? I propose to reconstruct his reasoning in three steps. The first step amounts to stressing the following epistemological point: although quantum mechanics tells us that there is *physically* no sharp distinction between the microscopic system and the apparatus, we can and must *mentally think* such a distinction in order to express unambiguous information concerning the microscopic system. To distinguish mentally the microscopic system and the apparatus is a fundamental requirement for the unambiguous formulation and communication of the information gained from a measurement: "the description of quantum phenomena requires a distinction in principle between the objects under investigation and the measuring apparatus by means of which the experimental conditions are defined" (Bohr 1963, 78). In this quotation, Bohr supports that this distinction has to be made "in principle." This is an a priori demand for knowledge to be expressed and communicated. Without such a distinction, it is not possible to state any knowledge concerning the microscopic system itself. Camilleri and Schlosshauer (2015, 76) express this point as follows: "[Bohr] regarded the condition of isolation to be a simple logical demand, because, without such a presupposition, an electron cannot be an 'object' of empirical knowledge at all."

Let us emphasize that, due to individuality, the microscopic system isolated mentally by making this system/apparatus distinction cannot be considered as an object existing in itself. As Cuffaro (2010, 310–11) expresses it: "In my description of a classical object ... I am able to distinguish the object 'as it really is' from the object 'as it appears.' But this is not possible for atomic phenomena. Although we must make some 'subject-object' distinction—some 'cut' in what we observe—it is an arbitrary cut—one in which the interaction between apparatus and object cannot be disentangled from our description of the object."

The second step in Bohr's reasoning consists in remarking that, by contrast, within the frame of classical physics, there is no problem in distinguishing the system under study from the apparatus. The interaction between them can be neglected or we can determine the contribution of this interaction and subtract it by a calculation in order to determine the state of the system in isolation: "the edifice of classical physics [relies on] the assumption that the interaction between the object and the tools of observation can be neglected or, at any rate, compensated for" (Bohr 1963, 59). According to Bohr, the concepts of classical physics are relying on such a distinction. Or as Faye (1991, 140) puts it, "The object-instrument distinction is ... an inherent feature of the use of classical concepts." For instance, space-time coordination corresponds to the location of an object in space and time, where the object is tacitly considered as being distinct from any apparatus we may use to determine this location in space and time.

The third step in Bohr's reasoning can be expressed as follows: any unambiguous description concerning a microscopic system relies on the distinction made mentally between this system and any measurement apparatus; since such a distinction is fundamentally foreign to quantum mechanics which implies individuality, we have no choice than making use of the concepts of classical physics which tacitly rely on such a distinction.

Note that there was a debate between Bohr and Schrödinger on the question as to whether we can create new "quantum concepts" suited for the description of the microscopic systems. As Chevalley (1991, 81–8) explains it, Bohr rejected the idea that such quantum concepts are possible because quantum mechanics does not provide us with a new "intuition" or "representation" of microscopic systems; this theory breaks with any intuitive representation of a system insofar as it does not allow us to conceive it as an isolated system in space and time (which is a consequence of individuality). Or as Honner (1987, 63) puts it: "There are no such 'quantum concepts' which can be drawn from quantum object themselves, since a sharp separation between object and observing system is rendered impossible by the indeterminacy relations."

Note that Bohr's thesis concerning the necessity of making use of classical concepts has been interpreted by von Weizsäcker (1952, 85) as having a Kantian dimension: the description of the measurement outcomes by means of classical concepts is a "condition of the possibility" of "every actual experiment known to us." Nevertheless, like above concerning the impossibility to apply a classical physical description, Bohr's epistemology differs in some respect from Kant's approach. For, the necessity of making use of classical concepts results from the lack of suitable concepts yielded by quantum mechanics due to individuality; and, in turn, individuality has been brought to light through an empirical inquiry, a posteriori and not a priori.

The system/observer distinction

In other parts of his writings, Bohr puts forward a slightly different requirement for a description to be objective, and hence, to be unambiguous for communication: such a description must rely on a *clear distinction between the system and the observer*, or more precisely, between the system under study and the observer who performs a measurement on this system and describes it on the basis of this measurement: "The description of atomic phenomena has ... a perfectly objective character" if "no explicit reference is made to any individual observer," which implies that "no ambiguity is involved in the communication of information" (Bohr 1963, 3); "any appeal to the observing subject ... would hinder unambiguous communication of experience" (7).

Since the measurement apparatus can be viewed as a material expansion of the observer, the system/apparatus distinction may be confused with the system/observer distinction. However, the two distinctions are related to two distinct issues, which Bohr not always made clear. The first issue is how to make the system/apparatus distinction so that we are able to consider the system in isolation and express knowledge about it. As seen in the previous section, Bohr's solution to this issue consists in supporting his view that classical concepts are necessary, because they rely on the required system/apparatus distinction.

The second issue is how to provide a description of the system that does not depend on the observer, to prevent this description to be flawed by subjectivity. There is an issue here because the dependency of the measurement outcome on the apparatus, due to the individuality of the system and the apparatus, may suggest that the outcome is dependent on the observer himself insofar she/he is the one who makes the choice of the apparatus. Bohr's solution to this second issue is quite different from his solution to the first one. It amounts to arguing that any subjectivity in the description of the measurement outcome can be avoided merely by including the description of the apparatus in the very description of the measurement outcome: "the whole experimental arrangement must be taken into account in a well-defined description of the phenomena" (Bohr 1958, 90); we must "take into consideration the conditions under which the experience has been obtained" (Bohr 1963, 78).

According to Bohr, there is also a relation between the subjectivity problem and the requirement of unambiguous communication: a description of the measurement outcome that can be communicated without ambiguity must be devoid of subjectivity. As a consequence, describing the experimental setup not only enables avoiding any observer dependency, that is, any subjectivity, but also is a necessary condition for the measurement outcome to be communicated without ambiguity. In Bohr's (1963, 92) words: "the characteristic new feature in quantum physics is merely the restricted divisibility of the phenomena, which for unambiguous description demands a speci-fication of all significant parts of the experimental arrangement." Ultimately, although he did not write it explicitly in this manner, Bohr identified two necessary conditions for unambiguous communication: the use of classical concepts and the description of the whole experimental setup.

Is there a relation between Bohr's thesis and complementarity?

One can wonder if there is a relation between Bohr's thesis of the necessary use of classical concepts and complementarity, which lies at the core of the epistemology he developed in the frame of microphysics. Let us recall briefly the meaning of comple-mentarity. As Murdoch (1987, 61) and Faye (1991, 142) have pointed out, comple-mentarity is based on two notions: mutual exclusiveness and joint completion. Indeed, two measurement outcomes are complementary if they are obtained under mutually exclusive experimental conditions, and if together they provide a complete description of the microscopic system under study (see, for instance, Bohr 1958, 19; and 1963, 4). Mutual exclusiveness can be viewed as the negative component of complementarity, while joint completion as its positive component. Mutual exclusiveness is a new feature of microphysics which imposes a *limitation* on the use of classical concepts to describe the measurement outcomes: an experimental setup which allows us to apply a classical concept (e.g., space-time coordination) prevents us from applying another classical concept (e.g., momentum and energy conservation principles). As for joint comple-tion, it is a novel feature of the framework composed of the classical concepts: two descriptions of measurement outcomes making use of different classical concepts can be combined so as to "exhaust all conceivable knowledge about the object" (Bohr 1963, 4).

As emphasized by Folse (1985, 114), Bohr's initial concern when speaking of complementarity was to show that space-time coordination and the momentum and energy conservation principles are "two modes of description" that can no more be "applied simultaneously to the same object," as it was the case in classical physics, but can be combined in a "complementary fashion" when considering two different situa-tions. In other words, complementarity is Bohr's solution to overcoming the problem of the applicability of classical concepts in the frame of microphysics.

On the one hand, there is indeed a relation between Bohr's thesis and complemen-tarity insofar both are dealing with the application of classical concepts. On the other hand, they are independent: Bohr's thesis states a requirement for the unambiguous communication of the knowledge yielded by a measurement on a microscopic system, while complementarity is a new conceptual framework proposed by Bohr as a solution to overcoming the problem of the impossibility of applying simultaneously different classical concepts. Accordingly, Bohr's thesis can be considered and discussed on its own (which we are doing in this chapter) independently of the endorsement or rejection of complementary.

Summary

Bohr's thesis is based on at least four conceptual distinctions. They have to be taken into account, I contend, in order to properly understand the meaning of this thesis. Let me sum up the main points discussed above:

a) *The distinction between "classical concepts" and "classical physical description"*: Bohr claimed that all measurement outcomes must be described either by means of the "classical concept" of space-time coordination or by the "classical concept" of causality, but not that they must be receive a "classical physical description" implying the simultaneous application of these concepts (which is forbidden in microphysics);

b) *The distinction between the description of the measurement apparatus and the description of the measurement outcomes*: Bohr's thesis concerns primarily the description of the measurement outcomes; this description has to made by means of classical concepts; Bohr also argued that the measurement apparatus and the permanent mark must be described by means of classical concepts; but this is required to ensure the possibility to describe the measurement outcome (i.e., to interpret the permanent mark) by means of classical concepts;

c) *The distinction between the system and the apparatus*: to be objective, information on a microscopic system obtained from a measurement must be communicable without ambiguity; in turn, unambiguous communication implies a distinction to be made mentally between this system and the measurement apparatus; such a distinction is in contradiction with the feature of individuality of quantum mechanics, so that we have no other choice than making use of the concepts of classical physics which tacitly relies on such a distinction;

d) *The distinction between the system and the observer*: the system/observer distinction is another distinction not to be confused with the system/apparatus distinction; to avoid subjectivity in the description of the measurement outcome (which depends on the choice of the observer concerning the apparatus being used), the description of the measurement outcome must also include the description of the apparatus.

4. Bohr's thesis and decoherence theory

After having tried to clarify Bohr's thesis of the necessary use of classical concepts, let us discuss its relation to decoherence theory. Is this thesis rejected by decoherence theory? Or, on the contrary, does it make a contribution to the problem tackled by decoherence theory?

Bohr's thesis in the frame of today's physics

Before investigating these questions, let us consider three ingredients involved in Bohr's thesis that might be viewed differently in the frame of today's physics. First, this thesis makes reference to "classical concepts." Although this expression is not often used by today's physicists, this is the case for the examples of so-called classical concepts given by Bohr: localization in space and time, and the conservation principles for momentum and energy. Therefore, with respect to these examples, the formulation of Bohr's thesis in terms of "classical concepts" may still appear meaningful today.

Second, the justification of Bohr's thesis relies on the notion of "individuality" (or "wholeness") which is assumed to be a feature of quantum mechanics. No physicist makes use today of this expression. Nevertheless, Bohr's notion of individuality can be viewed as a precursor to today's talk of entanglement. "Entanglement" is a feature of any state vector associated to a system composed of two subsystems having interacted (such as the microscopic system and the apparatus): this state vector cannot be expressed as a simple tensor product of state vectors associated to each subsystem; there is no well-defined state vector associated with each subsystem; the state vectors of the two subsystems are "entangled" (initially, Schrödinger [1935], who introduced the word, was speaking of the "entanglement of predictions" concerning both subsystems). Like individuality, entanglement means that two systems having interacted cannot be conceived as being physically distinct. For this reason, Bohr's justification of his thesis might still be supported today.

A third ingredient associated with Bohr's thesis to be reconsidered is the claim that, to avoid ambiguity and subjectivity, the "whole experimental arrangement" has to be described. Today, this requirement appears to be excessive. When they communicate to each other the measurement outcome that has been obtained, physicists can merely specify the observable that is measured. Of course, the knowledge of the experimental arrangement is needed to determine the observable. But, once this observable has been determined, there is no need to mention the experimental arrangement anymore. Nevertheless, the fact that this claim is somehow obsolete does not affect Bohr's thesis. Indeed, this claim is neither a premise nor a consequence of Bohr's thesis, but rather an independent claim, which shares with it a motivation, namely, the search of the conditions for an unambiguous communication of the measurement outcomes.

Is Bohr's thesis rejected by decoherence theory?

Camilleri and Schlosshauer (2015) have provided a detailed and compelling analysis of this question leading to a negative answer: "Our analysis suggests that, contrary to claims often found in the literature, Bohr's doctrine is not, and cannot be, at odds with proposed solutions to the dynamical problem of the quantum–classical transition that were pursued by several of Bohr's followers and culminated in the development of decoherence theory" (abstract).

To argue for this answer, they put forward two points. First, Camilleri and Schlosshauer (2015, 82) stress that Bohr's thesis concerns a very different problem from the one addressed by decoherence theory: "Bohr's epistemological thesis concerning a functional description of experiment should be distinguished from the efforts, which gathered momentum in the 1960s, to supply a dynamical explanation of the emergence of classicality from quantum theory." To put it differently, Bohr's thesis is related to the problem of how to describe the outcomes of measurement performed on microscopic systems (what we labeled "problem B" in Section 2), whereas decoherence theory is intended to contribute to the quantum-to-classical transition problem ("problem A"). Camilleri and Schlosshauer are insisting on the idea that Bohr's thesis is an "epistemological thesis" (79). One can reproach them for suggesting thereby that problem A, addressed by decoherence theory, is an ontological problem, by contrast with problem B, addressed by Bohr. Yet, as pointed out above (in Section 2), problem A might be tackled either according to an epistemological or an ontological perspective. Besides, the fact that Bohr's thesis addressed another problem than the one addressed by decoherence theory does not imply, strictly speaking, that this thesis is not challenged by decoherence theory.

So let us turn to the second point made by Camilleri and Schlosshauer. Putting forward the epistemological nature of Bohr's thesis is a way to stress that Bohr did not claim that measurement apparatuses *are* classical objects from an ontological point of view. The requirement that they be *described* by means of classical concepts is an epistemological requirement. Bohr never claimed that they could not be described by means of quantum mechanics. As Camilleri and Schlosshauer (2015, 79) put it, "While Bohr insisted that we must employ the concepts of classical physics to describe whatever part of the system we have designated to function as a measuring instrument, it is always possible to give a quantum-mechanical description of the apparatus in its entirety." If Bohr was claiming that measurement instruments *are* classical objects, he would be in contradiction with decoherence theory which describes the instruments by means of quantum mechanics. It is true, as Camilleri (2009, 295) writes in a former paper, "In providing a quantum-mechanical treatment of the measuring instrument, Wootters and Zurek departed from Bohr's insistence on the primacy of classical concepts." However, Bohr did not claim that the measurement instrument *cannot* be described by means of quantum mechanics. He claimed it *must* be described by means of classical concepts *if* we want to infer from it unambiguous information concerning the system under study.

Note that Bohr (1963, 3) himself attempted to provide a physical justification for the possibility of describing the measurement apparatus by means of classical concepts. He held that "measuring instruments [are] rigid bodies sufficiently heavy to allow a completely classical account of their relative positions and velocities." Nonetheless, this justification does not appear convincing. Bohr does not explain why heaviness implies the possibility of applying classical concepts. This is an objection made by Everett in 1957 (see Osnaghi, Freitas, and Freire 2009, 106; and Camilleri and Schlosshauer 2015, 79).

There remains a possible way of viewing decoherence theory as challenging Bohr's thesis that is not discussed by Camilleri and Schlosshauer in their 2015 paper. Decoherence theory does not resolve completely the quantum-to-classical problem. If we consider this problem from an epistemological point of view, decoherence theory only yields improper mixtures, which is due to the fact that the interference terms (i.e., the non-diagonal terms of the reduced density operator associated to the studied system and the apparatus $S+A$), although practically insignificant, are never strictly equal to zero (Bächtold 2008a, b). This raises the following question: is it legitimate to make use of classical concepts to describe the measurement apparatus and the measurement outcomes? We will come back to this question in the next section.

Does Bohr's thesis make a contribution to the problem tackled by decoherence theory?

One can wonder if Bohr's thesis does not indirectly contribute to the quantum-to-classical transition problem (problem A). Let us come back to the argument underlying this thesis: the measurement outcomes must be described by means of classical concepts because these are the only available concepts relying on the division between the system under study and the measurement apparatus; this division has to be made because otherwise it is simply not possible to express unambiguous information concerning the system. Now, the fact that the concepts needed to describe the measurement outcomes without ambiguity are those developed in the frame of *classical physics* is a contingent fact. In this respect, the classical character of these concepts is not fundamental. Bohr's main point, one may argue, is that we need to make a mental distinction between the system and the apparatus, and that this distinction must be involved in the description of the measurement outcomes. This distinction, rather than the use of classical concepts, is an a priori condition of possibility of the very description of the measurement outcomes.

Following Bohr's line of reasoning, we could support that this distinction must be introduced a priori in the formalism of quantum mechanics if this formalism is intended to make predictions concerning a microscopic system. As a matter of fact, the distinction between the microscopic system under study and the apparatus is made from the beginning in the frame of any quantum description of the measurement as well as in the frame of decoherence theory. Bohr does not provide us with a satisfactory reason, from the point of view of the theory, for the *possibility* of making this distinction (see the previous section). Nevertheless, he argues for its *necessity* from an epistemological point of view: to make this distinction has to be considered as a necessary condition for the statement of unambiguous knowledge of the microscopic system.

If the "classical concepts" are defined as those concepts implying the system/apparatus distinction, it may appear legitimate after all, *from an epistemological point of view*, to make use of these concepts to describe the measurement outcomes. From this point of view, the fact that decoherence theory does not lead to a complete solution to the quantum-to-classical transition problem (since it yields only improper mixtures) should not be considered as an objection to Bohr's thesis. One can reverse the way to consider the problem.

Epistemologically speaking, the system/apparatus distinction is first: it sets what has to be derived by the theory, namely, a description of the measurement outcome implying a well-defined distinction between the microscopic system and the apparatus.

In this regard, Bohr's thesis stands as a *regulative idea* (to take Kant's words) for the physicists' research dealing with the quantum-to-classical transition problem. It leads us to restate this problem in another manner: how can we account for the transition between the quantum description, characterized by the entanglement of the state vectors associated with the microscopic system and the apparatus, and the description of the measurement outcome implying a well-defined distinction between the microscopic system and the apparatus?

5. Conclusion

This chapter is intended to show that Bohr's thesis of the necessary use of classical concepts can shed light on the problems related to the quantum-classical cut. To understand this contribution one has first to clarify the thesis: what it states exactly and how it is justified. Four conceptual distinctions were put forward. The two first distinctions are important to understand the actual meaning of Bohr's thesis: (a) it does not state that all measurement outcomes must receive a "classical physical description," that is, that they must be described by means of the theories of classical physics; they must be described in terms of "classical concepts," that is, by the concepts that have been developed, for contingent reasons, in the frame of classical physics; (b) the thesis concerns primarily the description of the measurement outcomes and not the description of the measurement apparatus (as some authors suggest); the description of the measurement apparatus in terms of classical concepts, according to Bohr, is only a requirement for the possibility of describing the measurement outcomes with classical concepts. The third distinction helps us to understand Bohr's justification of his thesis: (c) the concepts of classical physics are the only available concepts relying on the mental distinction between the microscopic system under study and the apparatus—this distinction being a necessary condition for the unambiguous formulation and communication of the knowledge gained about the microscopic system by means of a measurement. The last distinction concerns another necessary condition: (d) the description of the system must not depend on the observer, which can be avoided, in Bohr's view, by specifying explicitly the experimental setup used to perform the measurement.

The discussion of these four points leads us to identify the system/apparatus distinction as being essential in Bohr's proposal for dealing with the two problems associated with the quantum-classical cut. Couldn't we describe the measurement outcomes in microphysics by means of quantum mechanics (problem B)? Bohr does not claim the contrary. However, according to his thesis, any description of the measurement outcomes must rely on a well-defined distinction between the system and the apparatus; and this distinction may appear foreign to quantum mechanics, which is characterized by the entanglement of the state vectors associated to interacting systems. The quantum-to-classical problem (problem A) is not directly addressed by Bohr's thesis. Nevertheless, this thesis provides us with another way to look at this problem: the challenge is to explain the transition, not from an alleged quantum world to an alleged classical world, but rather from a description implying the entanglement of the systems under consideration to a description implying a well-defined distinction between the microscopic system under study and the apparatus.

References

Bächtold, M. 2008a. "Five formulations of the quantum measurement problem in the frame of the standard interpretation." *Journal for General Philosophy of Science* 39: 17–33.

Bächtold, M. 2008b. "Interpreting quantum mechanics according to a pragmatist approach." *Foundations of Physics* 38: 843–68.

Bächtold, M. 2008c. "Are all measurement outcomes 'classical'?" *Studies in History and Philosophy of Modern Physics* 39: 620–33.

Blanchard, P., et al. (eds). 2000. *Decoherence: Theoretical, Experimental, and Conceptual Problems*. Berlin: Springer.

Bohr, N. 1931. "Maxwell and modern theoretical physics." *Nature* 128: 691–92.

Bohr, N. 1934. *Atomic Theory and the Description of Nature*. Cambridge: Cambridge University Press.

Bohr, N. 1935. "Can quantum-mechanical description of physical reality be considered complete?" *Physical Review* 48: 696–702.

Bohr, N. 1958. *Atomic Physics and Human Knowledge*. New York: J. Wiley & Sons.

Bohr, N. 1963. *Essays 1958–1962 on Atomic Physics and Human Knowledge*. New-York: Interscience.

Bub, J. 1997. *Interpreting the Quantum World*. Cambridge: Cambridge University Press.

Camilleri, K. 2009. "A history of entanglement: Decoherence and the interpretation problem." *Studies in History and Philosophy of Modern Physics* 40: 290–302.

Camilleri, K., and M. Schlosshauer. 2015. "Niels Bohr as philosopher of experiment: Does decoherence theory challenge Bohr's doctrine of classical concepts?" *Studies in History and Philosophy of Modern Physics* 49: 73–83.

Chevalley, C. 1991. "Introduction & glossaire." In *Physique atomique et connaissance humaine*, N. Bohr. Paris: Gallimard, 19–140 and 345–567.

Cuffaro, M. 2010: "The Kantian framework of complementarity." *Studies in History and Philosophy of Modern Physics* 41: 309–17.

D'Espagnat, B. 1995. *Veiled reality: An analysis of present-day quantum mechanical concepts*. Reading, MA: Addison-Wesley.

Faye, J. 1991. *Niels Bohr: His Heritage and Legacy*. Dordrecht: Kluwer Academic Publishers.

Folse, H. 1985. *The Philosophy of Niels Bohr: The Framework of Complementarity*. Amsterdam: North Holland.

Giulini, D., et al. (eds). 2003. *Decoherence and the Appearance of a Classical World in Quantum Theory*, second revised edition. Berlin: Springer.

Honner, J. 1987. *The Description of Nature: Niels Bohr and the Philosophy of Quantum Physics*. Oxford: The Clarendon Press.

Joos, E., et al. (eds). 2003. *Decoherence and the Appearance of the Classical World in Quantum Theory*, second edition. Berlin: Springer.

Kauark-Leite, P. 2012. *Théorie quantique et philosophie transcendantale: dialogues possibles*. Paris: Hermann.

Mittelstaedt, P. 1998. *The Interpretation of Quantum Mechanics and the Measurement Process*. Cambridge: Cambridge University Press.

Murdoch, D. 1987. *Niels Bohr's Philosophy of Physics*. Cambridge: Cambridge
University Press

Osnaghi, S., F. Freitas, and O. Freire Jr. 2009. "The origin of the Everettian heresy." *Studies in History and Philosophy of Modern Physics* 40: 97–123.

Psillos, S. 1999. *Scientific Realism: How Science Tracks Truth*. New York: Routledge.

Scheibe, E. 1973. *The Logical Analysis of Quantum Mechanics*. Oxford: Pergamon.

Schrödinger, E. 1935. "Die gegenwärtige Situation in der Quantenmechanik." *Die Naturwissenschaften* 23: 807–12, 823–8, 844–9.

Van Fraassen, B. 1991. *Quantum Mechanics: An Empiricist View*. Oxford: Clarendon Press.

Von Weizsäcker, C. F. 1952. *The World View of Physics*. Chicago: The University of Chicago Press.

Zurek, W. 1981. "Pointer basis of quantum apparatus: into what mixture does the wave packet collapse?" *Physical Review D* 24: 1516–25.

Zurek, W. 1982. "Environment-induced superselection rules." *Physical Review D* 26: 1862–80.

Zurek, W. 1991. "Decoherence and the transition from quantum to classical." *Physics Today* 44: 36–44.

12

Individuality and Correspondence

An Exploration of the History and Possible Future of Bohrian Quantum Empiricism

Scott Tanona

1. Introduction

In this contribution I offer an interpretation of Bohr's philosophy grounded in both the history of his work in early quantum theory and developments in interpretation post-Bohr. I consider Bohr's philosophical writings to be insightful but also often vague and occasionally impenetrable, and in the end incomplete as interpretation. I nevertheless believe we have a lot to learn from Bohr if we put his writings in the right context. I attempt to provide such a context and thereby develop a story about what we can take away from Bohr. I then offer an outline of an approach to interpreting quantum mechanics that I believe updates Bohr. While I aim to make this account accurate with respect to history and to Bohr's writings, I will not focus on historical or interpretive detail and instead will aim for general account that, I hope, fits details that can be found elsewhere.[1] I hope also that it is of some interest independent of whatever it adds to our understanding of Bohr.

The picture of Bohr I aim to paint is empiricist but not anti-realist. I believe that Bohr strove to advance early quantum theory by trying to infer atomic structure directly from empirical phenomena rather than developing speculative theoretical models and hypotheses that might account for those phenomena. While he certainly ventured into speculative territory, he was always concerned to ground his models in phenomena. Bohr felt keenly the epistemological gap introduced by his 1913 model of the atom. That model, while it addressed the instability of any would-be classical model of the atom, nonetheless did not offer empirical access to everything we might want to learn about the structure of the atom besides the stationary states—and worse, took away access we thought we had. Bohr's correspondence principle was meant to fill that gap. Through its use Bohr developed an appreciation for the assumptions and theoretical background being used to gather and analyze empirical data, and he became aware of a disconnect between the developing quantum theory and much of the theory and assumptions used in generating and interpreting empirical data. The history I try to briefly fill in here is a story of Bohr's attempt to maintain enough of the classical framework for gathering data as he could and to fit it to the new quantum theory (or vice versa) so that we could better determine how to build a model of atomic structure from that empirical data.

The end result is that Bohr developed a philosophical stance in which the job of the physicist, and later the interpreter of physics, was to connect the hidden properties of the things being investigated to the more mundane, more accepted, more classical, and ultimately more direct properties evinced in experimental instrumentation. Applying that method to experimental results of 1925 he concluded that distant events were "coupled" and occurred as unified wholes rather than as distinct separate events describable by their own local physics. This recognition of nonseparability or "individuality" stayed with him and formed critical background to his interpretation of quantum mechanics.

Bohr's approach to understanding quantum mechanics—which he viewed as a culmination of the use of the correspondence principle in the old quantum theory—maintained the empirical approach of inferring hidden quantum properties from corresponding, classically described empirical phenomena. His approach was at odds with the apparent nonseparability of phenomena because that nonseparability implied that the measuring instruments used to produce the empirical results were themselves part of the overall quantum phenomenon, and so apparently were not suitable as independent epistemological starting points for interpreting quantum formalism from the outside in.

The picture Bohr developed to address those tensions, with its emphases on the use of classical language, the physicality of measurement, and the resultant complementarity of different measuring contexts, was influential but ultimately incomplete, and today it seems outdated. A survey of modern interpretations of quantum mechanics is not likely to include a truly Bohrian interpretation. While the legacy of the "Copenhagen interpretation" is large, that beast is more mythical than real, and the name is often misapplied to refer to a standard von Neumann collapse interpretation or something closer to Heisenberg's views (see Howard 2004). Presentations of Bohr's views often serve more historical purpose than any other, and are used as either foil or fodder for the views of those invoking them. Complementarity itself is unlikely to have a real interpretive role in any viable interpretation, though it might be included as a description of quantum phenomena.

This state of the representation of Bohr's views is understandable. Bohr did not present his views in a systematic, comprehensive fashion but instead elaborated them historically, arguing for their naturalness in light of how they developed. It is difficult to reconstruct a comprehensive modern interpretation from those writings. This difficulty is partly due to vagueness in his language, some of which may have come by way of intended generality. It is also partly due to what we have learned since Bohr developed his views and how our standards for interpretation have changed since then. One would be hard-pressed today if one attempted to provide a new interpretation of quantum mechanics without saying explicitly how the interpretation addressed the measurement problem and indicating where the interpretation stood with respect to interpretive and physical principles such as the collapse postulate, the eigenvalue-eigenstate link, the completeness of the formalism, and the possibility of alternative dynamics. It's not clear how to interpret Bohr in these terms. But even if we cannot reconstruct a modern interpretation from Bohr, I believe we can continue to clarify some of the elements of Bohr's philosophy, that we can learn from updating Bohr in a careful way, and that we may find alternative ways to frame the problem of interpretation that may bear some fruit.

Instead of trying to faithfully recover a comprehensive philosophy of quantum mechanics from Bohr, I will aim only to identify and describe key elements of his approach and to try to explain them in the context of their historical development. I will then move to the question of what we can take away from Bohr today. I will try to present a more modern approach that is built on those elements but also makes use of some of what we have learned since then, especially about decoherence. I hope that that part of the chapter will be in line with the spirit of his approach, though I will not promise fidelity to everything Bohr said. That extension of his ideas will not form a complete interpretation of quantum mechanics, and in particular will not satisfactorily address the measurement problem. I will indicate why I don't think that is necessarily a failure.

2. History: The old quantum theory

Bohr's approach to the development of a theory of atomic structure was bottom-up, data-first. It was based as much as possible on making definitive inferences from described phenomena rather than developing models that might explain phenomena. It is easy to view his 1913 model of the hydrogen atom as a product of theoretical speculation, a hypothesis that avoided the instability of a classical orbital atomic model almost by fiat by quantizing angular momentum to derive stationary states, and then also fit the empirical data. Indeed, Bohr sometimes introduced the model by starting with postulates of stationary states and energy relation for transitions (e.g., 1918, 1965). However, the discovery process began with the empirical phenomena of atomic spectra.

Starting from the Balmer series $v = \dfrac{K}{n''^2} - \dfrac{K}{n'^2}$ and interpreting it with the quantum energy-frequency relationship $hv = E_1 - E_2$ (which to Bohr did *not* imply light quanta) Bohr (1920, 23) *inferred* the existence of distinct energy states of value $E_n = -\dfrac{Kb}{n^2}$: "The spectra do not give us information about the motion of the particles in the atom, but only a knowledge of the energy changes in the various processes which can occur in the atom. From this point of view the spectra show the existence of certain definite energy values corresponding to certain distinctive states of the atoms." This quotation suggests that the justification of the model was data-driven. Bohr's claim that the spectra "show" us something about atomic structure is typical. He does not merely say that the hypothesis of stationary states explains or would explain the spectra. The inference is more a deduction from phenomena (in the sense of Harper [1990]) than a best explanation of the phenomena. Bohr remained confident of this inference and did not doubt the existence of stationary states even after the later developments in quantum theory and alternative explanations from Schrödinger.

Both the discovery process and the evidential justification required a starting description of the phenomena that could be established independently and anteced- ently of any new hypotheses about atomic structure. This requirement was not trivial. The radical nature of the new theory increasingly threatened to undermine existing descriptions of a lot more than atomic structure. The light-quantum idea, in particular, would require radical rethinking of interference and diffraction of radiation. Because the inference to stationary states depended on having an accurate description of the spectra to start from, Bohr's approach required that the new theory not require revolu- tionary re-descriptions of empirical phenomena.

His later insistence on the primacy of classical language stems from this early work and its continuation in the old quantum theory from 1913 to 1925. Atomic spectra are optical phenomena, patterns seen in the frequencies, wavelengths, intensities, and polarizations of radiation emitted from excited atoms. While some of the structures of those patterns might be described without wave interpretation (e.g., leaving the meaning of radiation frequency aside and looking just at the right-hand side of Balmer series equation above), for Bohr the patterns could not be separated from their method of production via the use of instruments such as spectrometers and the interpretation of those instruments' readings with wave theory. The possibility of particulate light quanta did not sway Bohr (1965, 14):

> In spite of its heuristic value, however, the hypothesis of light-quanta, which is quite irreconcilable with so-called interference phenomena, is not able to throw light on the nature of radiation. I need only recall that these interference phe- nomena constitute our only means of investigating the properties of radiation and therefore of assigning any closer meaning to the frequency which in Einstein's theory fixes the magnitude of the light-quantum.

Bohr's approach aimed to infer atomic structure from spectra by relying on parts of classical theory to interpret and extend empirical results. The development of this approach after 1913 led to his famous *correspondence principle*. According to classical electrodynamics, an electron in an orbit will continuously emit radiation tied to the motion of the electron, with frequencies of the radiation equal to the Fourier compo- nents of the electron oscillation. That theory would have let us infer atomic properties of electron motion from emitted radiation had that theory been correct. The discon- tinuous orbits of Bohr's 1913 model ruled out the classical mechanism of the continu- ous emission of radiation. This model came at a cost. The apparently instantaneous or near-instantaneous emission process could not account for the extended wave proper- ties of the radiation and could not even predict properties of the radiation other than frequency. Worse, we could no longer hope to infer properties of electron motion in the atom from the spectra. This problem led Bohr to search for a method for inferring further atomic structure from empirical phenomena.

The correspondence principle was inspired by the discovery that in the limit of high quantum number the frequency of the radiation emitted in a transition from a stationary-state orbit agreed with the frequency of the radiation that would have been emitted classically in that same orbit (if somehow it could have done so without decay). While light seemed to be emitted only intermittently in transitions between discontinuous stationary states of the electron, Bohr maintained that the properties of the emitted light were still related in some way to the motion of the electron in the orbit in the stationary state. More specifically, he maintained that the association in the limit between a Fourier component of motion and the frequency of radiation emitted from a transition also applied in transitions in lower quantum numbers, even though the association did not lead to the same quantitative agreement.

The correspondence principle relied on the agreement in the limit but the principle itself was a claim about relations or correspondences between three different components of the theory and data: motion of the electron within a stationary state, transitions between stationary states, and emitted radiation. The idea was that for each transition that occurred, there was a component of motion in the stationary state that determined both the probability of that transition and the properties of the radiation emitted in that transition.

Bohr used the presumption of these relations to bridge two gaps in the theory. First, there was the epistemological gap between the empirical phenomena and the atomic model: specifically, between the patterns in the wave properties of the spectra and any properties of electron motion except the transitions. That epistemological gap existed because there was a second, causal gap *within* the theory between the transitions and what was presumably occurring in the stationary states before and after the transitions—something like electrons revolving around the nucleus (though not emitting radiation). Bohr (1920) used classical theory to try to bridge both the theory-data epistemological gap and the within-theory causal gap. By bridging those gaps he hoped to use spectral evidence to infer atomic structure by establishing the right "connection between the spectrum and the atomic model" (26).

The approach therefore posited two *correspondence relations* that would fill those respective gaps. The *within-theory* relation posited a correspondence between motion in the stationary states and transitions between stationary states, more precisely between the Fourier components of motion and different possible transitions: "We shall assume that even when the quantum numbers are small the possibility of *transition* between two stationary states is *connected* with the presence of a certain harmonic component in *the motion of the system*" (Bohr 1920, 28; emphasis added).

That presumed within-theory correspondence was then associated with another, more important *theory-data* correspondence between motion in the states and the atomic spectrum: "On account of the general correspondence *between the spectrum of an atom and the decomposition of its motions into harmonic components*, we are led to compare the radiation emitted during the transition between two stationary states with the radiation which would be emitted by a harmonically oscillating electron on the basis of the classical electrodynamics" (Bohr 1920, 51; emphasis added).

Together these relations generated the *between-theory* agreement, an empirical correspondence of a different kind in which the empirical content of the quantum theory approached classical theory so that there was a "correspondence between the frequencies determined by the two methods" (Bohr 1920, 26).

Bohr did not always take care to distinguish these different aspects of his correspondence approach and he tended to refer simply to "the" correspondence principle.
He sometimes talked about correspondence in terms of the agreement in the limit and
he used other terms that were consistent with thinking of it in terms of the limit, for
example, by describing it as a "rational generalization" of classical theory. But the correspondence principle was not about that agreement itself. It was rather about those
within-theory and theory-data associations which were responsible for the in-the-
limit agreement.[2]

These within-theory and theory-data relations were only "correspondences," not
identities and not direct equivalencies between values. But while the connection
was not strict, Bohr thought that by following an approach of following "the" correspondence principle that assumed there were these connections or correspondences
to be found, one could use the agreement in the limit and other more qualitative or
approximate relations to bridge both the gap in the theory and the epistemological gap
between the theory and the empirical data. Bohr used the correspondence principle
successfully, though not without criticism (e.g., Sommerfeld 1923, 275), to tie things
such as the intensities of the emitted radiation to both the probability of transition and
the motion in the stationary state.

While the connection between the elements was not always clean—in the old quantum theory it was precise only in the limit—Bohr (1924, 224) thought that it established firm conclusions grounded directly in empirical phenomena:

> It is possible, *from the empirical rules* governing the remarkable limitation of the
> combination of spectral terms, *to draw conclusions* as regards further details of the
> orbit of the outer electron.

> The above mentioned conclusions may ... be considered as *empirical deduc-
> tions from the spectral evidence*, providing us with information as to the details
> of the interaction between electrons in the atoms.... it appears possible, by pro-
> ceeding in this manner, to arrive from experimental evidence on spectra in a
> rational way step by step at a classification of the orbits of the electron in the atom.
> (Emphases added)

The main relevance of the correspondence principle for his later interpretative
work was the epistemological legacy of this empirical approach of inferring otherwise hidden properties (atomic structure) from independently described observable
data (properties of emitted radiation). The observable phenomena and their empirical description had epistemological priority. The requirement that the new quantum
theory agree with applications of classical theory was needed not merely because
of the desire to retain past successes, but more importantly to enable the approach
of developing new theory from existing, repeatable, and antecedently described
phenomena.

Despite some significant successes, the approach ran into significant difficulties,
especially around the question of determining which motions, in which orbits, were
associated with transitions. Since in low quantum numbers the motions in successive
orbits were significantly different—unlike what was found in the limit of high quantum

number—it was a lot harder to apply the transition-motion and the motion-spectrum correspondence relations to find appropriate general functions relating them.

There were deeper conceptual challenges, too. One was an apparent inconsistency between the continuous wave-nature of the emitted radiation and the discontinuous quantum transitions. The light in the spectrum obeyed wave theory and its frequency and other wave properties were measured with classical techniques. But light emitted in a short burst should not have a precise frequency—instead of an infinitely extended pure-frequency wave, a wave pulse is essentially a combination of such waves with a spread in frequency reciprocally related to the width of the pulse. The light of the spectrum had fairly precise frequencies, in seeming contradiction with the apparently discontinuous transition.

By 1924 these and other problems were mounting. Bohr, with Kramers and Slater (BKS), worked on a last-ditch effort to account for the interaction of light and matter that preserved classical notions of space and time. The BKS theory was the result of taking the correspondence principle very seriously. BKS posited that radiation was not really emitted in the transition but instead was somehow emitted continuously in correspondence with electron oscillation within an orbit, just as the classical theory suggested, but now with no energy loss until a transition. This continuous emission was consistent with experimental evidence that outside fields disrupted spectra as if they perturbed the electron motion—this seemed to imply a classical emission mechanism in association with regular electron motion.

Since the stationary states did not match the expected motion, BKS posited "virtual oscillators" to stand in for the missing electron oscillations, which turned out to be neither starting, nor ending, stationary state motion, but something in-between. The virtual oscillators caused the continuous emission of radiation and probabilistically caused transitions between stationary states. BKS were explicit that this was a continuation of the correspondence program of inferring structure from spectra: "On the correspondence principle it seems nevertheless possible, as it will be attempted to show in this paper, to arrive at a consistent description of optical phenomena by connecting the discontinuous effects occurring in atoms with the continuous radiation field" (Bohr, Kramers, and Slater 1924, 785).

This new conception of the interaction of light and matter was both radical and conservative. It was conservative in its retention of classical electrodynamics, which reintroduced classical explanations of empirical phenomena such as the precision of the spectral frequencies. It was radical because of the consequences of that retention. Radiation, while still causally efficacious, was more epistemologically distant, "virtual" in that its presence would not always be felt (it induced transitions only probabilistically) and did not directly carry energy. This virtual radiation was emitted and absorbed by virtual oscillators that, while associated with real electrons, themselves had no known direct physical presence and had properties that reflected not only the existing stationary state orbits but also the possible stationary state orbits the electron might transition to. Most radical was the idea that energy would be conserved only statistically, as changes in atomic stationary states could occur on one end of the emission or absorption of virtual radiation without it occurring on the other. According to the theory, a virtual oscillator emits virtual radiation whether or not a transition occurs in the atom. The radiation spreads across space in three dimensions and hits atoms in all directions, inducing stationary state elevations with probabilities in proportion to its intensity. The virtual radiation emitted by one atom might cause "absorption" changes in stationary states in multiple other atoms, possibly even without a corresponding "emission" change.

It was not long before challenges to the theory, especially from Einstein, raised doubts in Bohr's mind, and not much longer before the BKS theory died. The final moment came with announcement of the 1925 Bothe-Geiger experiments (Bothe and Geiger 1925) that showed that emission and absorption events in distant atoms are "coupled." The Compton effect showed coincidences between the timing of detection of the recoil electron and the emission of a secondary electron from the scattered radiation that were too close and numerous for the statistical BKS theory. But despite the obvious failure of the theory as a whole, the BKS approach lived on in two important ways. First, while the BKS ontology and statistical energy conservation died, the formalism formed the basis of matrix mechanics. Kramers continued to work on the problem of finding the right expression of atomic oscillation to connect to the spectra. It is possible that the BKS theory, having broken the identification of the radiation-causing electrical oscillator with the motion of an electron in an orbit, made it easier to address that problem a little more abstractly. Kramers's (1924a, b) success was a critical stepping-off point for Heisenberg in his development of matrix mechanics, which essentially did away with modeling electronic orbits and was built just on transition probabilities.

Second, Bohr attempted to reconcile his continued interpretation of light in wave terms, based on the success of classical electrodynamics and its use in optical measuring instruments, with the newly demonstrated coupling of energy changes in distant atoms. Bohr (1925, 204) thought that this coupled matching of emission and absorption events and especially the directionality of the coupling detected in the Compton-Simon experiments (Compton and Simon 1925) could not be causally determined by the properties of the radiation as it traveled through space, at least not if the radiation traveled in any way consistent with the wave behavior it otherwise demonstrated in optics: "No space-time mechanism seemed conceivable that permitted such a coupling and at the same time achieved a sufficient connection with classical electrodynamics, which has been successful to such a great extent in describing optical phenomena."

Bohr's conclusion was that the existence of the coupling meant that the whole phenomenon of radiation emission, travel, and absorption was an "indivisible" or "individual" whole that could not be broken down into intermediate component steps. That is, the initial transition and emission are not fixed at some earlier point in time, which then cause a later absorption and second transition. Rather, the properties of the initial transition and emission are determined only when the absorption occurs. Causality in such quantum events is not a local process traceable step-by-step in time through space, but occurs as a unified whole across space and time.

Bohr must have concluded something along these lines because he rejected the light quantum, at least as a general, stable feature of radiation. The wave nature of radiation meant that at the moment of emission of radiation there could not be an already-fixed unit of energy traveling in a fixed direction. While the light quantum would account for the coupling of distinct emission and absorption events, it would not account for interference and diffraction effects, upon which so much of the empirical data rested.

The same sort of reasoning is what led to the radical BKS: if radiation is being emitted in extended form that must occur at least somewhat independently from the instantaneous or near-instantaneous transitions; so transitions must be only probabilistically related to the emission of radiation; then since the emission and later, distant absorption events are connected by the radiation as the causal intermediary, those emission and absorption events must be only probabilistically related, too. To find that they are strictly coupled meant that something else besides a simple causal intermediary of the traveling radiation must tie those events together.

Bohr's concerns about how literally to take his use of classical theory had been growing. The difficulty of reconciling the strict coupling with the wave mechanism led Bohr to question how to interpret the wave mechanism. He worried about "difficulties to our ordinary space-time descriptions of nature ... presented by the simultaneous understanding of interference phenomena and a coupling of changes of state of separated atoms by radiation" (April 21, 1925, letter to Geiger, in *Bohr's Collected Works*, ed. Stolzenberg 1984, 79).

The Bothe-Geiger results and Compton-Simon results settled it: "The ... coupling between changes of state in distant atoms by radiation excludes the possibility of a simple description of the physical events in terms of intuitive pictures.... if the coupling should really be a fact, ... we must take recourse to symbolic analogies to a still higher degree than before" (May 1, 1925, letter to Born, in *Bohr's Collected Works*, ed. Stolzenberg 1984, 85).

Hence the Bothe-Geiger and related results led Bohr to give up the search for a universal, local, traditionally causal theory. Instead Bohr began to think more abstractly about both traditional wave-theoretic descriptions of the spread of radiation through space and ascriptions of classical particle properties to that radiation. Realistic interpretations of neither sets of concepts seemed possible any longer.

Questions about the difficulty of interpreting partial uses of classical theory only grew after Heisenberg's development of quantum mechanics. But instead of questions about developing theory, the questions now were about how to interpret a theory that Bohr thought was the culmination of the correspondence approach. Heisenberg had built on Kramers's dispersion work associated with BKS (sans virtual oscillators) and the resulting matrix mechanics encapsulated the correspondence between radiation properties and properties of the orbital mechanics, but now without any orbital mechanics. The elements of the matrices stood in place of all that Bohr meant to uncover with the correspondence principle—they were the hidden structures that corresponded to the emitted radiation. However, while they were now able to accurately predict and account for emitted radiation properties and more, they could not be interpreted realistically, at least not easily, and certainly not in terms of particle motion in orbits.

One might think then that the quantum formalism could replace the correspondence principle. And it did in one sense: Bohr no longer discussed the correspondence principle as a tool for theory building and no longer stressed the need to find agreement with classical theory in the limit. However, because the matrix elements did not obviously represent realistic "hidden" properties behind the observable aspects of the phenomena, there remained deep questions about what one could infer about quantum systems from observations.

3. Legacy of the history

We can take away several things from this history. Bohr worked in an empirically driven attempt to infer atomic properties from phenomena. This approach required an independent and antecedently available description of phenomena from which to work. Bohr maintained that the empirical evidence we had for atomic structure was based largely on atomic spectra and that those spectra were optical phenomena, which we measure and interpret in terms of classical wave concepts. He remained convinced that to use the spectra as evidence for atomic structure we therefore had to determine how those extended wave phenomena were connected to the apparently discontinuous phenomena in the atom. To find that connection Bohr put significant effort into trying to find a natural continuation of classical electrodynamics. The extreme version of this approach was to posit a mechanism that retained a classical conception of the spread of radiation through space and time and maintained that space-time propagation of radiation was caused by (and caused) electrical oscillations just as in classical theory, but at the cost of the connection between those oscillations and stationary state transitions. The Bothe-Geiger experiments significantly narrowed the range of possibilities for making the connection between spectra and atomic properties. As Bohr would later describe it, the space-time account of radiation traveling through space could not be entirely reconciled with the close pairing of the "causal" mechanism that paired an initial stationary state transition and emission event with a distant and later absorption event and second transition. But he thought that the spatial wave-like behavior of the emitted radiation could not be denied, at least not in optical instruments, and that it still was a critical component of measurement in spectra.

Given both spatial-extended propagation of radiation and the distinct coupling and emission and absorption in distant atoms, Bohr concluded that the emission and absorption formed a connected "individuality." To say that Bohr fully recognized entanglement as we understand it today would of course be too strong, but it is not too strong to say that Bohr concluded that the properties of emission and absorption were nonseparable, in the sense that they were not independently describable with separate states dynamically related in time. It appeared to him that radiation is in some sense emitted and spreads out in space and that in that spread it must potentially touch many different atoms. But it only transfers energy to one, which then must end that propagation and ensure that it will not transfer energy to any other atoms. The "beginning" and "end" of the quantum process are inexorably tied together and the radiation spread in the middle somehow ends up being contained within this particular coupling. As it does not spread to and affect other atoms, the spread cannot literally be true.

This understanding was consistent with the mechanism of transitions between stationary states. If the transition were understood as a traditional causal process, the probability of transmission and the properties of what would be emitted would depend only on the initial state. But the probability of transition and the properties of the radiation emitted depended not only on what state the electron was in, but also on what possible states the electron could end in and moreover what one it actually ended in. A common complaint about the transitions from friends and critics of the theory alike had been that the electron seemed to need to know what its final orbit would be before it could emit the correct frequency, and Sommerfeld said that the theory seemed more teleological than causal (Kragh 2012, 165–8). Bohr (1965, 16) recognized this: "The nature of the radiation sent out from the atom is not determined only by the motion of the atom at the beginning of the radiation process, but also depends on the state to which the atom is transferred by the process."

Presumably those merely possible states did not apply a force on the particle when it was in the initial state, and so the emission was a function of both beginning and ending states, not only of properties local to the electron in a stationary state orbit.

Bohr's conclusion was that the process was an integrated whole and causality was a property of the whole emission process, not something that drove the process step by step. The relatively spatially distant state also determined what would be emitted, in a process that could not be described in any neat space-time picture. Likewise, the "causal" process of the transfer of energy between distant atoms could not be captured by the "space-time" process of the spread of radiation that he thought surely was still happening in some sense.

Bohr thought that quantum mechanics incorporated this individuality of phenomena and later realized that it must apply also to the instruments used in the investigation of phenomena. But even after this realization, Bohr's approach retained the correspondence approach of inferring atomic properties from empirical phenomena through correspondence relations between them. This developed into the idea that a measurement of non-observational properties of a system is a process of establishing a "correspondence" between those system properties and some properties of a measuring instrument:

> We must recognize that a measurement can mean nothing else than the unambiguous comparison of some property of the object under investigation with a corresponding property of another system, serving as a measuring instrument, and for which this property is directly determinable according to its definition in everyday language or in the terminology of classical physics. (Bohr 1939, 19)

Since the quantum formalism describes those systems as a nonseparable whole, this correspondence between the object properties and the separately described, empirically accessible instrument properties must occur partly outside that formalism. Bohr then had to deal with a central tension that quantum mechanics presented to him: the conflict between the apparent nonseparability of systems and the need to establish empirical access to part of a system.

4. Elements of Bohr's philosophy of quantum mechanics

While Bohr never gave a complete interpretation of quantum mechanics, he did provide a variety of principles and conceptual tools that addressed interpretive problems. These might provide a basis for a more complete interpretation. Here I will provide brief analyses of several important concepts and interpretive elements. These analyses will be informed by the above history but will not try to capture the history of further development of his views over time. I intend that these analyses will accurately represent Bohr's general views.

General aspects

It can be easy to think of Bohr as an instrumentalist or other sort of anti-realist who denied the reality of the ascriptions we make and rejected our ability to infer properties behind observables. After all, he declared that our ability to communicate the properties of a system was a prerequisite for ascribing properties, that that communication depended on our ability to first establish appropriate measurement conditions, that quantum mechanics was an "abstract" formalism, and that we ought not interpret formalism outside a context in which we could "unambiguously" communicate measurement results. The picture of complementarity he developed tied our descriptions of a system to the experimental context in which they were made, and it denied that object properties are determined independent of a context of measurement. Together these do seem to paint an instrumentalist or anti-realist picture.

While I will not fully argue this here, I do not think that's an accurate portrayal of Bohr. The point of his empirical, bottom-up approach in early quantum theory was to infer properties behind the observable phenomena—a realist's project. I don't think he ever gave up on that project, even if it had to evolve post Bothe-Geiger and post-quantum mechanics. For example, even in what might be his most positivistic-sounding work—his reply to Einstein, Podolsky, and Rosen (EPR)—there are clear and strong, though restricted, realist sensibilities.[3] In their argument that quantum mechanics was incomplete, EPR gave conditions for ascribing reality to quantum mechanical symbolic representations (Einstein, Podolsky, and Rosen 1935). In his response, Bohr did not generally reject the ability to make inferences to unobservable properties. Instead he suggested that we have learned that our abilities to ascribe reality to hidden phenomena are constrained and that such ascriptions are limited to the context of a particular type of experimental arrangement that sets the conditions for those ascriptions. One reading of Bohr's response is that he totally missed the point, and that while EPR were worried about whether there are "elements of reality" that are not included in the limited list of what we can say in quantum mechanics, Bohr responded only by pointing out that there is a limited list of what we can say in quantum mechanics. I think the best way of interpreting Bohr is that he thought that those limitations influenced the system as a whole—though he did not clearly state how—so that they were not mere limitations on language but also affected the overall system.

The general view can be made by way of analogy to relativity. Analogies to relativity are common but often brief in Bohr's work. With relativity we learn that any ascription of position and momentum is relative to some reference frame and that those properties are not absolute; rather, they are aspects of phenomena as they occur relative to a reference frame. Similarly, we learn something about properties in general from quantum mechanics: we learn that properties are not absolute but are aspects of phenomena as they occur in a measurement context. According to this view, complementarity is something we discover from quantum mechanical structure when we learn that there is no description of physical reality that is not measurement-frame dependent. But just as relativity does not imply anti-realism about the underlying events or the reality of the space-time measures relative to a particular reference frame, similarly quantum mechanics does not imply anti-realism about the underlying entities or the reality of the property measures relative to a quantum reference frame.[4]

Another factor that recommends a general qualified realism is his view on formalism. Bohr has been described as being not too comfortable with abstract mathematical formalism (e.g., Beller 1992, but see also Dresden 1987). Certainly abstract mathematical arguments do not play much of a role in his interpretive work. But he was not a skeptic or anti-realist about the formalism. He apparently thought that the formal quantum theory applied broadly, perhaps universally, and that use of the formalism allowed us to infer properties from measurement observations. He did caution that the formalism was "symbolic" and abstract, to be interpreted only when one could align them with classically describable properties, and that our ability to do that was limited: "The essentially statistical nature of this account [is] a direct consequence of the fact that the commutation rules prevent us to identify at any instant more than a half of the symbols representing the canonical variables with definite values of the corresponding classical quantities" (Bohr 1939, 14–15).

However, while he was concerned about the limited ability to associate all of the property variables with definite, classical values, this did not imply anti-realism about the formalism. His was instead a cautious realism informed by his bottom-up empirical approach, a picture of nature that admitted a reality represented by the formalism but said that our relationship with the systems we interact with exceeds the partial view we will get of them from our own limited perspectives.

Individuality and nonseparability

Nonseparability is central to Bohr's views, but unfortunately—though perhaps understandably—is not well explicated by him. While it was not until at least the Schrödinger (1935) cat paper that we had a nice characterization of entanglement and some of the puzzles it poses in measurement, it seems clear that Bohr had some understanding of it.[5] Nonseparability is prominent in the very beginning of the Como lecture in which he first introduced complementarity, though it is also mixed in with a number of other features of quantum mechanics:

> [The essence of the quantum theory] may be expressed by the so-called quantum postulate, which attributes to any atomic process an essential discontinuity, or rather individuality, completely foreign to the classical theories and symbolized by Planck's quantum of action. This postulate implies a renunciation as regards the causal space-time co-ordination of atomic processes. Indeed, our usual description of physical phenomena is based entirely on the idea that the phenomena concerned may be observed without disturbing them appreciably... . Now the quantum postulate implies that any observation of atomic phenomena will involve an interaction with the agency of observation not to be neglected. Accordingly, an

independent reality in the ordinary physical sense can neither be ascribed to the phenomena nor to the agencies of observation. (Bohr 1928, 580)

In the first sentence Bohr substitutes "individuality" for "discontinuity." While he relates these two terms, with this substitution he also recognizes features of quantum phenomena that are broader than the existence of a finite, smallest unit of exchange. The broader issues recall the conclusion he reached after the Bothe-Geiger experiments, in which he recognized a "coupling" between beginning and ending events that was inconsistent with any account of a space-time path of causal influence by radiation from one atom to another. The broader issues also include the reasons we are forced to reject the ideal of observing a system without disturbing it. The connection between individuality and discontinuity might seem to imply that the issue we face is that there is a limit to how small we make that interaction between system and "agency of observation." But later in the Como lecture Bohr takes pains to reject Heisenberg's claim that uncertainty is caused by disturbance in the form of unmeasured momentum exchange. Bohr says that that sort of disturbance would only interfere with our ability to know what happened to a system being measured. That might be unfortunate but would not account for the deeper issues we actually run into.

The indication that the issues are deeper comes in the last line: "an independent reality in the ordinary physical sense can neither be ascribed to the phenomena nor to the agencies of observation." Bohr was not denying realism here. He was denying *independence* between the systems being investigated and the systems being used to investigate them. That is, while the ordinary ontology in our conception of "reality" consists of independent objects, the truth is that that ordinary concept of independent reality does not generally apply to systems in quantum interactions. Quantum phenomena involving multiple systems and multiple degrees of freedom are generally integrated wholes with properties that are not reducible to separate properties of the component subsystems. Dynamically, changes to a system in the beginning of a quantum process, say an atom emitting radiation, are not fixed except in virtue of the changes that occur at the end of a process. The "individuality" of these processes means they cannot be broken down into separate, distinct, fixed, intermediate steps.

Bohr related the "quantum postulate," Planck's constant, discontinuity, the "quantum of action," and "individuality" all with the denial of "independent" reality between objects or their behavior. He wasn't exactly explicit where one concept began and where one ended, or whether in the end they reduced to the same new feature discovered in quantum theory. But he was at least sometimes clear about nonseparability, especially in measurement. Bohr's (1949, 210) clearest phrasing declared an "impossibility of any sharp separation" between objects: "[The] impossibility of any sharp separation between the behaviour of atomic objects and the interaction with the measuring instruments which serve to define the conditions under which the phenomena appear [is a] straightforward consequence ... of a minimum quantum of action."

The "individuality" or holistic nature of quantum phenomena meant that the total phenomena could not be fully captured by descriptions of the parts, and that according to the true state of the composite system of measured systems and the agents and instruments used to measure them, even the ascription of separate, individual systems with independent properties is at best a partial accounting of reality.

Measurement, classical language, and external conditions

One of the most important elements of Bohr's view was the necessity of using classical language. This sounds like a linguistic commitment, but the requirement is more than linguistic and has its roots in two aspects of Bohr's theoretical work. The first was the empirical tradition of his work, in which making inferences about hidden structure required an independent starting point. The second was the epistemological requirement these starting points be "unambiguous," intersubjective, and communicable.

Bohr repeatedly described measurement in correspondence or correlational terms. The task of measurement was to put object system properties into correspondence with measuring system properties that meet the above criteria, so that the observation of a measuring system in one state can be tied to one of the properties of the object system:

> We must recognize that a measurement can mean nothing else than the unambiguous comparison of some property of the object under investigation with a corresponding property of another system, serving as a measuring instrument, and for which this property is directly determinable according to its definition in everyday language or in the terminology of classical physics. (Bohr 1939, 19)

The view expressed here is a continuation of his earlier correspondence approach. The approach had to change in the context of quantum mechanics, as nonseparability posed a challenge to the method of establishing correspondence between object and instrument properties. A measurement interaction is a holistic phenomenon in which there is no real division between object and measuring system. So how are we to independently access the results of the measuring instrument and communicate its results in ordinary or classical terms? We must be able to treat the measuring system as if it has independent properties accessible to us, but it is not clear how to do so given nonseparability.

Despite the fact that the use of ordinary or classical language was foundational principle of Bohr's approach, Bohr was not very clear about either the physical conditions for imposing a classical description or really even what a classical description is. The "classical" properties Bohr seemed most concerned about were described broadly in terms of "space-time" and "causal" frameworks, associated with the respective abilities of locating a system in space and time and identifying transfers of momentum and energy. Applying those frameworks, or even specific properties associated with them, means neither that we assign the measuring system a full classical state nor that all classical physics applies to it. At the very least, it's clear that "classical" implies an unambiguous description. One condition for that is that there be no superpositions of the relevant properties.[6] An instrument or measurement intermediary that remained in some superposition of the states meant to correspond to the measured object system states could not be used to convey the object state.

The "unambiguous" criterion would also mean that there be no questions about symbolic meaning for the relevant properties. The concern with symbolism, for Bohr, was about interpretation of the formalism. A symbol that stands in a similar place in quantum mechanics as a symbol in classical theory cannot simply be interpreted in the same way. For example, wave formalism in quantum mechanics should not be interpreted as literal waves. From a classical perspective, the presence of waves implies the possibility of observable interference and diffraction, which are not automatically presented in quantum mechanical phenomena. Bohr had already rejected the idea that radiation spreading out according to quantum dynamics implied that we can think of that in terms of actual waves. Thus the presence of a "symbolic" wave in the formalism would not unambiguously indicate the presence of a real wave. However, once one has observed interference effects, one can safely infer the existence of waves and use the term unambiguously.

There is little doubt that Bohr thought of these unambiguous "classical" descriptions in terms of the concepts of classical physics, though he never specified how to differentiate the essential characteristics of a classical concept from the aspects of that concept carried along by its role in the entirety of the associated physics. In any case, it meant that there were at least limited inferences one could make using at least parts of classical physics to the system once one had ascribed a classical property to the system.

Bohr (1935, 701) did not think this approach of relying on classical concepts for description of observations was unique to quantum mechanics:

> The singular position of measuring instruments in the account of quantum phenomena ... appears closely analogous to the well-known necessity in relativity theory of upholding an ordinary description of all measuring processes, including a sharp distinction between space and time coordinates, although the very essence of this theory is the establishment of new physical laws, in the comprehension of which we must renounce the customary separation of space and time ideas.

That is, despite the fact that relativity radically reenvisions space and time, frame-dependent measures retain their classical roles. Within the context of the inertial framework, certain applications of those concepts can be applied and the normal kinematics work relative to that frame. What changes is the way we translate those measurements to other frames of reference.

The use of classical concepts, then, is to describe the part of the total composite system that will reflect measurement observations relative to a measurement context. In Bohr's (1939, 28) terms, when we describe the measuring system with classical concepts we are describing the "external" conditions of the otherwise individual phenomenon: "The main lesson [is] ... the necessity of describing entirely on classical lines all ultimate measuring instruments which define the *external conditions of the phenomenon*, and therefore keeping them outside the system for the treatment of which the quantum of action is to be taken essentially into account" (emphasis added).

These outside conditions allow us to assign an "external" state to the measuring instrument. Despite the fact the pure quantum state applies to the composite system of object and instrument, we ascribe to the instrument a "classical" state in virtue of its relationship to the lab and its interactions with the rest of the environment, including us. For example, the lab experiment might let us look at a pointer, or a dot on a screen, or the motion of a diaphragm, and unambiguously associate those visible results with some aspect of the instrument's quantum state. These additional interactions between the instrument, the lab, and ourselves, together with the criteria for their use, set the "external conditions" for the measurement. The external conditions include both physical processes and the theoretical framework we use to interpret them to provide empirical content and allow us to understand, identify, and come to learn about the results in the measurement subsystem. Without these external conditions, it would not make sense to even assign the entangled instrument a state separate from object.

This idea was expressed by Bohr in response to EPR. Bohr (1939, 21–2) said that we cannot ascribe a definite state to a subsystem of a nonseparable system if the composite system is isolated:

> The [EPR] paradox finds its complete solution within the frame of the quantum mechanical formalism, according to which no well-defined use of the concept of "state" can be made as referring to the object separate from the body with which it has been in contact, until the external conditions involved in the definition of this concept are unambiguously fixed by a suitable control of the auxiliary body.

This statement expresses the fact that subsystems in nonseparable states do not have their own pure quantum states. It also claims that it only makes sense to ascribe a subsystem a state in certain conditions and that these conditions are "external," not grounded in the quantum state of the composite system.

The externality of these conditions and the general reliance on classical concepts is potentially problematic. Bohr did not believe there are inherently different kinds of systems. Rather, he stated that we have a "free choice" to describe a system quantum mechanically or classically: "in the system to which the quantum formalism is applied, it is of course possible to include any intermediate auxiliary agency employed in the measuring process" (1939, 23). But he also claimed that free choice is restricted, limited to "a region where the quantum-mechanical description of the process concerned is effectively equivalent with the classical description" (1935, 701). We cannot ascribe classical properties unless the subsystems will be "effectively" classical according to the correct quantum mechanical description.

Again, what makes a system effectively classical, how close that has to be to our actual classical descriptions, and so on are not specified precisely by Bohr. But we do see at least that there are physical conditions for our application of classical concepts. Epistemological needs drive the requirement of using outside descriptions of phenomena but the need for accuracy restricts their use. Since the correct quantum mechanical description of the measuring instrument includes the object and the actual quantum mechanical interactions in the external conditions, the physical requirements for their use involve all of them.

One of the physical requirements can be stated in terms of the measurement subsystem. In addition to the above general requirement that we have (as yet-undefined) "effective" classicality, the way Bohr (1939, 23–4) thinks about measurement requires that the classical state we attribute to the measurement apparatus via the external conditions must closely align with the quantum state of that system: "Since … all those properties of such agencies which, according to the aim of the measurement, have to be compared with corresponding properties of the object, must be described on classical lines, their quantum mechanical treatment for this purpose will be essentially equivalent with a classical description."

The picture of measurement we get is something like this: Measurement is a process of both physical interaction and empirical access. A measurement apparatus plays a critical role in the center of the two. Previous physical interactions with the object of interest align states of the measurement system with states of the object, but make the resultant composite state nonseparable. In the right conditions, there can be further interactions with the measurement system that we use to get information about the system. In measurement, these additional interactions are external conditions that we do not model quantum-mechanically. These external interactions must support unambiguous communication about results and in virtue of this must be able to be described, and then must actually be described, in classical terms. These descriptions must also align with or in some sense correspond to the quantum mechanical state of the measurement apparatus, though of course they will not reflect the true quantum state of the system. In sum, the state of the object must correlate with the state of the measurement system, which must correspond to classical descriptions we give it, which allow us to unambiguously communicate it. Physical interactions ground the correlations between the object and the instrument and between the instrument and its environment and observer. Measurement requires those interactions but also requires that we treat one set of those correlations different from the other, so that we may separate the object-instrument interaction from the external conditions of that interaction.

Physicality, open systems, and measurement frames

Bohr's discussions about complementarity stated that different experimental arrangements produce different behavior in the object system. Our mere use of classical language does not magically create different behavior. That point may seem obvious, but it reinforces that measurement has physical requirements and physical consequences.

Bohr emphasized the concrete physicality of measurement and described measurement apparatuses in material nuts and bolts terms—sometimes literally, for example, giving descriptions of the double slits in interference experiment as either bolted down hard to the lab or attached to the lab via springs.[7] These physical descriptions emphasized the old-fashioned classical normalcy of the measurement apparatuses. They also indicate for us that the choice of what sort of object behavior one wants to measure dictates both which interactions must occur with the instruments and the relation those instruments must have to the rest of the lab.

How a measurement apparatus is connected to the rest of the lab influences not only the role it can play in providing for unambiguous communication but also what its interaction with the object will be and what the influence on the object properties will be. In a two-slit experiment in which the particles are sent through an initial one-slit diaphragm to produce a coherent wave that will pass through a two-slit diaphragm, an interference pattern will be produced if the diaphragms are both bolted down to the same lab frame, but no interference pattern will be visible if the first diaphragm is put on springs to see which way the particle was deflected. The concrete physicality with which Bohr sometimes discussed these emphasized that the object system is not interacting just with a single measuring instrument. The object is really interacting with a whole lab (e.g., momentum transferred to a bolted-down diaphragm is passed to the lab).

The relationship to the lab has both a strong analogy and strong disanalogy with special relativity. The analogy is that measurement results are relative to a measurement context similarly to how space-time measurements are relative to an inertial frame of reference. Indeed, fixing a measurement context is sometimes the same as fixing a frame of reference: "To measure the position of one of the particles can mean nothing else than to establish a correlation between its behavior and some instrument rigidly fixed to the support which defines the space frame of reference" (Bohr 1935, 699–700). Bohr thought that there were similar requirements for measurements of non-space-time variables. The relativity of space-time occurs because of the constancy of the speed of light and the role of light in identifying space-time coincidences. The relativity of quantum claims comes from individuality and the fact that systems must be open in order to be observed. Since establishing a frame of measurement requires interaction, the system must be open: "a closed system ... according to the view presented here ... is not accessible to observation" (Bohr 1928, 586).[8] The object system must interact with a measurement system and the measurement system must further interact with the lab frame. Bohr claimed that quantum dynamics strictly speaking only applies to closed systems. If the object system were closed it would evolve according to the unitary dynamics but would not be observable, and its description would have "symbolic" but not empirical meaning. Putting it into one measurement context versus another, with one set of open interactions versus another, gives it different empirical meaning. The same holds for the object-plus-instrument system together: if they are isolated their description has symbolic but not empirical meaning, and they themselves must be put into specific open interactions to have their intended empirical meaning.

In relativity we might say that the space-time interval in some sense represents non-relative physical facts about the relation between two events, but that a space-time interval has different empirical meaning in different frames. Similarly we might say that the quantum state of a composite system represents non-contextual physical facts about the relation between subsystems, but that a quantum state has different empirical meaning in different measurement contexts.

A disanalogy with relativity is that while in relativity the frame does not affect the actual events (objective space-time "coincidences will not be affected by any differences which the space-time co-ordination of different observers otherwise may exhibit" [Bohr 1928, 580]), in quantum mechanics we learn that the measurement system is really part of a larger composite system of object plus apparatus, intermediaries, and lab. The lab frame is the context that provides the external conditions that align the symbolic quantum properties with classical descriptions of observations. But because the measurement frame is part of the phenomenon, a choice of which properties to measure is a choice of interaction that will lead to a specific composite state. One choice of measurement context sets up a different total interaction than another and a different open system than another. The different properties evinced reflect not just an observer's view of independent phenomena but rather part of the phenomenon itself.

Uncontrollability, disturbance, and neglect

A key element to Bohr's views is that measurement affects or "disturbs" measured systems in "uncontrollable" ways. Bohr's views about disturbance may have changed over time (see Fine and Beller 1994), but his view was never that measurements disturb existing properties. He did say that measurements involve "uncontrollable" exchange of momentum and energy: "Any unambiguous use of the concepts of space and time refers to an experimental arrangement involving a transfer of momentum and energy, uncontrollable in principle, to fixed scales and synchronized clocks which are required for the definition of a reference frame" (Bohr 1958, 72).

Despite the language about transfer of momentum and energy, Bohr did not think of this transfer in simple disturbance terms, even in the introduction of complementarity. His discussion of the gamma-ray microscope rejected Heisenberg's apparent analysis that uncertainty is a result of disturbance of the particle from the radiation that hits it: "[A] discontinuous change of energy and momentum during observation could not prevent us from ascribing accurate values to the space-time co-ordinates, as well as to the momentum-energy components before and after the process" (Bohr 1928, 583).

Uncertainty is a function not only of the change in momentum and energy but also of features of the microscope, including the size of its aperture, which we can use to constrain the change in momentum: "The reciprocal uncertainty which always affects the values of these quantities is, as will be clear from the preceding analysis, essentially an outcome of the limited accuracy with which changes in energy and momentum can be defined, when the wavefields used for the determination of the space-time co-ordinates of the particle are sufficiently small" (Bohr 1928, 583).

"Defined" might again be read epistemologically, but given the above account of individuality of quantum systems under measurement, a non-epistemological reading is proper. The size of the aperture is the "end" of the process that "began" with the scattering of radiation of the particle, but those are coupled interactions that form an individual whole.

Bohr (1958, 73) later explicitly cautioned against misinterpreting the word "disturbance" as an effect on preexisting properties:

> One sometimes speaks of "disturbance of phenomena by observation" or "creation of physical attributes to atomic objects by measurements." Such phrases, however, are apt to cause confusion, since words like phenomena and observation, just as attributes and measurements, are here used in a way incompatible with common language and practical definition.

Disturbance and even creation of properties might be thought of as causal events due to the influence of separate, outside influence. But Bohr stressed that any apparent disturbance reflects features of the individual phenomenon that covers all the systems involved. The "observed" "phenomena" cannot be described or understood in terms just of the object system. They are rather holistic features of an entire experimental arrangement: "On the lines of objective description, it is indeed more appropriate to use the word phenomenon to refer only to observations obtained under circumstances whose description includes an account of the whole experimental arrangement" (Bohr 1958, 73).

This clarification was made years after Bohr's initial description of complementarity and reflects Bohr's more mature views. However, it is consistent with his analysis of the microscope. The microscope need not have entered into his calculation if he had not had specific reason to include it, as he had just shortly before in that lecture described the reciprocal wave relations that would have applied directly to the gamma ray by itself and in interaction with the particle. That he did not talk just about the gamma ray and the particle is consistent with his statements that outside observation symbolic properties do not have empirical meaning, that the task of measurement is to tie object properties to corresponding instrument properties in empirical contexts, and that such composite systems are individual wholes whose properties are not determined except in the context of the whole experiment.

In any case, it seems clear that any simple notion of disturbance was never part of Bohr's picture.[9] The key to understanding "uncontrollable" exchanges, then, is to see them in light of the tension inherent in treating a nonseparable state as part of an open system that allows empirical access to properties of one of the parts of the system through the correlations in the state. While an "individual" interaction does not fix a definite state to the system, the choice of measurement context partly determines the state of the object: "The individuality of the typical quantum effects finds its proper expression in the circumstance that any attempt of subdividing the phenomena will demand a change in the experimental arrangement introducing new possibilities of interaction between objects and measuring instruments which in principle cannot be controlled" (Bohr 1949, 210).

Note again the recognition that composite systems are not generally separable: to subdivide a system requires doing something else to it, physically altering it by entangling it with other systems.

The interaction with the agency of measurement such as the microscope is not just an interaction between the microscope and the gamma ray. It is a whole phenomenon of interaction between the particle, the gamma ray, and the instrument. The entire interaction is not completed piecemeal. Rather the value of the momentum transfer between the gamma ray and the particle is not determined before the gamma ray has passed into the microscope. The change in momentum in the particle depends also on the change in the momentum of the microscope at the end. This is the "impossibly of neglecting the interaction with the agency of measurement," read in the context of Bohr's holistic interpretation of quantum phenomena. The interaction with the agencies of measurement partly determine object properties, not through inducing state collapse or anything like that, but by putting the object into a specific total system that changes the behavior of the object.

Uncontrollability occurs because the effects of these other entanglements are very much not negligible: the "impossibility of neglecting the interaction with the agency of measurement means that every observation introduces a new uncontrollable element" (Bohr 1928, 584). This uncontrollability is not about inability to measure definite exchanges but rather comes via a lack of "determination" due to the fact that the external conditions outside the open system are themselves entangled with the system. An interaction between a system and a measuring system (and possibly further measuring systems) will proceed according to standard quantum dynamics when the system is closed. But while the measuring instrument is in a state in which it can serve as a measuring instrument (rather than an object system), the system is open and standard quantum dynamics does not apply to the interaction between object and instrument alone. That does not mean that an interaction has not happened according to standard quantum dynamics. It means that the standard dynamical interaction that occurred is broader and involves the external conditions as well, but also that we do not have empirical means to determine how that total interaction proceeded and we do not have empirical access to the total resultant composite state.

The "impossibility of neglecting" the interaction means that we cannot observe a system without entangling it with the instrument and lab frame. "Uncontrollability" comes in when we make a measurement in which the interaction is "not to be neglected" but then we actually do neglect it because we do not have empirical access to it. We do not have empirical access to the total interaction and resultant state because in order to have empirical access to the object itself, we set up the system to be open. In Bohr's (1939, 19) words, the classical external conditions that allow empirical access necessitate that we neglect (for epistemological reasons) the quantum (entanglement) effects that are otherwise not to be neglected (i.e., real and of consequence): "The necessity of basing the description on the properties and manipulation of the measuring instruments on purely classical ideas implies the neglect of all quantum effects in that description, and in particular the renunciation of a control of the reaction of the object on the instruments more accurate than is compatible with the [uncertainty relation]."

When we establish and then describe a correlation between instrument properties and object properties, we must, by necessity, leave out some aspects of some of the interactions, leave out other interactions entirely, and definitely not include the full description of the total quantum state which includes our whole lab, and so on. The open "external conditions" help us observe and describe one aspect of that interaction, but they cannot help us observe and describe all aspects. While we can control the interaction to force correlations between some object and system properties, we cannot control it to force others at the same time.

Two features combine to generate this uncontrollability: the quantum-mechanical entangling "interaction that cannot neglected" and the way we proceed in measurements to give descriptions that "neglect all quantum effects" because they are partial and do not include all the involved systems. Unfortunately Bohr does not give a general theoretical account of precisely how or why uncontrollability is a consequence of these two features. It is clear, however, that it is reflected in complementarity.

Complementarity and indeterminacy

It is easy to view complementarity as something we bring to the interpretation of quantum mechanics on the basis of outside philosophical principles. Bohr did venture into descriptions of more general features of complementarity in nature (e.g., 1933, 1938). However, complementarity is better characterized as a *consequence* of physical and philosophical interpretive elements described above.

In particular, Bohr called complementarity a consequence of imposing the measurement distinction between object and instrument when they are not distinct according to "the quantum postulate." Complementarity "appears as an inevitable consequence of the contrast between the quantum postulate and the distinction between object and agency of measurement, inherent in our very idea of observation" (Bohr 1928, 584). We know from above that because of the openness required in measurement, any empirical ascription of properties is partial. Complementarity and indeterminacy reflect the trade-offs that occur when we make one partial description over another.

We know there is at least some epistemological component to complementarity, and Bohr (1928, 584) often described particular trade-offs in epistemological terms, for example, when "the determination of its momentum always implies a gap in the knowledge of its spatial propagation." Sometimes the epistemological trade-offs run deeper than mere lack of knowledge, for example, when "the fixation of its position means a complete rupture in the causal description of its dynamical behavior" (584). A "rupture" in our causal description of the particle is not merely an epistemological loss. A rupture means that there is no causal story to be told—there is no measurement context in which properties of the system are aligned with external conditions sufficient for describing momentum exchange, and so on. We know from the above that different measurement contexts create different entangled systems. The choice of one system over another is not merely a choice about what we can learn.

Complementarity is about trade-offs in our choices in setting up different measurement conditions. The choices we have are not entirely ones of language or description, of which information we choose to get out of a system once it is set up. The choices that matter for complementarity are the choices between the contexts themselves.[10] Our freedom of choice comes *before* the experiment, in our choice of experimental arrangement: "We are, in the 'freedom of choice' offered by the last arrangement, just concerned with a discrimination between different experimental procedures which allow of the unambiguous use of complementary classical concepts" (Bohr 1935, 698).

Bohr here (and elsewhere) draws attention to complementary *concepts*. The focus on concepts can make it seem as if complementarity is merely a conceptual phenomenon. But according to the above, classical concepts are employed in the context of an experimental arrangement. Complementarity is more fundamentally about those experimental contexts. The choice of measuring system is meant to align object properties with unambiguous instrument properties describable in classical terms. To be able to shift our description or attention to different measurement results, we must establish different measurement context with different external conditions, which require different physical interactions.

A more complete statement: Complementarity is a phenomenon concerning our ability to meet epistemological measurement criteria when gaining empirical access to quantum systems, in which, due to the nonseparable nature of many composite quantum systems, we face unavoidable trade-offs in the use of different experimental arrangements that set up different measurement contexts. The different measurement contexts allow the alignment of different unambiguous external conditions with the quantum system and generate both different total composite states and different partial substates for that system by putting the system into different open interactions.

The above statement indicates that complementarity is due to the nonseparability of entangled quantum states and the openness required for unambiguous measurement. The interpretative elements above indicate that the ascription of classical properties to a measuring instrument is at best partial, that a choice to establish one such partial relationship rather than another creates different total system states with different object substates, and that information about the differential effect on the system is in some sense lost to the measurement frame with which it interacts. They do not express what those effects are, which concepts are incompatible, and which information is lost.

Bohr generally used two methods for addressing such questions. First and foremost, he used particular examples, most famously in his discussions with Einstein about the two slit experiment and the clock on a spring (Bohr 1949). But these particular examples might best be thought of as consistency tests. If you thought you could cheat quantum mechanics and measure complementary properties, you find that there is something preventing that. But the something might be anything, with little principled guidance for what will matter. In his discussions with Einstein he seemed always to find something that showed incompatibilities in measurements, but finding them took the mind of Bohr, sometimes thinking overnight. There seems to be no general method for identifying what about one arrangement is incompatible with or leads to indeterminacies in another.

The other way the mutual incompatibility was expressed was via the indeterminacy relations and the commutation relations. Though how precisely they generate incompatibility in measuring circumstances is not made explicit, that they play this role seems to be expressed in passages like these:

> In particular is the essentially statistical nature of this account a direct consequence of the fact that the commutation rules prevent us to identify at any instant more than a half of the symbols representing the canonical variables with definite values of the corresponding classical quantities. (Bohr 1939, 14–15)

> ... as soon as we want to know the momentum and energy of these parts of the measuring arrangement with an accuracy sufficient to control the momentum and energy exchange with the particle under investigation, we shall, in accordance with the general indeterminacy relations, lose the possibility of their accurate location in space and time. (Bohr 1949, 215)

> The quantum mechanical formalism ... represents a purely symbolic scheme permitting only predictions, on lines of the correspondence principle, as to results obtainable under conditions specified by means of classical concepts. It must here

be remembered that even in the indeterminacy relation ... we are dealing with an implication of the formalism which defies unambiguous expression in words suited to describe classical physical pictures. (Bohr 1949, 210–11)

Complementarity between pairs of concepts and their arrangements occurs at least in accordance with the indeterminacy relations and possibly as a consequence of them or the underlying commutation rules. But since the indeterminacy relations and commutation relations are symbolic and complementarity is a relationship between different measurements contexts that establish correspondence with classical concepts in open experimental arrangements, the symbolic formula do not obviously explain the incompatibility of those measurement contexts because they do not describe them. These relations are consequences of the formalism, but they are formal and so need to be interpreted to get their meaning in a particular context. How to do so is not so clear. To interpret them is, of course, to align the symbols with classical descriptions of observable properties. But if differences between measurement contexts prevent simultaneous correlations, then some of the symbols cannot be given meaning while others are. So how are we to understand them?

Since those relations are part of the formalism we might imagine that they apply to closed quantum systems, but then they tell us nothing about open experimental settings in which we apply classical concepts and so do not explain the incompatibility of those experimental arrangements. If they are meant to directly say something beyond the behavior of quantum systems in isolation, then they conflate quantum behavior with the partial classical concepts we apply to that behavior in specific measurement contexts and make quantum mechanics incoherent. In neither case is it clear how the formalism explains the partial results in measurement contexts. If, however, the incompatibility of experimental arrangements is meant to explain those relations, we are still left with questions about the impossibility of an experimental context that allows more than partial correspondence and the production of the specific trade-offs in those partial accounts. Either way, there seem to be significant open questions about how to complete the account.

5. A tentative updating of Bohr

Decoherence

The above left a number of unanswered questions and incomplete explanations. I believe we can go some way to filling in those gaps by incorporating what we now know about decoherence. Decoherence has sometimes been suggested to undermine Bohr's views about classical physics (e.g., Bacciagaluppi 2012) because it indicates how classical behavior may emerge from purely quantum systems and so not deserve priority. I suggest here that epistemological priority of classical descriptions is not undermined by decoherence but that decoherence actually fills in gaps in Bohr's views, especially regarding the role measuring instruments play in recording definite results.[11]

Decoherence is a phenomenon of two parts. The first part consists of certain kinds of physical interaction with the environment. Models of environmental interaction suggest that systems that are not isolated from their environment (e.g., radiation, dust particles) can become entangled with that environment in a way that correlate system states $\{|o_i\rangle\}$ such as position with environmental states $\{|e_i\rangle\}$ that become very nearly orthogonal: $|\psi_T\rangle = \sum_i c_i |o_i\rangle|e_i\rangle$ where $\langle e_i|e_j\rangle \approx 0$ for $i \neq j$. The relevance of this environmental interaction is that if we measure the system by itself, without measuring the environment states, it will look as if it has "decohered," that is, lost ability to display interference effects. If it had been in a superposition of states $\sum_i c_i |o_i\rangle$ that then became correlated with orthogonal environmental states in a total state $|\psi_T\rangle = \sum_i c_i |o_i\rangle|e_i\rangle$, in interactions on the system alone it will appear as if it is no longer in a superposition of the states it had been in. Models of environmental interactions suggest that ordinary objects very quickly become entangled with their environment and often appear decohered in the position basis. Actually every bipartite system can be written as a sum of some jointly orthogonal bases for both systems, so the point is really that environmental interactions will often generate a state that is nearly orthogonal in "good," "classical" bases.

The second part of decoherence is what describes these appearances. This part is not a physical interaction but a prescribed technique for describing subsystems of larger systems. We represent pure states $|\psi\rangle$ of the total system with the density matrix $\rho = |\psi\rangle\langle\psi|$ and model the behavior of the subsystem by taking the partial trace of the total system state, summing over the other degrees of freedom, to get a "reduced" density matrix for the system. Consider a composite system of two systems A and B. If $|\psi\rangle$ is the state of the composite system in the tensor product space of the two systems with bases $\{|a_i\rangle\}$ and $\{|b_i\rangle\}$, then the reduced density matrix for system A is $\rho_A = \sum_i \langle b_i|\rho|b_i\rangle = \sum_i \langle b_i|(|\psi\rangle\langle\psi|)|b_i\rangle$. Assume $|\psi\rangle = \sum_j c_j |a_j\rangle|b_j\rangle$. Then $\rho = |\psi\rangle\langle\psi| = \sum_{j,k} c_j c_k^* |a_j\rangle|b_j\rangle\langle a_k|\langle b_k|$. The resulting reduced density matrix produces the same predictions for an observable O on just the subsystem A as the (here assumed pure) state of the total system does for the observable $O \otimes I$ on the total system. If $\{|b_i\rangle\}$ is orthonormal, then $\rho_A = \sum_{j,k}\left(\Sigma_i\langle b_i|b_j\rangle\langle b_k|b_i\rangle\right)c_j c_k^* |a_j\rangle\langle a_k| = \sum_i |c_i|^2 |a_i\rangle\langle a_i|$, which is diagonalized in the a basis. This is the feature that matters for decoherence. If a system, any system, has become entangled with another system, any system (not necessarily the environment), then even if the first system started in a superposition of a states, if the interaction entangled the system such that the a states are correlated with orthogonal b states, then measurements on the first system will not show any interference effects of those a properties. That is, the system will no longer act as if it is in a superposition of a states, but will rather appear as if it is in a classical mixture of a states. We get "effective" collapse. Even though no collapse has occurred, measurements on just the first system will make it look as if a collapse has occurred.

Evidence that the system has not *really* collapsed can be found either by making measurements directly on the total system or by comparing results of measurements on the A system with results of measurements on the B system. That evidence will appear in the form of possibly nonlocal "spooky" correlations. This is what so-called quantum erasers (e.g., Englert, Scully, and Walther 1999) often really do—they do not really erase what happened but recover entanglement that appeared to have dissipated or collapsed. Indeed one conclusion one might reach from quantum erasers is that measurement does not really involve true collapse (e.g., see Tanona 2013).

Environmental decoherence seems to be a real enough phenomenon, but there are several problems with turning to this phenomenon to help with interpretive problems. The first is that environmental decoherence is only approximate—typical interactions have only *nearly* orthogonal environmental states correlated with good states like position. Second, at best decoherence is only *for all practical purposes*. If one could find the right environmental systems and perform the right experiments, one would find the missing correlations and recreate the interference that was apparently lost. Third, decoherence itself is seemingly justified only via a collapse interpretation. Using the partial trace to represent a subsystem of an entangled system is justified because it gets the measurement statistics right, and the claim that the subsystem will look decohered is a claim about what the subsystem will look like when it is measured. But measurement statistics are based on the Born rule, which describes results upon measurement, which of course in the standard interpretation is what you get upon collapse. So trying to use decoherence to avoid collapse does not really work. Fourth, it may justify the assignment of a particular state to a subsystem if you have already identified a subsystem, but it will not do anything to justify one factorization of the total system into subsystems over another. Fifth, decoherence only mimics classicality in a statistical sense. It models the subsystem as a mixture that is interpretable as a classical mixture of different possible definite states. But it does not then account for specific results, for example, the electron going this way rather than that. If one needs from an interpretation not just an account of loss of visible superposition/interference effects but also an explanation for how and when a system went from a superposition of properties to a definite property, then relying on decoherence is not enough.

If one is not looking for decoherence to perform interpretive magic, these may not be such issues. I suggest we can supplement Bohr's approach with knowledge about decoherence without expecting it to do magic.

Decohering correspondence

I now turn to the task of putting together these various features of Bohr's expressed views in a contemporary context. This will be a rough sketch, given with no pretense of providing a complete interpretation, but with the hope that it can provide some insight and serve as a platform for further exploration. We will start with no collapse, no separate measurement dynamics, and no split of the world into quantum and non-quantum systems. Everything obeys basic quantum-mechanical laws, and there is nothing in principle special about measurement interactions—they are governed by the same unitary Schrödinger dynamics as other interactions. These principles are consistent with Bohr's recognition of nonseparability and his lack of collapse talk. The universality of quantum mechanics is consistent with Bohr's insistence that in a description of experiment in which we are treating an instrument outside our quantum description, we could widen our description to include it (though only at the expense of either forgoing empirical access to it or introducing another experimental arrangement). With the standard dynamics, processes often end up producing entangled states. Without collapse entanglement does not disappear, but it dissipates.

Measurements create object-instrument states correlated in certain properties of interest. If one is interested in the object system's A property (basis $\{|a_i\rangle\}$), then a "good" measurement will evolve the object and measuring system as follows: $\left(\sum_i c_i |a_i\rangle\right) \otimes |m_{ready}\rangle \to \sum_i c_i |a_i\rangle \otimes |m_{Ai}\rangle$.[12] If the object system had started in a definite A state, then all the c_i but one will be 0. If the object system started in a superposition of A states, the measurement interaction will have produced an entangled state in which neither subsystem has its own separate pure state, and there is in general not even any preferred factorization of the total system into specific subsystems. Had a property B where $\sum_i c_i |a_i\rangle = \sum_i d_i |b_i\rangle$ on the same object system been measured instead, the interaction would have proceeded as follows:

$$\left(\sum_i d_i |b_i\rangle\right) \otimes |m_{ready}\rangle \to \sum_i d_i |b_i\rangle \otimes |m_{Bi}\rangle.$$

For the above interactions to serve as measurements, additional criteria must be met. Focus on a measurement of property A. The $\{|m_{Ai}\rangle\}$ must be interpretable classically. The $\{|m_{Ai}\rangle\}$ at least must be orthogonal. They must also correspond to "classical" properties of an object we can observe or otherwise verify and communicate independent of the quantum description. The relationship between the $\{|a_i\rangle\}$ and the $\{|m_{Ai}\rangle\}$ must also be able to be given a classical description. Making that relationship meet that classical description likely involves other intermediaries and a lab serving as, among other things, a frame of reference, which means the total system is actually more complex than so far indicated. Moreover, to be visible requires further interaction, for example, simply with light. The measurement context then includes a great variety of other degrees of freedom that will then also be correlated with the object and measurement states. These additional degrees of freedom must serve to help establish, or at least not interfere with, that initial desired correlation. If the measuring instrument was suitably chosen for repeated unambiguous observations and communication, then at least some of the environment properties that will be correlated with the measurement result states will be orthogonal and themselves either be visibly distinct or lend themselves to making some other properties visibly distinct. For simplicity, labeling all this simply under a single environment variable, ideally the total system looks something more like this: $\sum_i c_i |a_i\rangle \otimes |m_{Ai}\rangle \otimes |e_i\rangle$. Also ideally, the basis $\{|m_{Ai}\rangle\}$ and the setting should be such that the $\{|e_i\rangle\}$ will be orthogonal and further environmental interactions will not select a different basis, so that the reduced state of the object and measuring system $O \otimes M$ will itself be diagonalized in the measurement basis $\{|a_i\rangle \otimes |m_{Ai}\rangle\}$ and interpretable as a mixture of effectively collapsed states. Under these conditions the combined system will appear as a mixture of states that reflect measurements of O influencing M in straightforward ways. Either it had value a_1 for A, which led M to have property m_{A1}, or it had value a_2 for A, which led M to have property m_{A2}, and so on. In this condition properties A and M_A can be thought of in classical terms, and the connection between them can also be thought of in partly classical terms. How to interpret the properties of the composite $O \otimes M$ system if the environmental states $\{|e_i\rangle\}$ are not orthogonal (or other of the idealized descriptions above do not obtain) is more complicated. Whether we can talk of idealized ascription of classical properties when the underlying quantum structure does not precisely support it is a topic that will have to wait for further analysis. In any case the orthogonality of the $\{|m_{A1}\rangle\}$ guarantees that the reduced state of the object system will be diagonalized in $\{|a_i\rangle\}$, that is, interpretable as a mixture of a states, which are then interpretable in the classical terms associated with the interaction that correlated it with the measurement states.

According to this view a measurement interaction produces three distinct states to be applied to the systems and interpreted: the total nonseparable quantum state of the composite system, the reduced states of the object and the measurement instrument, and the classical states such as "pointer here" and "pointer there" that are ascribed to the measurement instrument. Between these states are two correlations or "correspondence" relations. The first is the physical correlation between entangled states of the object system and the measurement system, or, alternatively, between the components of the reduced states ρ_A and ρ_{MA}. This correspondence is caused by an appropriate measurement interaction and can be modeled completely within the quantum formalism. The second correlation is between the respective states of the measuring instrument, m_{A1} and m_{A2}, and so on, and external descriptions of those, for example, "pointer here" and "pointer there." The physical interactions (e.g., we actually look, or the environment "records" the result) and the theoretical framework for describing these in unambiguous terms are not included in the quantum state but will be described outside the system, in the external conditions, ascribing "classical" properties via measurement result. These observations of the measuring system themselves could of course be modeled quantum mechanically, in which case then we would be back to a closed model without measurement. So the "classical" correlation comes about via the open system interactions. Without these interactions we could not treat the system $O+M$ as if it were classical.[13]

The measurement interaction puts the system into a new entangled state in which the reduced state predicts properties quite different from what would have been predicted before the interaction. If the system had been in a definite B state that was a superposition of A states, the measurement of A would effectively collapse the object system into A states, at least from the perspective of anyone looking just at that system. The system would now appear as if it were a mixture of A states where it would not have before. Repeating the same measurement on identical systems will show the frequencies of different A values according to the probabilities. Had B properties been measured, different interactions would have occurred, with different reduced density states. An open system with "classical" interactions with the instrument that decohered the object and instrument in the corresponding basis will not also decohere the object and system in an incompatible basis. For continuous variables, the indeterminacy relations will indicate trade-offs between the degrees of decoherence.

The appearances we get via measurement are only partial because every system we observe is open. We can only ascribe states relative to a specific measurement context defining that open system. A measurement context is a lab frame that can establish a measurement interaction between an object and a measurement system. The context must also establish interactions between the measurement system and "lab" that allow information flow from measurement system and alignment of "external" classically described relations according to the above criteria. Momentum exchange between instrument and lab will correlate their properties differently than where there is not, and different such contexts will decohere (or not) the systems differently.

The ascription of definite but unknown *A* states is relative to the measurement context in which one will be looking at just the object and instrument and not the rest of the environment. If one were to step back before actually interacting with the system or instruments and instead interact with the total entangled system of object plus instruments plus lab, it would no longer be appropriate to ascribe definite *A* states to the object system. The measurement interactions set up correlations that allow us to interpret definite measurement results with object properties. Open interactions with the instrument allow us to get information about the results. The results will be stable in a context that selects them, and so the instrument should be set up so that the lab interactions do that. The ability to ascribe a definite state to the object requires the good measurement interaction as well as good reinforcing lab interactions, but it's the conceptual view that "neglects" or leaves out these other interactions that commits one to the ascription of *A* properties instead of something else. That ascription will be good only for that context or for extensions of that context to include other interactions that also neglect that same environment. For different choices of measurement frame and subsystem, different statistics will occur. None are complete because the reduced states are not complete, and none give complete information about the original state of the system. Nonetheless, they are as complete as they can get in any particular measurement context.

From the context of the measurement context, the additional environmental interactions are "irreversible" because they are not controlled for and we will not have the ability to measure them as well. The "end" of the measurement occurs when these additional interactions have decohered the *O+M* system in the bases of the interaction. The ending of a measurement is context-dependent. If the environmental actions are "irreversible" for all practical purposes, then generalizing out from this measurement context to other measurements on other degrees of freedom are not likely to uncover the "neglected" coherence, and so it's safe to treat the measurement as final. Note the difference between shifting to a second measurement context and the counterfactual question about shifting to an alternative measurement context. In many circumstances shifting to a second context will not "undo" the measurement by finding instead evidence of missing complementary properties. However, shifting to an alternative measurement context will potentially produce complementary properties.

As in relativity, the "relativity" of the classical state (different measurement contexts lead to different "appearances") is not dependent on actual conscious observer. Interactions of any kind on any of these subsystems alone will proceed as if the system were collapsed into its respective definite state, but interactions on the total system will not. Anything interacting with either just the object system or just the measurement system is operating within this established measurement context. From the perspective of this context, the object system has that property, although from the wider perspective it is actually a subsystem of a system in a nonseparable state.

According to this view, natural "measurements" are possible. Natural conditions such as environmental decoherence can establish all of the elements of a measurement context except the neglect. Then future interactions with the systems that *do not* interact with the decohering environment are operating within the natural "measurement" context.

The above sketch is rough and incomplete, but suggests how decoherence can ground Bohr's approach. It does not justify the principles on which the approach is based, but it suggests a way to fill in features of the account related to uncontrollability, conditions for applicability of classical concepts, the restriction of claims to a measuring context, why measurements affect objects, and the general phenomenon of complementarity.

6. Conclusion

Whether such a supplemented Bohrian approach might form the basis of an interpretation in a modern sense is another question. In conclusion I will briefly entertain prospects for doing this and for relying on decoherence for interpretation.

To turn the above into a more comprehensive interpretation will probably require that we reject the need for explaining why measurements have definite results at all. With an empiricist approach modeled on Bohr, we would rather, and annoyingly, assume the obvious—that quantum measurements have results—and build our interpretation from there. We would note that even though the processes that produce those results are not classical, we will know these results are limitedly describable with ordinary language and classical physics, because we will already have so described them as our starting point for interpretation. We would not then try to explain how quantum mechanics explains the results but rather ask how we should understand and interpret what quantum mechanics tells us about the world based on the empirical data.[14]

A Bohrian empiricist approach that takes measurement results as given may avoid the problems of justifying the use of decoherence for interpretive ends. Trying to explain (or explain away) collapse using decoherence is problematic because justifying the use of the partial trace is based primarily on the fact that it gets the Born statistics right, which means one is trying to explain why we get definite results with a technique that already assumes we get definite results. This approach would not try to explain collapse. Of course, even trying to explain it away can be a problem: if we assume no collapse and then use a technique based on getting the collapse statistics right it looks like it is sneaking it in. This is something we would have to worry about if we were trying to solve the measurement problem. But in a quantum-empiricist Bohrian approach the job of interpretation does not require explaining why measurements have definite results. The measurement problem, so posed, frames the question the wrong way. The job is to start with empirical phenomena, with its results, and to say how to interpret those results. So framed, the statistics of quantum mechanics are not the statistics "upon measurement." From a phenomena-first frame in which we start with the fact that we have measurement results, the statistics are the bridge between the formalism and the empirical facts. It is then not question-begging to use decoherence to tell us under what conditions we can infer properties from those facts and which properties we can infer from which conditions.

From this perspective, it was perhaps a mistake for the standard interpretation to have identified the collapse postulate as the mechanism in the formalism with empirical content. Here we might say that the partial trace is the primary mechanism in the formalism with empirical contact and not justify it via its connection with collapse. Collapse is then derivative of the partial trace and may be used as an idealization describing the "end" of a measurement when our description of our system is limited and does not include the environment. Collapse mechanics applied to an idealized such system will give us the same statistics as the partial trace. What has really happened is that we are looking at a part of a wider system, and from this subsystem's perspective the coherence has been distributed elsewhere.

Without either collapse or some other mechanism for turning indefinite states into definite ones, it might seem we have another gap that decoherence does not close. Again, however, this may not be the problem it seems. Bohr was not an instrumentalist or anti-realist but he did seem to deny that quantum mechanical states by themselves indicate definite property values. According to this view, we need not explain how a certain definite property value appears when it was not there before. Rather we explain how the states describe the appearance of properties in different contexts. Following the relativistic analogy, we do not deny the reality behind the appearances in reference frames but deny that the appearances in reference frames straightforwardly reflect reality. Quantum states then determine both how a system will evolve in time in a closed system and how the system will appear in interactions in an open system. Ascription of properties will not simply follow the eigenvalue-eigenstate link, as systems may be in eigenstates of an operator when no interaction is in place to align that property with another in a measurement context. But the state is not simply an instrumental predictor of measurement results to which we attribute reality either. We may think of states as reflecting real features of reality but by themselves not reflecting the relational perspective of those in a measurement context—those phenomena are features of the whole experimental context. The properties in the context of measurement are not of the same type as whatever properties are reflected by the pure state of the object system (or object system plus instruments) in isolation.

Compared to modern interpretations, these ideas seem close to a perspectival modal interpretation (Bene and Dieks 2002). Perhaps, then, there is hope for updating Bohr and allowing Bohrian insights to help ground a modern interpretation, though of course questions will remain about both how well they fit into a modern picture and how faithful any such updated account may be to the original. Even if this hope is unfounded, I believe Bohr provides tools for thinking about many of the standard issues and gives reasons to stop worrying about others. Following Bohr's quantum empiricism, we may not need to worry about collapse, the measurement problem, or the appearance and disappearance of properties, and we can think about ascription of properties to systems in a manner both realistic and grounded in experience—as long as we start with the fact that we do have results and make inferences from there.

Notes

1 For example, among other sources cited below, in Tanona (2002, 2004a, b, 2013).
2 Others (e.g., Jammer 1966; Hendry 1984; Beller 1999; and especially Darrigol 1992) note that the correspondence principle was more than the agreement in the limit, but the importance of this distinction is often underappreciated.
3 But see Beller (1999) and Fine and Beller (1994) for arguments to the contrary.
4 Unfortunately, this analogy is underexplored by Bohr and little work has been done on quantum reference frames (but see Bene 1997; Dickson 2004a, b, 2007), so it is not clear how specifically we should understand Bohr's claims about quantum measurement frames.
5 See Howard (1979, 1994) for a defense of this view.
6 Howard (1994) argues forcefully for the position that "classical" does not mean all classical, that it also means ascribing some classical properties to the object system, and that the best reconstruction of Bohr's prescription that to treat certain systems as classical is that we should find conditions under which they can be treated as mixtures rather than superpositions. The view presented here considers classicality in broader terms than Howard's but is indebted to Howard's approach.
7 See Dickson (2004a) for an account of both this physicality and its relevance to the fixing of reference frames.
8 While the use of the word "closed" is clear here and in many other of his discussions of physics, Bohr (1958, 73) sometimes also used the word "closed" to mean something more like *completed* or *finished*, as he was striving to account for physical conditions for unambiguous results, for example, "every atomic phenomenon is closed in the sense that its observation is based on registrations obtained by means of suitable amplification devices with irreversible functioning such as, for example, permanent marks on a photographic plate."
9 For that matter, neither was collapse. Attributions of either to Bohr may be partly due to the faulty idea that there was a unified Copenhagen interpretation (see Howard 2004).
10 Camilleri and Schlosshauer (2015) argue that Bohr, unlike Heisenberg, thought that this choice was not really free and that a particular experimental arrangement more or less fixes where to make the distinction.
11 Camilleri and Schlosshauer (2015) also argue that decoherence could play a role in an updated Bohrian approach. They also describe Bohr as taking an empirical approach—in their terms, he was an experimentalist—and provide a nuanced picture of his use of classical concepts. The role for decoherence they suggest is more limited than that presented here and they weren't trying to defend Bohr or actually provide an updated Bohrian approach, but their interpretation of Bohr and their argument that decoherence is not inconsistent with that interpretation seem compatible with the account presented here.
12 Among the many details that will not be addressed here will be different types of measurement.
13 See Camilleri and Schlosshauer (2015) for additional argument along these lines. They argue that environmental decoherence grounds Bohr's instance that we *must* use classical concepts to describe results by showing what Bohr only presumed—that we actually *can* use classical concepts accurately to describe those results.
14 Halvorson and Clifton (2012) suggest one possible way to move forward in a Bohrian spirit that similarly takes measurement results for granted and, following Howard (1994), treat Bohr's "classical" in terms of mixtures.

References

Bacciagaluppi, G. 2012. "The role of decoherence in quantum mechanics." In *The Stanford Encyclopedia of Philosophy* (Winter 2012 Edition), edited by Edward N. Zalta. http:// plato.stanford.edu/archives/win2012/entries/qm-decoherence/.

Beller, M. 1992. "The genesis of Bohr's complementarity principle and the Bohr-Heisenberg dialogue." In *The Scientific Enterprise*, edited by Edna Ullmann-Margalit, 273–93. Dordrecht: Kluwer Academic Publishers.

Beller, M. 1999. *Quantum Dialogue: The Making of a Revolution, Science and Its Conceptual Foundations*. Chicago: University of Chicago Press.

Bene, G. 1997. "Quantum reference systems: A new framework for quantum mechanics." *Physica A: Statistical and Theoretical Physics* 242 (3–4): 529–65.

Bene, G., and D. Dieks. 2002. "A perspectival version of the modal interpretation of quantum mechanics and the origin of macroscopic behavior." *Foundations of Physics* 32: 645–71.

Bohr, N. 1918. "On the quantum theory of line spectra, Part I: On the general theory." *Det Kongelige Danske Videnskabernes Selskab, Matematisk-fysiske Meddelser* 4 (1): 1–36.

Bohr, N. 1920. "On the series spectra of the elements." In *The Theory of Spectra and Atomic Constitution*, edited by A. D. Udden, 20–60. Cambridge: Cambridge University Press. Original edition, originally published as "Über die Linienspektran der Elemente." *Zeitschrift für Physik* 2: 423–69.

Bohr, N. 1924. "Theory of series spectra." *Nature* 113: 223–4.

Bohr, N. 1925. "Über die Wirkung von Atomen bei Stössen." *Zeitschrift für Physik* 34: 142–57. Translated in *Bohr's Collected Works*, ed. Stolzenberg 1984, 194–206.

Bohr, N. 1928. "The quantum postulate and the recent development of atomic theory." *Nature* 121 (supplement): 580–90.

Bohr, N. 1933. "Light and life." *Nature* 131: 421–3, 457–9.

Bohr, N. 1935. "Can quantum-mechanical description of physical reality be considered complete?" *Physical Review* 48: 696–702.

Bohr, N. 1938. "Natural philosophy and human cultures." *Nature* 143: 268–72.

Bohr, N. 1939. "The causality problem in atomic physics." In *New Theories in Physics*, 11–30. Paris: International Institute of Intellectual Cooperation.

Bohr, N. 1949. "Discussion with Einstein on epistemological problems in atomic physics." In *Albert Einstein: Philosopher-Scientist*, edited by P. A. Schilpp, 201–41. Cambridge: Cambridge University Press.

Bohr, N. 1958. *Atomic Physics and Human Knowledge*. New York: Wiley. Reprinted as *The Philosophical Writings of N Bohr, vol. 2: Essays 1932–1975 on Atomic Physics and Human Knowledge*. Woodbridge, CT.: Ox Bow Press (1987).

Bohr, N. 1965. "Nobel lecture: The structure of the atom." In *Nobel Lectures in Physics 1922-1941*, edited by V. F. Hess, 7–43. Amsterdam: Elsevier.

Bohr, N., H. A. Kramers, and J. C. Slater. 1924. "The quantum theory of radiation." *Philosophical Magazine* 47: 785–802.

Bothe, W., and H. Geiger. 1925. "Experimentelles zur theorie von Bohr, Kramers, und Slater." *Naturwissenshaften* 13: 440–1.

Camilleri, K., and M. Schlosshauer. 2015. "Bohr as philosopher of experiment: Does decoherence theory challenge Bohr's doctrine of classical concepts?" *Studies in History and Philosophy of Modern Physics* 49: 73–83.

Compton, A. H., and A. W. Simon. 1925. "Directed quanta of scattered x-rays." *Physical Review* 26: 289–99.

Darrigol, O. 1992. *From c-Numbers to q-Numbers: The Classical Analogy in the History of Quantum Theory, California Studies in the History of Science*. Berkeley: University of California Press.

Dickson, W. M. 2004a. "Quantum reference frames in the context of EPR." *Philosophy of Science* 71 (5): 655–68.

Dickson, W. M. 2004b. "A view from nowhere: Quantum reference frames and uncertainty." *Studies in History and Philosophy of Modern Physics* 35 (2): 195–220.

Dickson, W. M. 2007. "Non-relativistic quantum mechanics." In *Philosophy of Physics*, edited by Jeremy Butterfield and John Earman, 275–416. Amsterdam: North Holland.

Dresden, M. 1987. *H. A. Kramers: Between Tradition and Revolution*. New York: Springer-Verlag.

Einstein, A., B. Podolsky, and N. Rosen. 1935. "Can quantum-mechanical description of physical reality be considered complete?" *Physical Review* 47: 777–80.

Englert, B. G., M. O. Scully, and H. Walther. 1999. "Quantum erasure in double-slit interferometers with which-way detectors." *American Journal of Physics* 67: 325–29.

Fine, A., and M. Beller. 1994. "Bohr's response to E-P-R." In *Niels Bohr and Contemporary Philosophy*, edited by J. Faye and H. J. Folse, 1–31. Dordrecht: Kluwer Academic Publishers.

Halvorson, H., and R. Clifton. 2012. "Reconsidering Bohr's reply to EPR." In *Non-locality and Modality*, edited by Thomasz Placek and Jeremy Butterfield, 3–18. Dordrecht: Springer.

Harper, W. 1990. "Newton's classic deductions from phenomena." *PSA: Proceedings of the Biennial Meeting of the Philosophy of Science Association* 2: 183–96.

Hendry, J. 1984. *The Creation of Quantum Mechanics and the Bohr-Pauli Dialogue*. Vol. 14, *Studies in the History of Modern Science*. Dordrecht: D. Reidel.

Howard, D. 1979. "Complementarity and ontology: Niels Bohr and the problem of scientific realism in quantum physics." PhD Thesis, Boston University.

Howard, D. 1994. "What makes a classical concept classical?" In *Niels Bohr and Contemporary Philosophy*, edited by Jan Faye and Henry J. Folse, 201–30. Dordrecht: Kluwer Academic Publishers.

Howard, D. 2004. "Who invented the 'Copenhagen interpretation'? A study in mythology." *Philosophy of Science* 71 (5): 669–82.

Jammer, M. 1966. *The Conceptual Development of Quantum Mechanics*. New York: McGraw-Hill.

Kragh, H. 2012. *Niels Bohr and the Quantum Atom: The Bohr Model of Atomic Structure 1913–1925*. Oxford: Oxford University Press.

Kramers, H. A. 1924a. "The law of dispersion and Bohr's theory of spectra." *Nature* 113: 673–4.

Kramers, H. A. 1924b. "The quantum theory of dispersion." *Nature* 114: 310–11.

Schrödinger, E. 1935. "Die gegenwärtinge Situation in der Quantenmechanik." *Naturwissenshaften* 23: 807–12, 823–8, 844–9.

Sommerfeld, A. 1923. *Atomic Structure and Spectral Lines*, third English edition. New York: E. P. Dutton. Original edition, translated from the third German edition by Henry L. Brose.

Stolzenberg, K. (ed.). 1984. *Niels Bohr Collected Works. Vol 5: The Emergence of Quantum Mechanics (Mainly 1924–1926)*. Niels Bohr Collected works, edited by Erik Rüdinger. Amsterdam: North-Holland Pub. Co.

Tanona, S. 2002. "From correspondence to complementarity: The emergence of Bohr's Copenhagen interpretation of quantum mechanics." PhD dissertation, History and Philosophy of Science, Indiana University.

Tanona, S. 2004a. "Idealization and formalism in Bohr's approach to quantum theory." *Philosophy of Science* 71: 683–95.

Tanona, S. 2004b. "Uncertainty in Bohr's response to the Heisenberg microscope." *Studies in History and Philosophy of Modern Physics* 35: 483–507.

Tanona, S. 2013. "Decoherence and the Copenhagen cut." *Synthese* 190 (16): 3625–49.

13

An Everett Perspective on Bohr and EPR

Guido Bacciagaluppi

1. Introduction

One of the most noteworthy developments in the philosophy of quantum mechanics since the early 1990s—closely related to the rise of structuralist positions in philosophy of science—has been the renaissance of the Everett theory (for the state of the art, see, e.g., Saunders et al. 2010; Wallace 2012). Ignored in the 1950s and 1960s, notorious in the 1970s and 1980s for the outlandish claims of some of its more "popular" versions, for a long time it was deeply suspect to most philosophers of physics, appearing as it did as measurement-induced collapse with the added grief of all the unwanted components. The philosophical landscape changed dramatically in the 1990s and 2000s with the work of Simon Saunders, David Wallace, and others, who reconceptualized the Everett theory along structuralist-functionalist lines and assigned a central role to decoherence theory (vindicating in hindsight ideas developed by Hans-Dieter Zeh in the 1970s). With the decision-theoretic approach by Deutsch and by Wallace and the work on confirmation by Greaves and others providing further a serious candidate for the meaning of probabilities in the theory, the Everett theory at last gained the status of a well-worked-out approach to the foundations of quantum mechanics, alongside the more familiar de Broglie-Bohm and spontaneous collapse theories, and seen by many even as the approach of choice, for its lack of theoretical revisionism and the relatively effortless compatibility with relativity (on which more below).

Even more recently, historians and historically oriented philosophers of quantum mechanics have provided an analogous reevaluation of the figure and work of Everett himself (Byrne 2010; Everett 2012; Freire 2015). As I have had the opportunity to stress elsewhere (Bacciagaluppi 2013), Everett's own work is far richer than the reductive image philosophers of physics have inherited of it from the pre-1990s period, and far closer to our current understanding (even though it does not use the technical tools of decoherence)—even in some aspects of its approach to probabilities. One point that has emerged from this historical work and is particularly significant to this chapter is the rather tragic interplay between Everett and his theory and Bohr and his circle and the Copenhagen interpretation (see in particular Osnaghi, Freitas, and Freire 2009). We now know that Wheeler would not let Everett's thesis pass until it was stripped of any elements that might fail to get Bohr's blessing—which was repeatedly sought and refused—a train of events that ultimately contributed to Everett's abandonment of physics. Everett, for his part, while aiming to leave behind Bohr's "overcautious" stance, was perfectly satisfied that complementarity was in fact subsumed in his theory (see, e.g., Everett 2012, 18, 153, and 175).

In recent work with Elise Crull (Bacciagaluppi and Crull 2018), I have been reexamining Bohr's reply to the EPR paper (which is a central source for Bohr's ideas on complementarity), trying to clarify in particular Bohr's distinction between "mechanical disturbance" and other kinds of "influence" in his strategy of criticizing EPR's "criterion of reality." A helpful framework for our analysis is provided by Howard's (1994) discussion of Bohr's doctrine of classical concepts, originally published in the predecessor to this volume. However, as argued in particular by Beller and Fine (1994) also in the predecessor to this volume, Bohr's approach to the "problem of physical reality" appears to drive him toward a positivist position. In this chapter, after briefly summarizing the essentials of EPR and in somewhat more detail our previous analysis of Bohr's reply, I shall suggest that Everett's theory indeed succeeds in recapturing Bohr's ideas about complementarity as expressed in the reply to EPR, but in a realist framework that goes some way toward mediating between Bohr and Einstein.[1] Such an Everettian perspective also suggests which aspects remain problematic in Bohr's reply to EPR.

2. EPR in a nutshell

As is well known,[2] the EPR argument was a development of Einstein's photon-box thought experiment of 1930: a clockwork mechanism opens a box at a precise time, letting exactly one photon out. We can thus know the exact time of emission of the photon. But we can also weigh the box before and after emission, and using mass-energy equivalence measure also the energy of the photon. In his later recollections, Bohr (1949) sees the photon-box experiment as an attempt to criticize the *empirical* validity of the uncertainty relations. At the latest a few months later, however, it is clear that Einstein was thinking of the experiment as a critique of the alleged *ontological* implications of the uncertainty relations (that the energy and time of emission of the photon cannot be simultaneously well defined). In the terminology of EPR, the aim of the critique is not the correctness of quantum mechanics but its completeness, which was arguably Einstein's main concern all along. As Ehrenfest (1931) (in his inimitable style) explains in a letter to Bohr, Einstein did not doubt the validity of the uncertainty relations and had not devised the photon-box to that end, rather his actual interest in the thought experiment was that[3]

> it is interesting to be clear about the fact that the projectile [the photon], which is already flying about in isolation "on its own," must be ready to satisfy very different "non-commutative" [*sic*] prophecies, "without even knowing" which one of these prophecies one will make (and check).

The EPR paper published in May 1935 argues for the incompleteness of quantum mechanics by arguing for the existence of "elements of physical reality" not included in the quantum mechanical description. To do this, EPR need a sufficient criterion for the existence of such elements of physical reality, namely: "*If, without in any way disturbing a system, we can predict with certainty (i.e., with probability equal to unity) the value of a physical quantity, then there exists an element of physical reality corresponding to this physical quantity*" (Einstein, Podolsky, and Rosen 1935, 777; italics in the original).

EPR consider in particular two no longer interacting systems in a generic entangled state $\Psi(x_1, x_2)$, which they write in two alternative ways (their eqs (7) and (8)):

$$\Psi(x_1, x_2) = \sum_{n=1}^{\infty} \psi_n(x_2) u_n(x_1) = \sum_{s=1}^{\infty} \varphi_s(x_2) v_s(x_1), \tag{1}$$

or (I switch to Dirac notation for later convenience)

$$|\Psi\rangle = \sum_{n=1}^{\infty} |\psi_n\rangle |u_n\rangle = \sum_{s=1}^{\infty} |\varphi_s\rangle |v_s\rangle, \tag{2}$$

where the states $|u_n\rangle$ and $|v_s\rangle$ are, respectively, (normalized) eigenstates of quantities A and B on system I, and the states $|\psi_n\rangle$ and $|\varphi_s\rangle$ are (not necessarily normalized or orthogonal) states of system II. Depending on whether one chooses to measure A or B, system II is then left in a state corresponding to either a (nonzero) $|\psi_n\rangle$ or a (nonzero) $|\varphi_s\rangle$. They further note: "On the other hand, since at the time of measurement the two systems no longer interact, no real change can take place in the second system in consequence of anything that may be done to the first system" (779).

The reasoning is not spelled out in the paper, but even without invoking the criterion of reality it now follows immediately that neither of these wave functions can provide a complete description of the state of system II. At most they can be incomplete descriptions of the same real state of the system. Indeed, as pointed out by both Fine and Howard, this indirect reasoning was Einstein's preferred version of the argument, and he thought the main point had not come across clearly in the published paper.

In the paper, EPR then turn to a special case, in which they proceed to apply the criterion of reality to establish the existence of elements of reality. Their special case is given by the state

$$|\psi\rangle = \int_{-\infty}^{\infty} e^{ix_0/\hbar} |-p\rangle |p\rangle \, dp = \int_{-\infty}^{\infty} |x + x_0\rangle |x\rangle \, dx, \tag{3}$$

where $|p\rangle$ and $|x\rangle$ are the standard (improper) eigenkets of momentum and position: whenever the measurement of momentum on system I yields the outcome p, system II is left in the momentum eigenstate corresponding to the eigenvalue $-p$, and whenever the measurement of position on system I yields the outcome x, system II is left in the position eigenstate corresponding to the eigenvalue $x + x_0$.[4]

Judging by the different readings in the literature (see Whitaker [2004] for an overview), the form of EPR's reasoning is not entirely clear. The following, however, is a straightforward reading that fits very well also with Bohr's reply. The criterion of reality is a conditional, with a statement about possibility in the antecedent. The special example considered by EPR establishes this antecedent: it is indeed possible to measure position on system I apparently without disturbing system II in any way. Therefore, the consequent follows by *modus ponens*: there is an element of reality corresponding to system II's position. The same reasoning applied to momentum establishes that there is an element of reality corresponding also to system II's momentum. (Note that the reasoning nowhere requires such non-disturbing measurements to be actually carried out, only that it be possible to carry them out.) Since no quantum mechanical state includes values of both position and momentum in the description it gives of a system, the quantum mechanical description of reality is incomplete.

3. Bohr's reply revisited

In his reply published in October 1935, Bohr addressed the argument as presented in the paper, and appears to read it as suggested above. His strategy is to undermine the criterion of reality as containing an "essential ambiguity" when applied to this case, so that EPR's claim no longer follows. As Bohr explains, the ambiguity in the criterion of reality is that while there is no "mechanical disturbance" of the second particle when the first one is measured (i.e., there is no interaction with the second particle), there is "*an influence on the very conditions which define the possible types of predictions regarding the future behavior of the system*" (Bohr 1935, 700; italics in the original). According to Bohr it is this kind of influence—not mere "mechanical disturbance"—that EPR need to rule out to be able to apply the criterion.

Furthermore, such "influence" is a very general feature of quantum measurements ("an essential property of any arrangement suited to the study of the phenomena of the type concerned" [Bohr 1935, 697]), already captured by his "viewpoint termed 'complementarity,'" with the EPR example involving no "greater intricacies." And, indeed, what Bohr does is to discuss at length an example of a simple quantum measurement, a particle passing through a single slit in a diaphragm initially movable along a given direction, where the initial momentum along that direction of both the particle and the diaphragm are known. (Later in the paper Bohr models explicitly also the EPR state of two microscopic particles, but for the purpose of his argument, he has no need of it.) Passage through the slit narrows the uncertainty in the position of the particle, and an "exchange of momentum" takes place between particle and diaphragm. Note that the total momentum and the difference in positions are then known, as in the EPR case. Note also that the original "exchange of momentum" between particle and diaphragm is a "mechanical" interaction, of course, but we are not interested in retrodicting the initial state of the particle (which Bohr explicitly says could be a known eigenstate of momentum), but in what predictions can be made about results of further measurements. Such predictions can now be made by manipulating only the diaphragm, so that there is no longer any interaction with the particle, again analogously to the EPR case. Note finally that (as mentioned by Bohr in his initial summary of the EPR argument) all of the above applies equally well if we describe the situation using classical mechanics.

The difference between the classical and quantum mechanical descriptions lies in the limitations arising from the choice of manipulations on the diaphragm. Specifically, if we wish to predict the position or the momentum of the particle by manipulating the diaphragm, we need to be able to reconstruct the relevant aspect of how the particle affected the diaphragm when they originally interacted. Bohr (1935, 697) calls this "controlling the reaction of the object on the measuring instruments if these are to serve their purpose." If we can legitimately use the "idea of space location" in describing the diaphragm and its interaction with the particle, then by performing a position measurement on the diaphragm (immediately after the passage of the particle[5]), we can predict the result of a further position measurement on the particle; and if we can apply the "conservation theorem of momentum" in describing the diaphragm and its interaction with the particle, then by performing a momentum measurement on the diaphragm we can predict the result of a further momentum measurement on the particle.

Crucially now the different choices of how to complete the measurement preclude the applicability of one or other of these classical ideas (rendering "maximal" the set of measurements whose results are predictable with certainty). If we measure the position of the diaphragm, its momentum becomes completely uncertain, and we may no longer legitimately apply the conservation theorem. Thus we have "voluntarily cut ourselves off" from the possibility of reconstructing the exchange of momentum between particle and diaphragm. If we determine the momentum of the diaphragm, its position becomes completely uncertain, and we may no longer apply the idea of spatial location. Thus we have "voluntarily cut ourselves off" from the possibility of reconstructing the space-time coordination of particle and diaphragm.

This analysis of measurement applies equally well to Bohr's example as to the EPR example, as emphasized also by Pauli in a letter of July 9, 1935, in which he describes to Schrödinger the strategy of Bohr's reply:

> A pure case of A is an overall situation in which the results of particular measurements on A (a maximal set) are predictable with certainty. I have nothing against calling this the "state"—but even then it *is* the case that changing the state of A—i.e. that which is predictable of A—lies within the *free choice* of the experimenter even without directly disturbing A itself—i.e. even *after* isolating A. ... In my opinion *there is in fact no problem here*—and one knows the fact in question even without the Einstein example. (Pauli 1985, 420; emphasis in the original)

For instance, as Jammer (1974, 96–7) points out, Bohr's example is almost identical to another famous thought experiment—Heisenberg's γ-ray microscope. Indeed, in both cases one has two systems immediately after the interaction in a state of known total momentum and zero difference in position (or approximately so: in Bohr's case because of the finite width of the slit, in Heisenberg's case because of the finite size of the microscope). In the case of the microscope, the freedom of choice in completing the measurement had been explicitly discussed by Weizsäcker (1931), and put to crucial use both by Grete Hermann (1935) as the linchpin of her analysis of causality and complementarity in quantum mechanics, and by Heisenberg in his own draft reply to EPR from the summer of 1935 (instigated by Pauli and related to Hermann's analysis).[6] Note that while neither thought experiment is a fully generic example of quantum measurement, because the state after the interaction is maximally entangled (or approximately so), the main point remains the same even if the entanglement between system and measuring instrument, or any other auxiliary system (in modern parlance an "ancilla"), is not maximal: choosing to perform one or another of alternative projective measurements on the auxiliary system still results in different kinds of states of the system, just as in EPR's more general example of equation (2).

Let us take stock. As far as Bohr's quantum mechanical description of the particle and diaphragm goes, nothing is in dispute here. EPR accept the correctness of quantum mechanics, and, indeed, the idea that the choice of measurements performed on one system determines which quantum state we assign to another system is explicitly asserted in the EPR paper. Even the idea of some kind of "cutting oneself off" need not be an issue. For instance, if quantum mechanics is correct, it is not in dispute that when we determine the position of the diaphragm its momentum will become totally uncertain, in the empirical sense of becoming unpredictable. But even if in this empirical sense "cutting oneself off" should be utterly final and irreversible, all that Bohr has established so far is compatible with the idea that by choosing to perform, say, a measurement of position on the diaphragm, we cut ourselves off merely *epistemically* from gaining access to the momentum of the particle. The only advance on a naïve disturbance theory of measurement would be that instead of saying that measuring the position of the particle uncontrollably disturbs the momentum of the particle (thus conceding that it exists), we say that measuring the position of the particle uncontrollably disturbs our possibility of knowing the momentum of the particle (thus equally conceding it exists). Bohr (1935), however, emphasizes that "cutting oneself off" is not meant merely epistemically: he states explicitly that "we have in each experimental arrangement suited for the study of proper quantum phenomena not merely to do with an ignorance of the value of certain physical quantities, but with the impossibility of defining these quantities in an unambiguous way" (699).

Bohr's position can arguably be spelled out in more detail using Howard's analysis of Bohr's doctrine of classical concepts. Howard (1994) has suggested that Bohr in fact agrees with Einstein on the fundamental necessity of the separability of system and apparatus for an objective description of the system (and the apparatus), but that he sees the "finite interaction between object and measuring agencies conditioned by the very existence of the quantum of action" as genuinely destroying the possibility of such an objective description.[7] While separability can be regained within an observational context, the price to pay is that the resulting description crucially depends on the choice of the observational context, whereby different choices generally exclude each other. In this sense, what can be considered objective in each context is different, so that we have different descriptions that are objective but complementary to each other.

As a formal "reconstruction" (a carefully chosen word) of Bohr's intuitions, Howard suggests we represent the lack of separability in terms of entanglement, and the passage to an observational context as substituting for the entangled state an appropriate mixture. Specifically, in the EPR case:

- we can choose to apply to both systems the idea of spatial location by substituting the mixture

$$\int_{-\infty}^{\infty} |x + x_0\rangle\langle x + x_0| \otimes |x\rangle\langle x| dx, \tag{4}$$

which allows us to selectively exploit the position correlations in the EPR state; or we can choose to apply to both systems the conservation of momentum by substituting the mixture

$$\int_{-\infty}^{\infty} |-p\rangle\langle -p| \otimes |p\rangle\langle p| \, dp, \tag{5}$$

•   which allows us to selectively exploit the momentum correlations in the EPR state.

Unlike what one might have expected, in choosing an observational context Bohr does not treat the apparatus as classical and the system as quantum mechanical, but treats certain aspects of *both* system and apparatus as classical, at the expense of others. Note this is perfectly in line with our above remarks about the need to reconstruct certain aspects of the interaction between system and apparatus in order to make certain kinds of predictions about the system.

   In this light, the core of the disagreement between Bohr and Einstein is the idea that quantum mechanical entanglement genuinely represents a lack of separability at the level of physical reality, as opposed to a failure to capture in the quantum mechanical description the underlying separability at the level of physical reality. The question remains why Bohr insists on the former, but in the next and final section we shall abandon this line of enquiry. We shall switch our focus to the Everett theory, endorsing the claim that Everett's theory can provide a detailed reconstruction of Bohr's views on complementarity, and indeed a justification of his specific criticism of the criterion of reality. The Everettian perspective, however, also includes a notion of physical reality that is independent of any observational context, and will suggest a different criticism of Bohr's reply.

4.   Everett: The best of both worlds?

Everett takes seriously the idea of a wave function of the universe that evolves according to the usual Schrödinger equation. From the universal wave function (or indeed the wave function of any given system), which we shall write in Dirac notation as $|\Psi\rangle$, we can define two types of states for any subsystem. Marginal distributions and expectations for a subsystem II are given by the reduced *density matrix* $\mathrm{Tr}_1(|\Psi\rangle\langle\Psi|)$ of the system (the density matrix obtained by taking the partial trace over the remaining degrees of freedom, comprising system I). Instead, a *relative state* $\langle\psi|\Psi\rangle$ gives the conditional distributions and expectations for subsystem II, conditional on any state $|\psi\rangle$ of system I for which $\langle\psi|\psi\rangle \neq 0$ (Everett 2012, 97–103).

   In state (2) above, the reduced density matrix of system II can be written in two alternative ways as:

$$\mathrm{Tr}_1(|\Psi\rangle\langle\Psi|) = \sum_{n=1}^{\infty} |\psi_n\rangle\langle\psi_n| = \sum_{s=1}^{\infty} |\varphi_s\rangle\langle\varphi_s|, \tag{6}$$

or in fact infinitely many other ways. Each $|\psi_n\rangle$ is the state of II relative to $|u_n\rangle$ and each $|\varphi_s\rangle$ is the state of II relative to $|v_s\rangle$. However, there are infinitely many other ways of defining relative states for any given system. The usefulness of the concept is thus not immediately obvious. Everett's strategy for recovering standard quantum mechanics, however, is to consider systems ("servomechanisms") that have a complex enough structure that they can store (and maybe act upon) records of the relative states of other systems they have interacted with in certain (measurement-like) ways. The theory will recover standard quantum mechanics if it predicts the usual quantum statistics for "typical" memory sequences of these observer systems (where it is clear that Everett has in mind an analogy with classical statistical mechanics).[8]

The consensus among today's Everettians is that the tool for identifying stable structures within the universal wave function is the theory of decoherence, and the strategy to recover the quantum probabilities is by suitably generalizing Savage's representation theorem from classical decision theory.[9] Irrespectively of these details, when applied to measurement situations as the ones above, when the observer performs, say, a position measurement on the diaphragm, then relative to the observation of a particular value, the diaphragm and the particle will be in a relative state which is a product of the corresponding position states.[10] The relative state of the observer indeed records the corresponding state of the diaphragm and the particle, and this record is dynamically stable (insofar as guaranteed by continued decoherence interactions with the environment). At the same time, more than one record is present. This can be phrased in the language of worlds branching and/or observers splitting. More than one "copy" of me is present, each recording a different result and dynamically decoupled from the others. When I look at the result, I realize which world I am in, and what from an absolute perspective is a deterministic development of different decoupled components within the universal wave function appears from the relative perspective of my own state as an indeterministic process with associated collapse of the state of the system.

I now claim that the relative states of Everett's theory are perfectly suitable to describe not only the empirical phenomena of quantum mechanics, as just sketched, but also the way these phenomena are seen from Bohr's viewpoint of complementarity (at least as analyzed above).

Consider first of all the state of the universe (or of that part that interests us) after the particle and diaphragm have interacted, but before we have chosen to measure either the position or the momentum of the diaphragm. This state is

$$|\Psi\rangle = \int_{-\infty}^{\infty} e^{ip_0/\hbar}|x\rangle|x\rangle dx = \int_{-\infty}^{\infty} |p_0 - p\rangle|p\rangle dp \qquad (7)$$

(a state like (3) but with $x_0 = 0$ and some total momentum p_0), times a "ready" state $|R\rangle$ of our measuring device (and/or of ourselves). Relative to the state $|R\rangle$, the particle and diaphragm are in the maximally entangled state (7), and neither has a (pure) relative state of its own.

Consider next the two possible choices of measuring the position or the momentum of the diaphragm. Depending on our choice, the total state either evolves to one including records $|R_x\rangle$ of the position of the diaphragm or to one including records $|R_p\rangle$ of its momentum, that is, we have either

$$|\Psi\rangle|R\rangle = \int_{-\infty}^{\infty} e^{ip_0/\hbar}|x\rangle|x\rangle|R\rangle dx \mapsto \int_{-\infty}^{\infty} e^{ip_0/\hbar}|x\rangle|x\rangle|R_x\rangle dx, \tag{8}$$

or

$$|\Psi\rangle|R\rangle = \int_{-\infty}^{\infty} |p_0 - p\rangle|p\rangle|R\rangle dp \mapsto \int_{-\infty}^{\infty} |p_0 - p\rangle|p\rangle|R_p\rangle dp. \tag{9}$$

In the first case, relative to a state $|R_x\rangle$ of the recording device, the particle and diaphragm are in the relative state $|x\rangle|x\rangle$. In the second case, relative to a state $|R_p\rangle$ of the recording device, the particle and diaphragm are in the relative state $|p_0 - p\rangle|p\rangle$.

Note that under the evolutions in (8) and (9), respectively, the density matrix of the composite of particle and diaphragm evolves from $|\Psi\rangle\langle\Psi|$ to either

$$\int_{-\infty}^{\infty} |x\rangle\langle x| \otimes |x\rangle\langle x| dx \tag{10}$$

or

$$\int_{-\infty}^{\infty} |p_0 - p\rangle\langle p_0 - p| \otimes |p\rangle\langle p| dp, \tag{11}$$

that is, the appropriate mixtures (4) or (5) of Howard's analysis. Before we have looked at the measurement result, these mixtures have an ignorance interpretation for us (and in this sense they are classical mixtures): "we" (the relevant copy or ourselves) live in only one branch of the universal wave function, in which the state of the particle is either one particular $|x\rangle\langle x|$ or one particular $|p_0 - p\rangle\langle p_0 - p|$. And this disjunction is exclusive in a strong sense: if we ourselves had not yet interacted in any way (even without consciously observing the result) with the recording device, we could still undo its interaction with the diaphragm at least in principle ("quantum erasure") and change our mind between choice (10) or (11), but as soon as we do interact with the recording device, we split into different copies of ourselves, and as far as what is in our power, we have indeed irreversibly "cut ourselves off" from the possibility of measuring the other quantity instead. In principle (and with a nod to Wigner), some friend of ours who had not yet interacted with us could undo the interaction. In practice, however, as soon as a macroscopic diaphragm or measuring device are involved, decoherence kicks in, and the universal wave function branches in an effectively irreversible way into worlds containing positions and records of position or momenta and records of momenta.

Crucially, moreover, we can now see that cutting oneself off is not merely epistemic, in the following sense. Indeed, even though the original state $|\Psi\rangle$ can be written nonuniquely in the biorthogonal forms (7), that is, at this stage we can still choose to exploit either the correlations in position or the correlations in momentum, the triorthogonal decompositions in (8) and (9) are unique (Elby and Bub 1994). In the first case, there are no states of the recording device other than $|R_x\rangle$ for which the relative states of particle and diaphragm have product form (i.e., for any other states of the recording device, the relative state of the composite particle and diaphragm is entangled, and neither the particle nor the diaphragm have a relative state of their own); and similarly in the second case there are no states of the recording device other than

$|R_p\rangle$ for which the relative states of particle and diaphragm have product form. Thus, when we choose to measure, say, the position of the diaphragm, we gain access to one of the components $|x\rangle\langle x|$ of (10). But now, while the other position components are contained in the other worlds that have branched away from ours when we performed the measurement, there are no worlds containing any momentum components! In this sense, cutting oneself off from the momentum of the particle is to be understood in an ontic rather than an epistemic sense: there is nothing to be known by us or any of our counterparts about the momentum of the particle in our or, indeed, any other world.

Thus the Everett theory is able to rederive Bohr's critique of EPR's criterion of reality: the "physical reality" one can be ignorant about in a world does in fact depend on the choice of manipulation made by the observer without "mechanically" disturbing the system of interest, so that the EPR criterion of reality is not applicable to the EPR example. That is not the end of the story, however.

Indeed, the Everett theory does not only contain "physical reality" as relativized to an observational context and as represented by the relative states, but also an absolute notion of "physical reality" that remains undisturbed by "mechanical" interactions and that is represented by the reduced density matrices. In Bohr's example, irrespective of whether we choose to measure position or momentum on the diaphragm, the density matrix of the particle is and remains the maximally mixed state

$$\int_{-\infty}^{\infty}|x\rangle\langle x|dx = \int_{-\infty}^{\infty}|p_0 - p\rangle\langle p_0 - p|dp \qquad (12)$$

(an improper mixture, because in the Everett theory there is no collapse), and in this sense the position components $|x\rangle\langle x|$ and the momentum components $|p_0 - p\rangle\langle p_0 - p|$ are still there, in fact all of them, even when we choose to perform the other measurement. Thus, the quantum mechanical description of reality as given by a particular position or momentum state after the corresponding measurement is clearly incomplete. It is true that the resulting incompleteness does not support Einstein's conclusion that the quantum state is merely a statistical description, but the fact that the absolute state of the particle is and remains (12) is still important in the context of Bohr's reply to EPR, for the following reason.

Suppose Bob (Bohr) has performed a position measurement on the diaphragm. In so doing, he has gained access to some particular component $|x\rangle$ of the state of the particle. The other position components are in different worlds, and the momentum components would even require reinterference between the different worlds in order to be reconstituted. The state $|x\rangle$ that he assigns to the particle is the one that he needs to use in order to make any further predictions of the future behaviour of the particle, because it is the only component of its absolute state he is able to interact with if he goes on to make measurements on the particle (which of course in general requires entangling it with some new instrument or ancilla). If he performs a position measurement, he will indeed find that the particle is in the state $|x\rangle$, thereby confirming the continued applicability of the "idea of space location" he has used to attribute the state to the particle in the first place.[11] And if he performs a subsequent momentum measurement on the particle, the state $|x\rangle$ will split into its (own) momentum components, independently of whatever result he might have obtained had he originally performed a momentum measurement on the diaphragm. Thus, the Everett theory helps us see how the viewpoint of complementarity can indeed explain the results of any other measurements that Bob might perform.

But let us now switch to the actual EPR example with the two particles, and ask what if it is Alice (Albertine) who performs a measurement on the distant particle? She interacts with the undisturbed state (12) (with $p_0 = 0$, and suitable $x_0 \neq 0$), possibly even at spacelike separation to Bob's measurement on the nearby particle. (Recall Ehrenfest's words: the particle is "flying about in isolation 'on its own.'") Why should *his* cutting himself off from the other components of the state force *her* to interact only with the particular $|x\rangle$ he has access to (or its momentum components)? Consider especially that one of the strengths of Bohr's account is that Bob's choice takes place entirely locally, that is, it is because of something that happens with Bob at his own location that he is unable to interact any longer with all the components of the state (12) of the particle on Alice's side of the experiment.

The Everett theory of course has a ready answer: *nothing* forces Alice to interact only with one particular component $|x\rangle$. Indeed, she interacts with all of them, and if she also performs a position measurement, she splits accordingly into components that have each interacted with some particular $|x\rangle$, just as if Bob had not performed his measurement at all. Bob (more precisely, each of Bob's components) will only gain access to that component of Alice that has interacted with the $|x\rangle$ to which he himself has access; and similarly with Bob and Alice interchanged.[12]

Bohr's complementarity view does not have such an answer ready.[13] Short of non-locality (observational contexts are determined by the first measurement to be performed in some privileged reference frame) or solipsism (observational contexts are determined only by me), the best answer appears to be that what should count as "physical reality" depends holistically on the entire observational context, consisting of both Alice's and Bob's local contexts. One might ask, echoing EPR's conclusions, whether any "reasonable definition of reality could be expected to permit this" (Einstein, Podolsky, and Rosen 1935, 780), but it would at least be an answer that keeps to the letter (if possibly stretching the spirit of) Bohr's entreaty that "the unambiguous account of proper quantum phenomena must, in principle, include a description of all relevant features of the experimental arrangement" (Bohr 1963, 4).

Notes

I would like to thank Elise Crull, my joint work with whom forms the background for this chapter; Hans Halvorson, who suggested our names as contributors to this volume; and Jan Faye and Henry Folse for their invitation to contribute to this volume and for their patience in waiting for the result.

1 This is a sentiment shared by Everett (2012, 158) himself: "Our theory in a certain sense bridges the positions of Einstein and Bohr, since the complete theory is quite objective and deterministic ..., and yet on the subjective level, of assertions relative to observer states, it is probabilistic in the *strong sense* that there is no way for observers

to make any predictions better than the limitations imposed by the uncertainty principle."

2 For a classic summary, see Jammer (1974, sections 5.2, 6.,2 and 6.3), and for necessary correctives Fine (1986, chapter 3) and Howard (1990).

3 All translations are by Elise Crull and/or myself.

4 Note that such a state is not given by a Schrödinger wave function; indeed, it is even more singular than a Dirac delta function. For a rigorous treatment in the formalism of C^* algebras, see Halvorson (2000). Standard wave functions, however, suffice to describe approximate EPR states. The mathematical details will be inessential to our discussion of either the EPR argument or Bohr's reply.

5 This is somewhat a limitation of Bohr's example, but note that even in the EPR paper there are no considerations of dynamics: depending on how the EPR state evolves, the observable whose values we can predict with certainty on system II, and the measurement we need to perform on system I, will not necessarily be position and will depend on the time of the further measurement (as discussed by Schrödinger [1935, section 4]).

6 Hermann's essay (published in March 1935) argues that quantum mechanics is causally complete, but that causal chains can be only reconstructed in retrospect, thus severing the classical link between causality and prediction. It then goes on to provide a comprehensive analysis of complementarity from a neo-Kantian perspective. For a full translation, see Crull and Bacciagaluppi (2018). For Heisenberg's reply to EPR and its relation to Hermann's essay, see Bacciagaluppi and Crull (2009, 2018) and Crull and Bacciagaluppi (2011, 2018).

7 As Bohr (1928, 580) wrote in the Como lecture, "The quantum postulate implies that any observation of atomic phenomena will involve an interaction with the agency of observation not to be neglected. Accordingly, an independent reality in the ordinary physical sense can neither be ascribed to the phenomena nor to the agencies of observation."

8 For Everett's discussion of observer memories, see especially Everett (2012, 118–19 and 137), and for his ideas on typicality, see Everett (2012, 123–30, 190–2, 261–4 and 294–5).

9 See again Saunders et al. (2010, parts 1 and 3) and Wallace (2012, chapters 3 and 5). Note that Everett's servomechanisms return as "Information Gathering and Utilising Systems" (IGUSes) in discussions of decoherence, and that the mathematical core of the quantum representation theorem is related to that of Everett's own derivation of the Born rule.

10 Similarly, relative to a position state of the particle, the apparatus and observer will be in a corresponding product state of position and recording position; and relative to a position state of the diaphragm, the particle and observer will be in a corresponding product state of position and recording position. Note again that the EPR state itself is highly singular, but if one considers suitably approximate states and realistic measurements, all relative states will be well defined.

11 As Hermann (1935) would put it, he is indirectly checking the causal connection established relative to the chosen context of observation by the original interaction between the particle and the diaphragm (or the electron and the photon in the case of the γ-ray microscope).

12 If Alice performs a momentum measurement, things are only slightly more complicated: she will first split into components in which she registers a particular value $-p$; then, when she crosses the future light cone of Bob's measurement and

can interact with any records of his measurement, she will split again (recall that a paradigm case of an observer splitting is precisely whenever they interact with the records of a measurement), so that now she has split into components that have registered some $-p$ on particle II and some x on particle I. When Bob interacts with Alice, he will interact only with those components of Alice that have registered his result x, but splits further himself (like Alice) to gain access to a particular among these, characterized by some or other value of $-p$. (The quantitative explanation of the observed correlations follows from any account of probabilities in Everett.) For further elaboration of this spacetime formulation of Everett, see Bacciagaluppi (2002) (see also Wallace [2012, chapter 8]).

13 See Bacciagaluppi (2014) for a closely related criticism of QBism, the radical subjectivist approach to quantum mechanics developed by Fuchs (2010) (sometimes classed as a "neo-Bohrian" view).

References

Bacciagaluppi, G. 2002. "Remarks on space-time and locality in Everett's interpretation." In *Non-locality and Modality*, edited by T. Placek and J. N. Butterfield, 105–22. Dordrecht: Kluwer Academic Publishers.

Bacciagaluppi, G. 2013. Review of *The Everett Interpretation of Quantum Mechanics: Collected Works 1955–1980 with Commentary, by Hugh Everett III*, edited by J. A. Barrett and P. Byrne. *HOPOS* 3: 348–52.

Bacciagaluppi, G. 2014. "A critic looks at qbism." In *New Directions in the Philosophy of Science*, edited by M. C. Galavotti, D. Dieks, W. Gonzalez, S. Hartmann, T. Uebel, and M. Weber, 403–16. Cham: Springer.

Bacciagaluppi, G., and E. Crull. 2009. "Heisenberg (and Schrödinger, and Pauli) on hidden variables." *Studies in History and Philosophy of Modern Physics* 40 (4): 374–82.

Bacciagaluppi, G., and E. Crull. 2018. *"The Einstein Paradox": The Debate on Nonlocality and Incompleteness in 1935*. Cambridge: Cambridge University Press. In preparation (expected publication date 2018).

Beller, M., and A. Fine. 1994. "Bohr's response to EPR." In *Niels Bohr and Contemporary Philosophy*, edited by J. Faye and H. Folse, Boston Studies in the Philosophy of Science, Vol. 153, 1–31. Dordrecht: Kluwer Academic Publishers.

Bohr, N. 1928. 'The quantum postulate and the recent development of atomic theory." *Nature (Suppl.)* 121: 580–90.

Bohr, N. 1935. "Can quantum-mechanical description of physical reality be considered complete?" *Physical Review* 48: 696–702.

Bohr, N. 1949. "Discussion with Einstein on epistemological problems in atomic physics." In *Albert Einstein: Philosopher-Scientist*, edited by P. A. Schilpp, 199–241. Evanston, IL: The Library of Living Philosophers.

Bohr, N. 1963. "Quantum physics and philosophy. Causality and complementarity." In *Essays 1958–1962 on Atomic Physics and Human Knowledge*, 1–7. New York: Wiley, 1963. Originally published in *Philosophy in Mid-century: A Survey. Vol. I, Logic and Philosophy of Science*, edited by R. Klibansky, 308–14. Firenze: La Nuova Italia, 1958.

Byrne, P. 2010. *The Many Worlds of Hugh Everett III: Multiple Universes, Mutual Assured Destruction, and the Meltdown of a Nuclear Family*. New York: Oxford University Press.

Crull, E., and G. Bacciagaluppi. 2011. "Translation of: W. Heisenberg, 'Ist eine deterministische Ergänzung der Quantenmechanik möglich?'" http://philsci-archive. pitt.edu/8590/.

Crull, E., and G. Bacciagaluppi. 2018. *Grete Hermann: Between Physics and Philosophy*. Dordrecht: Springer. In proof (expected publication date 2018).

Ehrenfest, P. 1931. Letter to N. Bohr, July 9, 1931. Archive for the History of Quantum Physics, mf. AHQP-EHR 17 (in German).

Einstein, A., B. Podolsky, and N. Rosen. 1935. "Can quantum-mechanical description of physical reality be considered complete?" *Physical review* 47: 777–80.

Elby, A., and J. Bub. 1994. "Triorthogonal uniqueness theorem and its relevance to the interpretation of quantum mechanics." *Physical Review A* 49 (5): 4213–16.

Everett, H. III. 2012. *The Everett Interpretation of Quantum Mechanics: Collected Works 1955–1980 with Commentary*, edited by J. A. Barrett and P. Byrne. Princeton: Princeton University Press.

Fine, A. 1986. *The Shaky Game: Einstein, Realism, and the Quantum Theory*. Chicago: University of Chicago Press.

Freire, O. Jr. 2015. *The Quantum Dissidents: Rebuilding the Foundations of Quantum Mechanics (1950–1990)*. Berlin: Springer.

Fuchs, C. 2010. "QBism, the perimeter of quantum Bayesianism." http: arxiv.org/abs/ 1003.5209.

Halvorson, H. 2000. "The Einstein–Podolsky–Rosen state maximally violates Bell's inequalities." *Letters in Mathematical Physics* 53 (4): 321–9.

Hermann, G. 1935. "Die naturphilosophischen Grundlagen der Quantenmechanik." *Abhandlungen der Fries'schen Schule* 6: 75–152. Translated in Crull and Bacciagaluppi (2017).

Howard, D. 1990. "'Nicht sein kann was nicht sein darf,' or the prehistory of EPR, 1909–1935: Einstein's early worries about the quantum mechanics of composite systems." In *Sixty-Two Years of Uncertainty*, edited by A. Miller, 61–111. New York: Plenum Press.

Howard, D. 1994. "What makes a classical concept classical?" In *Niels Bohr and Contemporary Philosophy*, edited by J. Faye and H. Folse, Boston Studies in the Philosophy of Science, Vol. 153, 201–29. Dordrecht: Kluwer Academic Publishers.

Jammer, M. 1974. *The Philosophy of Quantum Mechanics: The Interpretations of Quantum Mechanics in Historical Perspective*. New York: Wiley.

Osnaghi, S., F. Freitas, and O. Freire Jr. 2009. "The origin of the Everettian heresy." *Studies in History and Philosophy of Modern Physics* 40 (2): 97–123.

Pauli, W. 1985. *Wissenschaftlicher Briefwechsel mit Bohr, Einstein, Heisenberg u. a., Band/ Vol. II: 1930–1939*, edited by K. von Meyenn, A. Hermann, and V. F. Weisskopf. New York: Springer.

Saunders, S., J. Barrett, A. Kent, and D. Wallace (eds). 2010. *Many Worlds? Everett, Quantum Theory, and Reality*. Oxford: Oxford University Press.

Schrödinger, E. 1935. "Discussion of probability relations between separated systems." *Proceedings of the Cambridge Philosophical Society* 31: 555–63.

Wallace, D. 2012. *The Emergent Multiverse: Quantum Theory according to the Everett Interpretation*. Oxford: Oxford University Press.

Weizsäcker, C. F. von. 1931. "Ortsbestimmung eines Elektrons durch ein Mikroskop." *Zeitschrift für Physik* 70: 114–30.

Whitaker, A. 2004. "The EPR paper and Bohr's response: A re-assessment." *Foundations of Physics* 34: 1305–40.

14

Niels Bohr and the Formalism of Quantum Mechanics

Dennis Dieks

1. Introduction

Bohr's writings on the interpretation of quantum mechanics, from his 1927 Como lecture to his articles from the 1950s, are conspicuous by the absence of detailed discussions of technical aspects of the quantum formalism. In this respect they are very different from the tradition of foundational work that has started in the early 1950s, with Bohm's hidden-variables scheme, and has led via Bell's theorem and similar results to present-day work on quantum information theory. The predominantly qualitative character of Bohr's arguments, together with his convoluted and sometimes confusing style, can easily create the impression of a rhetoric without substantial support in quantum mechanics itself, aimed at intimidating the reader (Cushing 1994; Beller 1999). That this impression is superficial and misconceives Bohr's perspective on the interpretation of quantum theory has recently been argued from various sides (Howard 1994, 2004, 2005; Faye 2014; Landsman 2006, 2007; Camilleri and Schlosshauer 2015; Zinkernagel 2015, 2016). The present chapter is an attempt to further contribute to this program of rehabilitation, without losing sight of possible weaknesses or open questions in Bohr's interpretation. We shall argue that the nontechnical nature of Bohr's writing should not be misconstrued: Bohr's views are more intimately connected to the mathematical structure of quantum mechanics than usually acknowledged and reflect salient features of it. In fact, Bohr's interpretation bears affinity to modern noncollapse interpretations of quantum mechanics.

The idea that Bohr's statements about the meaning of quantum mechanics have a direct link to the structure of the mathematical formalism may at first sight seem implausible. After all, Bohr himself appears to downplay the significance of this formalism: from the Como lecture to his latest writings Bohr emphasizes the merely "symbolic" character of the mathematical apparatus that is used in the new theory. This, together with the notorious fact that classically describable measuring devices and measurement outcomes play a central role in Bohr's interpretation may suggest that he saw the formalism as possessing only pragmatic value, as a tool for making predictions on the macroscopic level. This would imply that Bohr championed instrumentalism, perhaps even denying the existence of a quantum world at all—a position that has indeed not infrequently been ascribed to Bohr and used as a *reductio* of his views by some anti-Copenhagenists (Cushing 1994; Beller 1999; cf. Faye 2014).

Recent scholarship has questioned this attribution of instrumentalism to Bohr (Folse 1985; Howard 1994; Camilleri and Schlosshauer 2015; Zinkernagel 2015, 2016)—a line that we shall follow and elaborate. To start with, we need to take a closer look at Bohr's reasons for calling the formalism of quantum mechanics "symbolic." After this, we shall pay attention to the role of classical concepts, which in turn will lead us to complementarity and measurement. We conclude with a discussion of how Bohr's views compare to the mathematical formalism from a modern point of view.

2. The symbolic character of the quantum formalism

Already in the first section of his famous 1927 Como lecture Bohr (1928) typifies quantum mechanics as a "symbolic method." Characteristically, to argue his point he does not start from the newly developed Schrödinger wave mechanics or Heisenberg matrix mechanics, but discusses the older, simpler, and more qualitative ideas of de Broglie (though it should not go unnoticed that in section 5 of the Como lecture Bohr makes it clear that he considers de Broglie's conceptions as having been embedded and generalized in the new wave mechanics). De Broglie had proposed to associate waves with particles: if E denotes the energy and p the momentum of a particle, de Broglie postulates an associated wave with frequency v and wavelength a satisfying the relations $E / v = p\lambda = h$ (with h Planck's constant). Bohr now observes that according to these relations the phase velocity of the wave associated with a particle moving at speed v is given by c^2 / v, whereas the group velocity is equal to v. He comments: "The circumstance that [the phase velocity] is in general greater than the velocity of light emphasizes the symbolic character of these considerations. At the same time, the possibility of identifying the velocity of the particle with the group-velocity indicates the field of application of space-time pictures in the quantum theory" (sec. 2).

The contrast that Bohr notes here gives us a first clue about the criteria he has in mind when he qualifies the formalism as "symbolic." It is clear that a wave propagating with superluminal velocity cannot correspond to something physically real propagating in space—this wave must therefore be symbolic in the uncontroversial sense of being not physical but rather having the status of a mathematical tool coming from Fourier analysis. However, a *packet* of such mathematical waves, propagating with group velocity v, apparently *can* be thought of as representing something physically real (taking into account caveats and restrictions to be discussed!). Elaborating on the latter point, Bohr goes on to discuss the spreads (dispersions) in energy and momentum inherent in de Broglie wave packets and concludes that these determine "the highest possible accuracy in the definition of the energy and momentum of the individuals associated with the wave-field." We shall soon return to the meaning of this "accuracy in definition"; but an immediate and obvious conclusion to be drawn is that according to Bohr we are allowed to think of physical "individuals" associated with packets of de Broglie waves. Indeed, the Como lecture is quite explicit that the wave packets correspond to atomic and subatomic entities, like electrons. The notion that the lecture is meant to promulgate an instrumentalist interpretation of quantum theory according to which the whole formalism possesses only mathematical and no physical descriptive content is thus immediately seen to sit uneasily with the textual evidence.

It is true that Bohr immediately follows up his mention of quantum individuals with the warning that the applicability of familiar classical concepts to them is severely limited, in view of the "inaccuracy in definition" just mentioned. Energy and momentum and also position of a wave packet do not possess sharp values so that the usual description of particles, as given by classical mechanics, can at most have restricted validity. Only in limiting situations, comparable to the limit in which geometrical optics is able to replace wave optics, is it to be expected that we can recover a picture in which energy and momentum are well-defined together with sharp positions (sharp positions at each instant of time lead to "a well-defined space-time picture" in Bohr's terminology). Arguing from the simple example of a single particle, formally represented by a de Broglie wave packet, Bohr at this point already formulates the thesis that well-defined space-time pictures and "the claims of causality" (i.e., the instantiation of well-defined values of energy and momentum fulfilling conservation laws) are not compatible in the description of quantum individuals. As Bohr (1928, sec. 2) says,

> The content of the relations (2) [i.e. the qualitative uncertainty relations $\Delta t \Delta E = \Delta q \Delta p = h$ derived from a Fourier analysis of wave packets] may be summarised in the statement that according to the quantum theory a general reciprocal relation exists between the maximum sharpness of definition of the space-time and energy-momentum vectors associated with the individuals.... At the same time, however, the general character of this relation makes it possible to a certain extent to reconcile the conservation laws with the space-time coordination of observations, the idea of a coincidence of well-defined events in a space-time point being replaced by that of unsharply defined individuals within finite space-time regions.

This first discussion of the symbolic character of the quantum formalism at the very start of the Como lecture is illuminating and already clarifies several issues. Importantly, that the formalism is symbolic is apparently not meant to imply that the formalism remains completely silent about physical entities on the micro level or even denies their existence. Quite the opposite, Bohr explicitly tells us that the theory is about physical individuals like electrons and photons. But, second and certainly not less important, these "things" should not be thought of as objects describable with values of physical quantities like energy, momentum and position in the way we would expect on the basis of classical mechanics. In classical mechanics particles necessarily possess well-defined positions, momenta and energies at all instants—but for quantum individuals the attribution of both sharp space-time positions and sharp energy/momentum values is inconsistent with the formalism (namely with the mathematical properties of wave packets), says Bohr. These two groups of concepts (dynamical versus space-time concepts) are incompatible (in fact, "complementary," as to be discussed more extensively in a moment) and their applicability is reciprocally restricted by the uncertainty relations. At the same time, it *is* possible to think of individuals with "unsharply defined momentum/energy and space-time locations"; and in situations in which Planck's constant h can be considered insignificantly small this gives us an approximation to the classical picture. However, in general the incompatibility of p and q stands in the way of a vizualizable picture of the kind we are used to in classical mechanics and in everyday experience.[1]

So it turns out that the "symbolic character" of the formalism refers first and foremost to the notion that familiar looking mathematical quantities like p and q do not stand for the well-defined physical properties that we would expect them to represent—the relation between the mathematical formalism and the quantum world must be different. It does not follow from this that the quantum formalism deals only with macroscopic measurement outcomes and has nothing to say about features of the quantum world. Quite the opposite, Bohr clearly takes the symbolism to yield information about what we can say about quantum individuals like atoms, photons, and electrons. In fact, it is the formalism itself that shows us that such "particles" cannot be pictorially represented in the way of classical physics.

This reading of the meaning of "symbolic" is supported by the remainder of the Como lecture. Bohr briefly discusses matrix mechanics, and observes that p and q in this formulation of quantum mechanics are not numbers, but rather noncommuting matrices obeying the canonical commutation relation. As he says, "This exchange relation [i.e., the commutation relation] expresses strikingly the symbolic character of the matrix formulation of the quantum theory." The point is that matrices are operators and not numerical quantities, and therefore cannot directly represent values of physical quantities.

Bohr then turns to wave mechanics, of which Schrödinger had hoped that it would restore visualizability to atomic physics: Schrödinger aimed to relate his Ψ to physical waves existing in three-dimensional space. Bohr (1928, sec. 5) rejects this possibility of a vizualizable depiction and asserts that "wave mechanics just as the matrix theory … represents a symbolic transcription of the problem of motion of classical mechanics adapted to the requirements of quantum theory"; and "in the wave equation, time and space as well as energy and momentum are utilized in a purely formal way."

Bohr explains that he has three grounds for the latter claim. First, the Schrödinger theory represents momentum by a differential operator $p = -i\hbar\partial / \partial q$, in which the *imaginary quantity i* occurs. Second, "there can be no question of an immediate connexion with our ordinary conceptions because the 'geometrical' problem represented by the wave equation is associated with the so-called co-ordinate space, the number of dimensions of which is equal to the number of degrees of freedom of the system, and hence in general greater than the number of dimensions of the ordinary space." Third, Bohr notes that "Schrödinger's formulation of the interaction problem, just as the formulation offered by matrix theory, involves a neglect of the finite velocity of propagation of the forces claimed by relativity theory."

This third reason is rather surprising, as it belongs to another category of objections than the other two. The non-relativistic character of 1927 quantum mechanics certainly signifies that this version of the theory cannot be completely correct—it will need a relativistic generalization. But this in itself is not relevant for visualizability or for whether the mathematical quantities in the formalism could be potential descriptors of physical reality; classical mechanics has also been superseded by relativity, but this does not make the formalism of classical mechanics automatically "symbolic" in a philosophically interesting sense. One can of course say that the basic incorrectness of non-relativistic quantum mechanics implies that the world is different from what a literal reading of the formalism would suggest. Still, the other two arguments are conceptually different, and more interesting, because they bear on the representational capacity *in principle* of the mathematical symbols to capture elements of physical reality, and not on the predictive correctness of the theory.

That the three arguments are here lumped together as if they were of the same sort appears to betray a certain lack of philosophical finesse on Bohr's part, or a lack of interest in matters of this kind—it is indicative of a physicist's rather than a philosopher's attitude. In the practice of physics research it is quite common not to be too "nitpicking" about small conceptual differences of this sort. And then, of course, the Como lecture was meant for a physics audience—its text is full of detailed examples from contemporary experimental practice and theoretical explanations offered by colleague physicists. This accords with the general point that Bohr is here concerned, as a physicist, with questions about the validity, applicability, consistency, and content of the new physical theory. It would be a distortion to see Bohr's views as basically stemming from an a priori philosophical background: assumptions of physical common sense plus the abstract reasoning typical of theoretical physics suffice to understand his reasoning in the Como lecture and later writings.

The argument that imaginary quantities occur in the theory and that for this reason the theory must be considered symbolic appears as rather unsophisticated from a philosophical viewpoint as well, since imaginary numbers can be used very well to represent physically real quantities. Of course, one should not come under the spell of the everyday language meaning of "imaginary" and confuse it with the mathematical significance of the term; one should not suppose that $i = \sqrt{-1}$ cannot represent anything physically existing because of this connotation. Apart from the trivial point that it is a matter of conventional choice of scale whether measurement results are expressed in real or imaginary numbers, there is the practical example of physically real vector quantities that can be handled very conveniently with complex numbers, as is standard practice in classical electrodynamics.[2]

Nevertheless, in the case before us sense can perhaps be made of Bohr's conclusion. Indeed, suppose we have defined a scale for physical position q in the usual way, that is, by means of real numbers. Then the relation $p = -i\hbar\partial / \partial q$ seems to indicate that real values for q cannot go together with real values for p, assuming that the function that is being differentiated is real-valued. This would entail that p and q cannot be jointly defined and measured on the same type of number scale—something very strange from the viewpoint of classical physics and impossible to visualize. If on the other hand the wave function that is differentiated has an imaginary value itself (as is actually common in quantum mechanics), this wave is apparently of a completely different dimension than the physical quantity q, which is also impossible to incorporate into a classical picture. So the occurrence of $i = \sqrt{-1}$ may in the present context be associated with the incompatibility of physical quantities. The symbols representing these quantities in the formalism must therefore be symbolical in the same sense as before, namely, as defying vizualizability and literal interpretation in the sense of classical mechanics.

Bohr (1928) himself tells us that his second argument, about the dimensionality of configuration space, is the most important one: "*above all* there can be no question of an immediate connexion with our ordinary conceptions because ... the wave equation is associated with the so-called co-ordinate space." In other words, the Schrödinger wave in the case of a many-particle system cannot be a physical wave in three-dimensional space (which would be an "ordinary conception") since it "lives" in a high-dimensional mathematical space.

It should be noted that although this is an argument against wave function realism, it does not at all entail that the wave function cannot contain information about the micro world: a counterexample would be the description of a many-particles system (e.g., a gas) by means of one point in phase space, in classical statistical mechanics, which also uses many more than three dimensions but still tells us about the properties of particles in three-space. We can consistently deny the physical reality of phase space and still be realists with respect to particles. So we should not mistake Bohr's argument for the symbolic character of the wave function for an argument in favor of instrumentalism *tout court*.

In fact, the analogy with classical statistical mechanics shows that Bohr's argument is more complicated than it may first appear. The force of the argument in quantum mechanics derives from the fact that the situation here is fundamentally different from that in classical statistical mechanics: a quantum many-particles wave in configuration space cannot in general be decomposed into many single-particle three-dimensional waves, whereas a classical phase point in a high-dimensional phase space *can* be understood in terms of ordinary three-dimensional descriptions of individual particles. This difference is due to the possibility of superposing quantum waves in configuration space: the sum of any two waves is again a bona fide Schrödinger wave. Even if the two original waves are decomposable as a product of one-particle waves, the superposed wave will not be. This relates directly to the typical quantum feature of entanglement and the corresponding holistic character of the quantum description of many-body systems. Bohr does not mention the point in this form (entanglement), but we have to assume that he realized the consequences of the superposition principle when he argued that the wave function lives in configuration space.

Essentially the same arguments as used in the Como lecture can be found in Bohr's later work (although, perhaps significantly, the argument about the non-relativistic character of the theory disappears). In his famous "Discussion with Einstein on Epistemological Problems in Atomic Physics" Bohr (1949, 238) explicitly connects the *symbolic* character of the quantum formalism to its *non-vizualizability* when he speaks of "the use of not directly visualizable symbolism," and he extensively comments on the difficulty of thinking of quantum objects in terms of conventional physical attributes.[3] In his brief 1948 *Dialectica* article in which he summarizes his position we again read: "These symbols [viz. p and q] themselves, as is indicated already by the use of imaginary numbers, are not susceptible to pictorial interpretation" (314). In the same article Bohr freely speaks about atomic objects, electrons and photons; while emphasizing again the novel quantum problem of describing them in terms of the usual physical quantities.

We may conclude that Bohr's emphasis on the symbolic nature of the quantum formalism is intended to warn us against a literal classical interpretation with respect to the meaning of mathematical quantities like p and q. The quantum world should accordingly not be thought of as mirrored in a simple way by the mathematical formalism: micro properties cannot be read off from the mathematical symbols in the uncomplicated manner they can in classical mechanics. Bohr argues that the problems that we encounter when we try to construct a classical picture, while staying in conformity with quantum mechanics, are problems of principle. Essential structural features of the quantum formalism preclude its classical interpretation.

The arguments that Bohr adduces for non-vizualizability and the symbolic nature of quantum mechanics by the same token show that the structure of the quantum formalism provides us with information about the quantum world. For example, we saw that the spreads of wave packets according to Bohr indicate to what extent a space-time picture and/or a causal description are applicable, and therefore to what extent such descriptions are able to latch on to the physical world. Directly related to this, the canonical commutation relations in the mathematical formalism tell us that sharp values of position and momentum cannot coexist, and so inform us about a characteristic feature of the quantum world.

3. The indispensability of classical concepts

The symbolic character of the mathematical formalism of quantum mechanics in the sense just explained implies that we cannot rely on the usual physical interpretation of mathematical symbols like p and q. This raises the question of how we can interpret the formalism at all. The dilemma is that to endow the symbols with physical meaning we seem to need knowledge about what the quantum world is like, what kind of things are "out there," so that we can make the formalism correspond to these things. But it is quantum mechanics itself that has to provide us with information about the structure of the quantum world. However, the theory can only do so after it has made contact with physical reality, when it is an interpreted physical theory: a purely mathematical scheme cannot tell us anything about the physical world.

A deceptively similar problem played a major role in early-twentieth-century logical empiricism. The logical empiricists attempted to make it clear how purely mathematical calculi could be interpreted empirically, without the intrusion of metaphysical elements. They proposed to solve this problem via the postulation of "correspondence rules" (Carnap) or "coordinative definitions" (Reichenbach); the idea simply being that we need an already empirically interpreted and stable vocabulary through which symbols in the as yet uninterpreted theory can be infused with physical meaning. Given that it is essential for this logical empiricist enterprise to stay as close as possible to theoretically uninfected experience, it is an obvious choice to take an observation language with a minimum of theoretical baggage for the already interpreted and stable language. The interpretational basis should thus be close to our pre-theoretical

everyday language and should enable us to unambiguously refer to observable things
and the results of experiments (as in Carnap's "*Ding-Sprache*"). The various ways in
which this basic idea was elaborated by the logical empiricists, and the many ensuing
debates (verificationism, cognitive significance, instrumentalism versus realism, etc.),
are well known and need not detain us here.

The short description of the basic idea just given already suffices to make it understandable why the logical empiricists expected Bohr to be a close philosophical friend.[4]
But although part of Bohr's ideas about the interpretation of the quantum formalism can
be seen as a physicist's version of the correspondence rule programme, there are also
important differences. Bohr's approach was not motivated by a strict empiricism or by an
anti-metaphysical attitude, but by the problems—reviewed in the previous section—that
arise when we wish to give the usual classical interpretation to symbols like p and q.

Nevertheless, the task Bohr had to face bears a striking resemblance to the problem
of how to give empirical content to an uninterpreted formalism, as we are also here
dealing with a formalism that needs empirical interpretation. Moreover, the sought-for
interpretation should obviously make it possible to adequately describe what is happening in the laboratory. But in this description of the laboratory world essential differences start to appear between Bohr and the logical empiricists. Bohr does not strive for
a purified theory-less observation language (and even less for a reliance on sense data),
but simply accepts that the interpreted new theory should be a generalization of classical physics. The anchor points for this generalization are the descriptions we already
use in experimental practice: in the laboratory we speak about macroscopic objects
and measurement outcomes, but also about many not directly observable things, like
atomic particles, electric and magnetic fields, electric currents, and so forth. Bohr does
not at all protest against this and does not attempt to suppress or reconstruct this language. However, as we have seen, he *does* warn against the supposition that all aspects
of this "classical language" possess unrestricted validity, and in particular against the
uncritical joint attribution of traditional physical attributes when we refer to micro
entities. However, that the assignment of classical attributes works eminently well in
the case of macroscopic objects is an unavoidable fact of laboratory life and more than
two centuries of experience with classical physics. Quantum theory must certainly be
able to reproduce this actual state of affairs. This situation suggests a way for linking the
quantum formalism to experience, namely, via our already existing classical description of the macro world. As Bohr puts it in an often-quoted passage in his contribution
to the 1949 Einstein volume:

> It is decisive to recognize that, however far the phenomena transcend the scope
> of classical physical explanation, the account of all evidence must be expressed in
> classical terms. The argument is simply that by the word "experiment" we refer to
> a situation where we can tell others what we have done and what we have learned
> and that, therefore, the account of the experimental arrangement and of the results
> of the observations must be expressed in unambiguous language with suitable
> application of the terminology of classical physics. (209)

This passage has been deemed mysterious and has given rise to speculations about
why Bohr thought that classical concepts were essential. Thus, it has become one of the

sources of the belief that Bohr was an instrumentalist or a strict empiricist. It has also been interpreted as evidence that Bohr was influenced by Kantianism and saw classicality as an a priori precondition of the possibility of empirical knowledge; or that he anticipated the more modern philosophical idea that we are "suspended in language," in fact classical language.[5]

But motivations originating in—or strongly influenced by—philosophical doctrines need not be assumed here. As is clear from the wording of the just quoted passage, Bohr himself thought of his claim as simple and obvious. This is in agreement with our view that Bohr took the classicality of experimental practice (in the sense that classical concepts are adequate for the description of results and are in fact the only ones we possess to describe them) as an undisputed and familiar fact. This classical description is basically just the description in terms of everyday language, generalized by the addition of physics terminology, and it is the one we de facto use to describe our environment. Bohr (1948, 313) apparently found this observation so self-evident, almost tautological, that he sometimes called the applicability of classical concepts to experimental practice a "logical demand"; because, as he says, "the word 'experiment' can in essence only be used in referring to a situation where we can tell others what we have done and what we have learned."

This simple reading of Bohr's doctrine of classical concepts is confirmed, as we shall see, by the way Bohr further elaborated on the role of these concepts in his interpretation of quantum mechanics, and also by evidence coming from explanations that Bohr gave in private conversations. For example, in discussions with Heisenberg in 1927 Bohr already stated: "in spite of your uncertainty principle you have got to use words like 'position' and 'velocity' just because you haven't got anything else" (Camilleri and Schlosshauer 2015); and von Weizsäcker tells the anecdote that when the physicist Edward Teller was expressing doubts about the importance of classical concepts, at an afternoon tea at Bohr's Institute, Bohr briefly fell asleep and awakening merely replied: "Oh, I understand. You might as well say that we are not sitting here, drinking tea, but that we are just dreaming all this" (Drieschner 2014, 78).

The use of classical *concepts* does not involve a wholesale acceptance of classical *physics*. As Bohr says in the above quotation, it is the evidence provided by macroscopic measuring devices (pointer readings, light flashes, sounds, etc.) that is to be expressed in classical terms, "however far the phenomena transcend the scope of classical physical explanation." In other words, although we need kinematical concepts like p and q on the macroscopic level, we have to expect that the classical dynamical laws will fail to give us adequate explanations of the values that we actually find for these quantities.

Specifically, we know that macroscopic objects can be ascribed a position and a momentum; and Bohr does not hesitate to relate these to the q and p occurring in the quantum formalism when it is applied to macroscopic objects. We here again discern the significant difference between Bohr's concerns, which are primarily physical, and the philosophical concerns of the logical empiricists. Bohr is not interested in formal rules that determine the empirical content of terms like "position" and "momentum," and is not engaged in a conceptual purification programme à la Mach. He accepts the descriptive vocabulary of classical physics as soon as this becomes possible without contradictions, that is, on the macroscopic level, and uses this as his cue to achieve a generalization of classical physics that applies to atomic objects as well. As we shall see in the next section, this generalization will be realized not by formal definitions of microscopic attributes in terms of macroscopic measuring results (as might be expected on a logical empiricist analysis), but by partly taking over the "metaphysics" of classical physics while complicating it by adding a relational aspect, namely, a dependence on the experimental context (the famous doctrine of complementarity). The very first two sentences of section 1 of the Como lecture already announce this: "The quantum theory is characterized by the acknowledgment of a fundamental limitation in the classical physical ideas when applied to atomic phenomena. The situation thus created is of a peculiar nature, since our interpretation of the experimental material rests essentially upon the classical concepts" (Bohr 1928, 580).

However, Bohr's strategy seems to be immediately menaced by the objection that his own arguments for the symbolic nature of the quantum formalism forbid an identification of the quantum symbols p and q with the classical concepts of momentum and position, even in the macroscopic realm. At least, this problem arises if we consider quantum mechanics to be a universal theory, applicable both at the micro and macro levels. Now, it has often been claimed in the older literature, and is sometimes still said today, that Bohr assumed that the world consists of two ontologically different parts, on two sides of a dividing line, a "cut": a quantum world on the microscopic side of the cut, and a classical world, governed by classical laws, on the macro side.

But the idea that Bohr denied the universal validity of quantum mechanics is mistaken. Indeed, from his earliest writings on Bohr argues from the assumption that quantum theory *is* universal, in the sense of applicable to both micro and macro systems (Howard 1994; Landsman 2006, 2007; Camilleri and Schlosshauer 2015).[6] In the Como lecture the universality of quantum theory is not an explicitly addressed theme, but is implicitly assumed throughout the text. For example, already in the first section of the lecture Bohr (1928, 580) mentions that observations of atomic systems consist in physical interactions with "the agencies of observation," and that these interactions are to be treated quantum mechanically and, importantly, *symmetrically* so that "an independent reality in the ordinary physical sense [i.e., the classical sense in which each system has its own independent state] can *neither be ascribed to the phenomena nor to the agencies of observation*" (emphasis added). The lack of independence to which Bohr refers here is a typical quantum feature that we shall discuss further in the next section; the important point for our present purposes is that this quantum feature is ascribed to *both* parties in a measurement, to the micro system *and* the measuring device.[7]

In his 1935 reply to Einstein, Podolsky, and Rosen, Bohr is more explicit in his discussion of macroscopic devices, and employs quantum mechanics to deal with a macroscopic diaphragm in the experimental set-up that he proposes as an analogue to the EPR experiment. Interestingly, he mentions here that it may seem a problem that this implies using quantum mechanics for "some process to which the ideas of classical mechanics can be applied." But, he says, "if all spatial dimensions and time intervals are taken sufficiently large, this involves clearly no limitation as regards the accurate control of momentum" and "a purely classical account of the measuring apparatus … implies the necessity of allowing a latitude corresponding to the quantum-mechanical uncertainty relations in our description of its behavior" (Bohr 1935, 698). Bohr implies here that the quantum description remains ontologically primary, even when we use (as we must) classical concepts to describe macroscopic objects. The classical description is *approximate* in the sense that in principle we should accept that there are "latitudes" in the classical quantities (like p and q), in accordance with the uncertainty relations. However, these latitudes will normally play no role in our dealings with the macroscopic world because they are enormously small ("conditioned by h") compared to the usual values of macroscopic quantities, and they will therefore not stand in the way of a classical description. However, they certainly exist and should be taken into account *explicitly* (698, second column) when we discuss the consistency of the quantum mechanical scheme—*implicitly* they are always there, even when we use only classical terms and do not apply the quantum formalism.

Bohr elaborates this point in his *Dialectica* article and in his contribution to the Einstein volume (Bohr 1948, 1949). In the latter, in his famous analysis of experiments in which a particle goes through a hole (or two holes!) in a diaphragm—prompted by Einstein's arguments at the 1927 Solvay conference—Bohr comments on the possibility of controlling the momentum transfer between the particle and the macroscopic diaphragm:

> Here, it must be taken into consideration that the position and the motion of the diaphragm and the shutter have so far been assumed to be accurately co-ordinated with the space-time reference frame. This assumption implies, in the description of the state of these bodies, an essential latitude as to their momentum and energy which need not, of course, noticeably affect the velocities, if the diaphragm and the shutter are sufficiently heavy. (Bohr 1949, 215)

The essential point to note here is that Bohr mentions the quantum and classical descriptions of the diaphragm and shutter as *jointly* applicable. Moreover, the classical description, in terms of a well-defined velocity, is clearly conceived of as approximate (the latitude required by quantum theory will not *noticeably* affect the definite value of the velocity if the bodies are macroscopic in the sense of sufficiently heavy). The same sort of remarks can be found throughout Bohr's analyses of Einstein's thought experiments.

It is true that Bohr (1949, 221–2) also says that we should distinguish "between the *objects* under investigation and the *measuring instruments* which serve to define, in classical terms, the conditions under which the phenomena appear" (emphases in the original). But he follows this up with the declaration that

> all unambiguous use of space-time concepts in the description of atomic phenomena is confined to the recording of observations which refer to marks on a photographic plate or to similar practically irreversible amplification effects like the building of a water drop around an ion in a cloud-chamber. Although, of course, the existence of the quantum of action is ultimately responsible for the properties of the materials of which the measuring instruments are built and on which the functioning of the recording devices depends, this circumstance is not relevant

for the problems of the adequacy and completeness of the quantum-mechanical description in its aspects here discussed.

This shows that Bohr considered it a matter of course that from an ontological point of view macroscopic objects are basically quantum mechanical. He also identifies, in the just-quoted passage, what he sees as the conditions that have to be fulfilled, *from the viewpoint of the quantum description*, to make classical concepts applicable: effects must be amplified to the macroscopic level and must become practically irreversible. The conclusion must be that for Bohr the necessity of using classical concepts has a purely *epistemic* status: it has to do with our access to the world, by means of macroscopic devices that are described by common language (extended by classical physics). It certainly does not correspond to any ontological dividing line between quantum and classical.

This conclusion is decisively supported by Bohr's *Dialectica* paper, in which he summarizes the situation with comparatively great clarity. Bohr (1948, 315–16) explains:

> Incidentally, it may be remarked that the construction and the functioning of all apparatus like diaphragms and shutters, serving to define geometry and timing of the experimental arrangements, or photographic plates used for recording the localization of atomic objects, will depend on properties of materials which are themselves essentially determined by the quantum of action. Still, this circumstance is irrelevant for the study of simple atomic phenomena where, in the specification of the experimental conditions, we may to a very high degree of approximation disregard the molecular constitution of the measuring instruments. If only the instruments are sufficiently heavy compared with the atomic objects under investigation, we can in particular neglect the requirements of relation (3) [i.e., the uncertainty relation] as regards the control of the localization in space and time of the single pieces of apparatus relative to each other.

Once again, it should be noted that Bohr here speaks about *approximations*, and *neglecting* the quantum character of the macroscopic devices. Obviously, one can only neglect something if it actually exists. It follows that according to Bohr the classical descriptions of devices must in principle be applicable at the same time as their quantum descriptions.

A difference between the two descriptions is that our immediate epistemic access is restricted to the classical one, and that we therefore need this classical description to give physical content to the quantum formalism at all. When we thus use classical language to endow the quantum formulas with physical content, it turns out that we cannot transfer the usual classical pictures to the micro world: even when interpreted with the help of classical concepts, the formalism of quantum mechanics remains non-visualizable in the general case of latitudes in position and momentum that cannot be neglected. This is the background of complementarity, to be discussed in the next section.

We therefore side with Landsman (2006, 2007), who briefly remarks that the Bohrian privileged status of classical concepts should be considered as merely epistemic; and also with Camilleri and Schlosshauer (2015) who argue that there is no conflict between Bohr's doctrine of classical concepts and the present-day "decoherence research programme" according to which classical descriptions emerge as very accurate approximations when quantum mechanics is applied to open systems (like the objects of everyday experience).

Our conclusion also bears affinity to what Howard (1994) proposes, but there are differences of nuance. Howard argues that Bohr demanded a classical description only of a *part* of macroscopic measuring devices, and that a quantum description would be possible for the remaining properties; and, second, that the use of classical concepts can be understood from a more fundamental requirement of "objectivity." The first thesis suggests (perhaps unintentionally) that the classical part of a device cannot be given a quantum description at the same time; this would be false, as we think to have demonstrated. It *is* true, of course, that to give a quantum description of a device we need the physically interpreted formalism, and therefore some other part of the world to which we apply classical concepts unrestrictedly.

Howard's second thesis, about an underlying philosophical requirement ("objectivity") that should justify the privileged status of classical concepts, in our opinion imputes too much in the way of philosophical motivation and approach to Bohr. In our portrayal, Bohr should first of all be seen as a physicist, motivated by concrete physical problems of his days. That we need classical language is from this viewpoint a simple bare fact, given the actual physical situation we find ourselves in; it is not something in need of philosophical justification. According to Bohr the strangeness of the situation, and what has to be explained, is not that we can apply classical concepts, but rather that classical descriptions cannot be applied across the board. Quantum features of the world manifest themselves in the results of experiments in our laboratories and these results can evidently be described in classical language. So how can we understand that this basic classicality does not extend to the micro world, as demonstrated by wave particle duality (in the Compton effect, for example) and similar phenomena? The task Bohr saw before him was to incorporate this unmistakable micro non-classicality into one consistent whole together with the equally undeniable classicality of our macro descriptions.

According to this analysis Bohr viewed the world as fundamentally quantum mechanical: we live in a quantum world. This contradicts a recent construal of Bohr's ideas by Zinkernagel, who argues that Bohr was *not* a "quantum fundamentalist." According to Zinkernagel (2015, 2016), Bohr held that although all systems can be treated quantum mechanically, they cannot all be treated this way at the same time.

This in itself is not in conflict with what we said above about the epistemic necessity of describing *some* systems classically in order to have a pragmatic starting point for the treatment of other systems. But Zinkernagel (2015) rejects this epistemic interpretation, and puts an ontological spin on Bohr's statements. He objects to the idea that Bohr's requirement of a classical description of measuring devices is a pragmatic or epistemological demand and also rejects the view that according to Bohr the device is really a quantum mechanical system. For example, he writes:

A way to understand Bohr's requirement is that we need a reference frame to make sense of, say, the position of an electron (in order to establish with respect to *what* an electron has a position). And, by definition, a reference frame has a well-defined position and state of motion (momentum). Thus the reference frame is not subject to any Heisenberg uncertainty, and it *is* in this sense (and in this context) classical. This does not exclude that any given reference system could itself be treated quantum mechanically, but we would then need another—classically described—reference system e.g. to ascribe position (or uncertainty in position) to the former. I think Bohr's view can be summarized in the claim that any system may in principle be seen as and treated quantum mechanically, but not all systems can be seen as and treated in this way simultaneously. This reading of Bohr's viewpoint might be called restricted quantum universalism. (430)

But as we have seen in our earlier quotations of Bohr, Bohr does *not* maintain that a classical system *is* not subject to Heisenberg uncertainty; quite the opposite, Bohr claims that it is essential for consistency that this uncertainty exists in principle even in macroscopic objects, in the form of very small latitudes, although it is true that we should necessarily discard these latitudes in a classical *description* of the object (in which they make no sense). That we need another, classically described, system as soon as we wish to treat a macroscopic object quantum mechanically is true, but as we have seen this can be given a natural epistemological reading. Zinkernagel's ontological interpretation is therefore not well founded in what Bohr writes. What is more, the precise ontological content of his proposed "restricted quantum universalism" is obscure. It is completely intelligible that descriptions of physical systems can be context dependent; this is an epistemological point of whose validity we can easily become convinced even without any appeal to quantum mechanics. But what can it mean to say that the *ontological status* of an object is context dependent; that in one context it *is* classical, and in another context it *is* quantum mechanical, even though these two "modes of being" exclude each other? A systematic elucidation and elaboration of this idea would require a drastic change in basic ideas about ontology and it seems implausible to impute such intentions to Bohr, who hardly makes any explicit reference to ontology at all (whereas epistemological claims abound in his work).

4. Measurement and complementarity

From the very start of quantum theory it has been a leitmotiv that interactions between quantum systems have to be treated in a way that is different from what we are used to in classical physics—quantum theory can be said to have begun with the suggestion that interactions can only take place in the form of exchanges of finite portions of energy and momentum. This basic idea can be traced from Planck's derivation of the black body radiation law and Einstein's 1905 light quantum hypothesis, via Bohr's "old quantum theory" of 1913, to the research in which Bohr was engaged in the 1920s. This discreteness of interactions developed in Bohr's thought from a simple discontinuity in energy exchanges between well-defined physical systems to a fundamentally unanalyzable aspect of interactions and then to the notion that interacting quantum systems are so intertwined with each other that they form a whole that cannot be thought of as consisting of subsystems with their own states. As Bohr puts it in his 1948 concise position statement in *Dialectica*: "a wholly new situation in physical science was created through the discovery of the universal quantum of action, which revealed an elementary feature of 'individuality' of atomic processes far beyond the old doctrine of the limited divisibility of matter originally introduced as a foundation for a causal explanation of the specific properties of material substances" (313).

It is important to note that Bohr uses the term "individuality" in this passage (and in many similar ones) in a rather idiosyncratic way: although it does carry the standard meaning of referring to the individual, own nature of each single process or situation, the term equally refers to the *indivisibility* and the *unanalyzable character* of the whole (in accordance with the original meaning of "individuality"). This has not always been understood in the literature, with the result that some of Bohr's explanations have seemed mysterious.

Discreteness in interactions by itself obviously does not necessitate a renunciation of divisibility or analyzability. One can easily imagine a completely classical theory in terms of well-defined subsystems in which only discontinuous and finite energy exchanges occur, and even a classical theory in which such exchanges can be said to be "uncontrollable" is easy to conceive. To have an example of the latter, one might think of a scheme in which interactions follow purely probabilistic laws—this sacrifices the deterministic character of classical physics, but need not conflict with the classical principle that each of the parties in the interaction possesses its own independent state. In fact, a theory along precisely these lines was proposed by Bohr, Kramers, and Slater in 1924. The experimental disproof of this "BKS theory" appears to have played a major role in Bohr's conclusion that a picture in which interacting quantum particles each possess their own description cannot be valid (Howard 2005).[8]

In the 1927 Como lecture Bohr (1928, 580) states this conclusion right at the start as the very essence of the considerations that are to come: "Notwithstanding the difficulties which are involved in the formulation of the quantum theory, it seems, as we shall see, that its essence may be expressed in the so-called quantum postulate, which attributes to any atomic process an essential discontinuity, or rather individuality, completely foreign to the classical theories and symbolised by Planck's quantum of action." It is true that Bohr's formulation here (like in other places) is not as transparent as one would wish; but if one considers it in context and pays attention to Bohr's idiosyncratic use of "individuality," there is no doubt that Bohr is referring to the holistic nature of interacting systems. Any remaining hesitation on this point is removed when we read the further explanation that follows a few lines after the just-given quotation: "Now the quantum postulate implies that any observation of atomic phenomena will involve an interaction with the agency of observation not to be neglected. *Accordingly, an independent reality in the ordinary physical sense can neither be ascribed to the phenomena nor to the agencies of observation*"[9] (580; emphasis added).

In this quotation the mention of an *interaction* that should not be neglected may seem to *presuppose* separate entities with well-defined states—states that are subsequently disturbed by the interaction. This is obviously not intended: the second sentence in the quotation denies precisely this existence of independent states. Indeed, in several places in his writings Bohr warns us that when he uses phrases like "disturbing phenomena by observation," this should not be taken literally (Bohr 1948, 315; 1949, 237). So the "quantum postulate" should not be understood as the statement that energy is exchanged in discrete packets, and that there is a random disturbance of prior existing properties by this exchange, but rather as a metaphorical expression of the fact that objects entering into an interaction come to form one whole ("as *symbolized* by the finite value of *h*," as Bohr often puts it).

Bohr's terminology certainly is apt to confuse here, because it makes use of expressions that already possessed a meaning but now acquire a new, metaphorical sense within the new conceptual framework. Bohr's global approach and way of arguing are very characteristic here: he comes to the subject as a physicist, adapting his intuitions on the basis of new empirical facts, and wrestles to make these new intuitions clear by couching them in familiar terms. This is completely different from the attitude of a philosopher starting from a desire to hygienically regiment language.

In Bohr's (1935, 697) reply to EPR we again read:

> Indeed the *finite interaction between object and measuring agencies* conditioned by the very existence of the quantum of action entails—because of the impossibility of controlling the reaction of the object on the measuring instruments if these are to serve their purpose—the necessity of a final renunciation of the classical ideal of causality and a radical revision of our attitude towards the problem of physical reality. (Emphasis in the original)

No wonder that this passage has led to misunderstandings, as it appears to deal explicitly with discrete interactions and disturbances. However, such a "common-sense reading" would make it completely mysterious why these finite interactions would "entail the necessity of a final renunciation of the classical ideal of causality and a radical revision towards the problem of physical reality"—a theme already familiar from the Como lecture. The mystery is solved when a little bit further on the same page we find: "The impossibility of a closer analysis of the reactions between the particle and the measuring instrument ... is an essential property of any arrangement suited to the study of the phenomena of the type concerned, where we have to do with a feature of individuality completely foreign to classical physics." This brings us back to wholeness and individuality (in the sense of indivisibility and "the impossibility of a closer analysis"), which is clearly what Bohr has in mind even when he uses expressions like "finite interactions between object and measuring device."

Note how easily Bohr switches, in the just-given pair of quotations, from "the impossibility of controlling the reaction of the object on the measuring instruments" to "the impossibility of a closer analysis of the reactions between the particle and the measuring instrument." The first expression, when taken literally, does not indicate at all that a drastic revision of classical concepts is necessary: the impossibility of *controlling* an interaction does not imply anything about the status of the various parties in the interaction. However, the impossibility of any further theoretical *analysis* may very well have such conceptual consequences. For Bohr, however, there seems to be no significant difference between the two expressions, which reinforces our earlier diagnosis: Bohr speaks as a physicist, using intuitive common sense language, without being too anxious about conceptual hairsplitting.

There can be no doubt that the metaphorical reading of "uncontrollable disturbance" is the correct one. It makes sense in all of Bohr's publications on the interpretation of quantum mechanics and, moreover, it plays a pivotal role in grounding the doctrine of "complementarity." For example, in the Einstein volume Bohr (1949, 222) reviews once again experiments that illustrate the necessary "renunciation of the visualization of atomic phenomena imposed on us by *the impossibility of their subdivision*" (emphasis added), and draws the conclusion that "arguments of this kind which recall the impossibility of subdividing quantum phenomena ... reveal the ambiguity in ascribing customary physical attributes to atomic objects." He then goes on to point out that an *unambiguous* use of space-time concepts is only possible when classical concepts become applicable to one of the parties in the interaction, for example, when marks have formed on a photographic plate in a practically irreversible amplification process. This then forges a connection to the doctrine of classical concepts and leads to the famous principle that definite properties can only be ascribed to atomic objects in a classically describable measurement context.

In his reply to EPR Bohr (1935) emphasizes that the ambiguity in ascribing physical properties to atomic systems is not a matter of insufficient knowledge. Rather, "we have not merely to do with an ignorance of the value of certain physical quantities, but with the impossibility of *defining* these quantities in an unambiguous way" (699; emphasis added). This lack of definiteness is in principle also there in the case of the description of the measuring device itself: it is subject to the uncertainty relations as well, as we shall further discuss below. But when macroscopic devices are at stake the smallness of the "latitudes" comes to the rescue (as we have seen in Section 3): this makes them negligible and allows "a purely classical account of the measuring apparatus" (698). So in the case of these measuring devices and other macroscopic objects it is consistent that we make epistemic contact with quantum reality by means of ordinary perception and classical concepts. Measuring devices can be described by means of values of classical quantities and by extension the measured atomic system, whose properties are correlated with the properties of the measuring device (pointer positions, dark spots on a photographic plate, and so on) becomes unambiguously describable as well.

We thus have returned to the theme of the indispensability of classical concepts of the previous section: measuring devices, like all macroscopic objects around us, can and must be described classically. It is an immediate consequence of this that measurements necessarily have only one single outcome. Pointers can only have one position at a time, a light flashes or does not flash, and so on—this is all inherent in the uniqueness of the classical description. Because of this, Bohr's interpretation does not face the "measurement problem" in the form in which it is often posed in the foundational literature, namely, as the problem of how to explain—in the face of the presence of superpositions in the mathematical formalism—that there is only one outcome realized each time we run an experiment. For Bohr this is not something to be explained, but rather something that is given and has to be assumed to start with. It is a primitive datum, in the same sense that the applicability of classical language to our everyday world (see Section 3) is a brute fact to which the interpretation of quantum mechanics necessarily has to conform. An interpretation that would predict that pointers can have more than one position, that a cat can be both dead and alive, and so on, would be a nonstarter from Bohr's point of view. So the measurement problem in its usual form does not exist; it is dissolved. This does not mean that we cannot or should not inquire how this "uniqueness of the classically describable world" fits in with the formalism of quantum theory, of course. We shall consider this question of the relation with the formalism in the next section.

For now, we want to return to another central point in Bohr's interpretation, namely, the issue of how the classical description of the measuring device relates to complementarity, that is, the doctrine that concepts like "position" and "momentum" cannot be applicable at the same time. If measuring devices were purely classical, so that the classical description applied in all exactness, there would be no reason to assume that there is an incompatibility between certain concepts: it is a hallmark of classicality that all physical quantities are well-defined and possess sharp values simultaneously (in mathematical terms: that the quantities form a Boolean algebra). If this were strictly true of the measuring device, there would be no problem of principle in attributing sharp values also to all quantities of correlated micro systems.

We have already seen in Section 3 that Bohr denied that the classical descriptions on the macro level have a fundamental ontological import; macro systems are basically quantum mechanical. Still, in a famous part of his report about his discussions with Einstein, Bohr (1949, 219–23) may appear to suggest that a completely *classical* account of measuring devices suffices to explain that it is impossible to assign both position and momentum to the measured quantum systems, as the experimental set-up needed to measure one of these quantities is incompatible with the set-up necessary to measure the other. Thus, in order to measure position, one needs to fix all relative positions in the device by means of rigid connections, and the whole assembly should be bolted firmly to a common support. As Bohr comments, "in such an arrangement … it is obviously impossible to control the momentum exchanged between the particle and the separate parts of the apparatus." By contrast, if we want to experimentally study momentum exchange, "certain parts of the whole device must naturally be given the freedom to move independently of others." These statements (famously illustrated by drawings that plastically exhibit the mutual exclusiveness of motion and stationarity) might create the impression that Bohr is contending that even on purely classical grounds one cannot simultaneously fix both momentum and position of a micro object. But this interpretation would be mistaken.

If the classical description applied with full precision, even the momentum absorbed by a device firmly bolted to the ground (in a laboratory on Earth) would be well-defined and could in principle be determined. Certainly, one would need a measuring procedure of fantastic accuracy for this purpose—but this is only a practical objection. By the same token, even the fastest moving classical object possesses at all times a well-defined position. Bohr is perfectly aware of this and leaves no doubt, in the passages in question, that it is only the basic *quantum* nature of the measuring devices that is responsible for complementarity.

For the case of a moving diaphragm with a slit in it, this means that "there will always be, in conformity with the indeterminacy principle, a reciprocal relationship between our knowledge of the position of the slit and the accuracy of the momentum control."[10] Similarly, the impossibility of momentum control in the case of a stationary device is not a consequence of its classical stationarity, but is due to its quantum character. A very small latitude in position implies a very large quantum latitude (indeterminacy, "uncertainty") in momentum and it is only this that makes it impossible to control the exchanged momentum. The bolting to the ground (which is equivalent to giving the device a huge mass) only serves to guarantee that the latitude in velocity associated with the large latitude in momentum can be completely neglected, so that we have a truly stationary device, as described in classical terms.

In other words, the doctrine of complementarity in its relation to measurement contexts only makes sense if it is acknowledged that measuring devices, like macroscopic objects in general, are in principle subject to quantum mechanics, even if it is true that we need classical concepts to have epistemological access to them. This reiterates our earlier point that the use of common language, extended by classical physics, is an epistemological maneuver that does not imply any ontic divide.

That the structure of the quantum formalism, in spite of its "symbolic character," was a decisive factor in the formation of Bohr's thoughts about complementarity is something we have already seen in Section 2. In the Como lecture Bohr used the superposition principle and de Broglie wave packets to argue for the uncertainty relations, conceived as limitations on sharp definition, and based complementarity on this. In his reply to Einstein, Podolsky, and Rosen, in the Einstein volume, and in his *Dialectica* article Bohr explicitly applies the uncertainty relations to macroscopic objects in order to justify complementarity. As he puts it in the EPR paper:

In the phenomena concerned we are not dealing with an incomplete description characterized by the arbitrary picking out of different elements of physical reality at the cost of sacrificing other such elements, but with a rational discrimination between essentially different experimental arrangements and procedures which are suited either for an unambiguous use of the idea of space location, or for a legitimate application of the conservation theorem of momentum.... In fact, the renunciation in each experimental arrangement of the one or the other of two aspects of the description of physical phenomena—the combination of which characterizes the method of classical physics, and which therefore in this sense may be considered as *complementary* to each other—depends essentially on the impossibility, in the field of quantum theory, of accurately controlling the reaction of the object on the measuring instruments, i.e., the transfer of momentum in the case of position measurements, and the displacement in case of momentum measurements. Just in this last respect any comparison between quantum mechanics and statistical mechanics—however useful it may be for the formal representation of the theory—is essentially irrelevant. Indeed we have in each experimental arrangement suited for the study of proper quantum phenomena not merely to do with an ignorance of the value of certain physical quantities, but with the impossibility of defining these quantities in an unambiguous way. (Bohr 1935, 699; emphasis in the original)

This may not be the most transparent and elegant way to explain the situation, but still, all the important elements are here. Whether a given set-up is appropriate for fixing position or momentum of a micro object is not a choice up to us, but is decided by whether the instantiated quantum properties are either close enough to classical position or to momentum. According to quantum theory, these two possibilities are mutually exclusive. This is not due to some epistemic conundrum, but mirrors a feature of quantum reality: position and momentum cannot be sharply defined at the same time, not even for macro objects.

As we have seen in the beginning of this section, the "uncontrollable disturbance" and "individuality" that Bohr frequently mentions as being a necessary part of any interaction, in particular a measurement, are meant to refer to the holistic nature of the combined interacting system. Accordingly, Bohr says, in a measurement the "micro part" does not possess its own independent state. So the situation is not like the one in classical mechanics in which a particle can always—whether in a measurement situation or not—be characterized by its position and momentum values. This means that we can only attribute position or momentum to the micro system relative to the actual measurement context in which it finds itself. This then is the essential ingredient of the complementarity doctrine: a micro system, for example, an electron, can only be assigned a position or a momentum relative to a concrete context.

Given that even macroscopic objects—like measuring devices—are basically quantum mechanical, consistency requires that their properties should in principle be contextual as well. That implies that also their properties depend on the measurement context. Bohr does not discuss this in depth: he only remarks, as we have seen, that if objects are sufficiently big and heavy quantum effects can be neglected for all practical purposes; and that we should only consider the remaining very small latitudes in their classical properties if consistency requirements force us to treat macroscopic objects as explicitly quantum mechanical. As we shall see, more can be said about this and similar issues if we take account of the quantum formalism in a more detailed way than we have done, in the footsteps of Bohr, so far.

5. Bohr's interpretation and the formalism

We can conclude from our discussion up to now that the formal structure of quantum mechanics played a role in Bohr's interpretation that is more important than standard stories tell us. True, Bohr saw the formalism as "symbolic"; by this he meant that a visualizable classical interpretation was not possible. But he did not at all think that the formalism should be considered a mathematical calculus without physical content. Quite the opposite: the structure of the formalism gave Bohr the essential motivation for his doctrine of complementarity, via the uncertainty relations and the canonical commutation relations. Complementarity entails an account of what the physical world is like: it implies a "relativization" of the applicability of classical concepts to the quantum world and thus tells us how the physical structure of the micro world should be characterized. Moreover, the interpretation of the mathematically defined uncertainties as "indeterminacies in definition," "latitudes," in classical quantities made it possible for

Bohr to combine the epistemologically unavoidable classical descriptions of macroscopic objects in the everyday world around us with their inherent quantum character.

It remains true that Bohr appealed only to general features of the quantum formalism: all he used was some Fourier analysis of wave packets, the operator form of p and q and the canonical commutation relations, the superposition principle for waves, and a number of qualitative statements about the role of macroscopic masses and irreversibility in going to the classical limit. In addition, Bohr commented on the importance of the fact that wave functions of many-particles systems are defined in high-dimensional n-particles configuration spaces, and that these wave functions cannot be reduced to n three-dimensional waves. As we have seen, the latter is due to the validity of the superposition principle in the configuration space, which in general makes the total wave function non-factorizable. In present-day parlance, the n-particles wave function generally is an entangled state. It is now recognized that entanglement is one of the most characteristic features of quantum mechanics—if not *the* most characteristic feature, as famously suggested by Schrödinger (1935). Although Bohr *in effect* referred to entanglement as his most important ground for thinking that the formalism has a nonpictorial content, he did not go into further details and did not explicitly single out entanglement (or an equivalent notion) as a key mathematical feature of the formalism.

He did, however, single out the "individuality" and wholeness occurring in quantum interaction processes as an essential and revolutionary trait of the new theory. From a modern point of view this wholeness, according to which systems do not keep their own wave functions during an interaction, is a direct consequence of entanglement. When two systems—each in their own quantum state—start to interact, the Schrödinger equation (more generally: any unitary evolution equation) dictates that an entangled two-particles state is formed in which the component systems can only be described by "mixtures" (obtained by "partial tracing") instead of pure states (wave functions). So one is tempted to relate Bohr's emphasis on the vital importance of "individuality" in the interpretation of quantum mechanics to the pivotal role of entanglement in modern, more formal considerations. In particular, when Bohr speaks about the inextricable whole formed by measuring device and quantum system (vital for complementarity), this seems translatable into statements about the combined entangled state of device and object system that results from the interaction between them.[11] So one could argue, as Howard suggests, that Bohr was familiar with entanglement and recognized its importance in the formalism (Howard 1994, 2004, 2005).

There can be no doubt that Bohr indeed knew about consequences of entanglement, for example, from his discussion with Einstein, Podolsky and Rosen, and had even seen its formal representation in von Neumann's (1932) theory of measurement. Still, it seems less than certain that he actually thought about his "individuality" along lines suggested by the formal treatment of entangled states à la von Neumann. Bohr's way of speaking about "individuality" and wholeness always kept a flavor of the old quantum theory, appealing as it did to an interaction that cannot be analyzed and therefore blurs the distinction between the partners in the interaction. It is true that Bohr came to use this terminology in a metaphorical way, but he never made a transition to an explicit discussion of formal aspects of entanglement in this context. We must of course realize that nobody did so at the time, so the question of whether Bohr was really aware of the significance of entanglement may be too anachronistic to deserve a debate.

However this may be, I think it is safe to say that Bohr's "individuality" *actually* corresponds to entanglement in the formalism, and that by using this we may translate— or "reconstruct," as Howard (1994) says—Bohr's statements in formal terms, regardless of whether Bohr himself was thinking in terms of this precise formal correspondence.

With the help of this translation of "individuality" into "entanglement" a formal picture of Bohr's account of quantum mechanics can be constructed. To start, let us translate into mathematical terms Bohr's claim that two interacting quantum systems generally lose their own states and become part of one "individual" whole. In the Hilbert space formalism this is appropriately captured by the standard von Neumann form of a unitary interaction:

$$\left|S_0^1\right\rangle \otimes \left|S_0^2\right\rangle \rightarrow \sum_i c_i \left|S_i^1\right\rangle \otimes \left|S_i^2\right\rangle, \tag{1}$$

in which $\left|S_0^1\right\rangle$ and $\left|S_0^2\right\rangle$ are the initial states of system 1 and system 2, respectively, and in which the arrow represents the unitary evolution that results from the interaction. The right-hand side of eqn (1) is an entangled state, which for convenience has here been written in its Schmidt form (so that the states $\left|S_i^1\right\rangle$ and also $\left|S_i^2\right\rangle$ are mutually orthogonal for values of i that differ from each other). In this entangled state no single (pure) state can be assigned to either system 1 or system 2; the best one can do is to extract, by "partial tracing," a density operator for each of the two systems: $\sum_i \left|c_i\right|^2 \left|S_i\right\rangle\left\langle S_i\right|$ (for ease of notation the index 1 or 2 has been suppressed).

This then is the mathematical Hilbert space translation of Bohr's statement that the separate systems do not possess their own wave functions when interactions are going on; and of the claim that the whole system displays an "individuality" typical of quantum mechanics—"individuality" being represented by entanglement.

The interesting special case of course is the one in which we are dealing with a *measurement*, in the sense of an interaction between a quantum system and a macroscopic system in which an irreversible recording of an outcome takes place. As we have seen, this is the situation used by Bohr to inject physical content into the formalism: macroscopic devices can be characterized by sharp values of position, momentum, and other classical quantities and this characterization gives us epistemic access to the quantum world. In order to formalize this thought, we have to look for representations of interactions in which at least one of the systems can be described classically; that is, it should be represented in the formalism with quantum "latitudes" that are so small that the attribution of sharp classical values becomes possible without practical contradictions.

Bohr himself gives us the hint that we should think of big and heavy systems that are able to amplify micro differences in an irreversible way, so that well-defined measurement results arise. This accords well with the present-day consensus that *decoherence* processes are vital for understanding the classical limit of quantum mechanics. The core principle of decoherence is that systems which are sufficiently big will in ordinary circumstances be in continuous interaction with their environment, and that this interaction will wash out the quantum aspects of the behavior of the system. Typical interactions in our macroscopic environment (primarily electromagnetic interactions) are sensitive to the *positions* of the interacting objects and this has the consequence that macro objects, through the irreversible "decohering" interaction with their environment, will lose quantum coherence between different positions—which in turn means that they will behave as if they possessed a definite position at each moment. More precisely, measurement interactions involving classically describable devices have a form like:

$$|S_0\rangle \otimes |M_0\rangle \otimes |E_0\rangle \rightarrow \sum_i c_i |S_i\rangle \otimes |M_i\rangle \otimes |E_i\rangle, \tag{2}$$

where S denotes the micro system on which a measurement is being made, M is the macroscopic measuring device, and E stands for the decohering environment that is in continual interaction with the device. The idea is that because of the interaction with the environment, the $|M_i\rangle$ Hilbert space states will be (practically) mutually orthogonal and entangled with orthogonal environment states $|E_i\rangle$—this is the decoherence effect that washes out interference between different values of i. Furthermore, the states $|M_i\rangle$ will be characterized by macroscopic pointer position values with a very small indeterminacy (i.e., $\langle M_i|(X - \overline{X})^2|M_i\rangle$ vanishingly small in comparison to $\overline{X}$, with X the relevant position operator and $\overline{X}$ its expectation value). The irreversible interaction between the measuring device and the environment E keeps these practically classical position values intact, in the sense that sharp values will not be blurred over time. Finally, the interaction between S and M should represent a measurement of some observable, for example, momentum p. That means that different positions of the almost-classical pointer (on the dial of the device) should become correlated, by the interaction, to different momentum states of the system S (so that the Hilbert space states $|S_i\rangle$ of eqn (2) in this case should be momentum eigenstates).

This account is only a rough sketch of how macroscopic measurements are modeled, with the help of decoherence, in present-day quantum mechanics; Rosaler (2016) gives a more detailed account. The essential point is that classical structures can be identified in the formal quantum representation, in the sense that device states with almost classical properties naturally occur (i.e., states with practically sharp values of position together with practically sharp values of velocity, obeying classical equations of motion with high precision). This then is the point where the formalism allows contact with Bohr's epistemological demand that we must describe the world around us in classical terms. So the (almost-classical) state $|M_i\rangle$ of the measuring device represents the situation that we describe in classical terms with a pointer position "i," and we assign the correlated value of the measured quantity to the micro system S. On the micro level the indeterminacies (latitudes) can of course not be neglected, and the quantum formalism automatically fulfils the requirements of complementarity in the form of the uncertainty relations for canonically conjugated quantities. So quantities like position and momentum can never be assigned sharp values together. The nature of the measuring device and its interaction with the system, finally resulting in a state of the form of the right-hand side of eqn 2, fixes which properties can be assigned to the object (viz., those that are definite-valued in the states $|S_i\rangle$).

We have glossed over an important point however. In eqn 2, as in all equations representing final states after unitary interactions, we do not have *one* state of the measuring device, but all these states $|M_i\rangle$ figure together in the superposition at the right-hand side of the arrow. So in the Bohrian interpretation of the formalism that we just explained we seem to have committed the error of replacing the full superposition by just one term in it. In fact, this is the standard objection against claims that decoherence solves the measurement problem. In the quantum formalism with only unitary evolution (so no collapses) we always end up with superpositions after interactions, also in the case of decoherence (as can be verified by looking at eqn 2). But this means, so the objection goes, that the measurement problem, in the form of the question why there is only *one* outcome, has not been solved—and even cannot be solved.

However, this objection misconstrues Bohr's approach to the interpretational problem, as we have already explained in Section 4. Although Bohr takes the mathematical formalism of quantum mechanics seriously, as providing us with information about the micro world, he does not *start* from the formalism in his interpretation. The formalism can only tell us about properties of micro systems once it has made contact with physical reality, and this contact can only be established via identifications between symbols and physical quantities on the macro level, described in classical language. The starting point for attributing physical meaning is thus the classical description of laboratory practice—and here experiments certainly have only one result at a time. So the uniqueness of experimental outcomes is not something to be explained, but is something accepted from the outset. The task is to find suitable expressions in the formalism that can represent these unique outcomes, and in this way to endow the mathematics with physical content.

In other words, the mistake made in the objection is the supposition that the mathematical framework, with its superpositions, *already* possesses physical meaning before the empirical interpretation has even started, and that the occurrence of superpositions of states $|M_i\rangle$ must signal the joint physical presence of all results i. In the Bohrian approach that we are outlining this is putting the cart before the horse. The measurement problem in its usual form, coming from the urge to explain how superpositions can fail to correspond to multiple outcomes, does not exist for Bohr.

In order to make this Bohrian approach correspond to the mathematical formalism, we need the following interpretational move: in measurement situations in which the final formal state is of the form $\sum_i c_i |S_i\rangle \otimes |M_i\rangle \otimes |E_i\rangle$, as in eqn 2, the corresponding physical situation is that the pointer of the device points to *one* of the possible pointer positions i; the probability that a specific i instead of one of the alternative values is realized is given by $|c_i|^2$.

This interpretational maneuver is not only Bohrian, but in accordance with a wide class of non-collapse interpretations of quantum mechanics. In particular, modal interpretations employ an explicit interpretational rule of precisely this kind (Lombardi and Dieks 2014; Bub 1997; Dieks 1989a, b); but the many-histories interpretation and the Everett interpretation are also compatible with the idea that our observable world corresponds to one branch from the superposition, selected by decoherence (Rosaler 2016). We shall briefly return to this connection with non-collapse interpretations in the next section.

The proposal to interpret the superposition in terms of one single outcome bears similarity to Howard's (1994) "reconstruction" of Bohr. Howard reads Bohr as saying that the description of the measuring device by the superposition of eqn 2 can be replaced by one of the components in the density operator of M. According to Howard, Bohr proposes that we act "as if" the right-hand side of (2) were $|S_i\rangle \otimes |M_i\rangle \otimes |E_i\rangle$. But it is misleading to introduce such an "as if." It suggests that the full superposition by itself cannot have the desired meaning, and that Bohr therefore had to have recourse to a sleight of hand. But as we have just argued, the superposition does not have an empirical meaning independently of its interpretation via classically described experiments, so no replacement by another mathematical state is needed. We just have to interpret the formulas correctly.

One might get the impression that the interpretation of entangled states of object system plus a measuring device in terms of just one measurement result is a verbal move that is equivalent to accepting the projection postulate. If this were right, the interpretation would bring in the projection postulate via the back door, and would effectively violate what we have accepted as one of Bohr's principles, namely, that projection or collapse should not be seen as a dynamical evolution process, but rather as a way of efficient bookkeeping that takes into account available information about experimental outcomes. However, there is an important difference between the interpretational principle outlined above and the projection postulate. The projection postulate tells us that the quantum mechanical state, as defined in the mathematical formalism, changes in a measurement from an entangled state to a product state, namely, one of the terms occurring in the original entangled superposition. This is a nonunitary evolution that is impossible in non-collapse interpretations. In principle, this difference in the sorts of interaction that occur in the formalism can lead to a difference in empirical predictions. The reason is that if we do not have collapses, superpositions will always maintain themselves, and this entails the possibility in principle of interference between the "branches" defined by the individual product states; whereas in the case of collapses no other branches are left with which interference can take place. It is true that this possibility of interference is remote in the circumstances of ordinary experiments, and Bohr probably never considered such exotic consequences of quantum theory. Moreover, his emphasis on the irreversibility of measurements, which is vindicated by the modern decoherence program, effectively blocks the possibility of a recombination of branches in ordinary experimental practice. Nevertheless, experimental techniques have advanced enormously since Bohr's days, and it is not excluded that effects of unitarity will become accessible to empirical tests.

Given that modern experimental techniques have created possibilities that were unheard of in Bohr's days, there may also be ways to investigate other (though related) possible effects of unitarity on the macroscopic level. Think of a hermetically closed chamber, in which a quantum measurement takes place. According to the interpretation of entangled states that we have outlined, a macroscopic device inside the chamber will register one unique result, even though according to the unitary theory the state of quantum system plus device is a superposition. However, an external observer who performs measurements on this room from the outside may be able to find results that indicate that the formal description of the chamber with its contents by the superposition is the right one (Bene and Dieks 2002; Dieks 2016).

The scenario reminds one of the predicament of Schrödinger's famous cat, which is shut up in a sealed room together with a decaying atom that is able to activate an infernal machine. After some time, the macroscopic cat system plus its microscopic environment will be formally described by an entangled superposed state. According to the Bohrian interpretation of the formalism as we have discussed it here, the cat is either dead or alive after the experiment. However, an external observer who performs measurements on the closed room, and thus has no direct access to the cat, can in principle find out that the superposed formal "dead plus alive state" is correct. For example, this superposed state may be an eigenstate of the observable measured by the external observer, and this can be borne out by the results of experiments. Experiments with such "Schrödinger-cat states" have actually been performed during the past two decades. The results show that an external observer can indeed observe the existence of the superposition, and thus verify that no collapse has taken place.

We thus see that the relational character of Bohr's interpretation must even go further than he himself emphasized, if we take its strict relation to the mathematical formalism seriously. Bohr stressed that the physical state of a system (in terms of the physical properties that can be attributed to it) only becomes definite in a well-defined experimental context. Thus, an electron can be assigned a position in the context of a position measurement, or a momentum value in the context of a measurement of momentum; and these two property assignments never apply simultaneously. But now an additional relational aspect reveals itself: for an observer inside the room with the cat (in the Schrödinger experiment) the cat is either dead or alive, but for an outside observer this need not be the case. So properties not only need "defining circumstances" to make them definite, but can also vary from one observer to another in (admittedly far from common!) circumstances of the kind we just described (Bene and Dieks 2002; Dieks 2016).

Even in such very strange situations, classical terminology remains applicable, in accordance with Bohr's claim. Thus, the internal observer describes the cat as either dead or alive, and the external observer establishes that the pointer of his measuring instrument, with which he performs his measurement on the sealed room, adopts a definite position that indicates the result. But the prediction and explanation of these measurement results cannot be given along classical lines; it needs the full (non-collapse) formalism of quantum mechanics. This illustrates Bohr's (1949, 209) earlier-quoted dictum that "however far the phenomena transcend the scope of classical physical explanation, the account of all evidence must be expressed in classical terms."

6. Conclusion

Although Bohr qualified the mathematical formalism of quantum theory as *symbolic*, he took it more seriously than usually acknowledged. He realized that physical meaning has to be injected into the formalism in order to make it a physical theory, and took the pragmatically and epistemologically well-justified position that this can only be done via contact with our world of experience, which we describe in common language enriched with terms from classical physics. However, the form of the formalism itself stands in the way of implementing a full classical interpretation across the board, extending to the micro domain. Classical pictures of micro systems are therefore impossible, according to Bohr, and this is responsible for the fact that the formalism is symbolic even after empirical interpretation, namely, in the sense that it does not support a vizualizable representation of the micro world—at least not in the sense of the pictures we normally associate with physical systems. This does not mean that according to Bohr we should opt for instrumentalism and forswear talk about electrons, photons, and atoms. Such quantum systems do exist, but their mode of being cannot be captured by describing them with *independent* properties and states. This is all made more precise by the doctrine of complementarity, as we have seen in Sections 2, 3, and 4. Significantly, structural features of the quantum formalism (superposition of waves, the canonical commutation relations) inspired Bohr decisively in this doctrine.

There can be little doubt that Bohr saw quantum mechanics as a universal theory, with an unrestricted domain of validity. It is also highly plausible that he did not think of the collapse of the wave function as a physical process, as a rival in the formalism of unitary Schrödinger time evolution. From a modern perspective, this places Bohr in the camp of non-collapse interpretations of quantum mechanics (Section 5).

The latter observation raises the (partly historical) question of why Bohr did not embrace attempts to formulate formally precise versions of non-collapse interpretations, in particular why he remained hostile to the "relative state" interpretation proposed by Everett (1957) in his dissertation. Significantly, Everett's supervisor Wheeler was convinced that Everett's work could be seen as a formal elaboration of Bohr's interpretation and started a correspondence about the subject with the Copenhagen group; Everett even visited Bohr on Wheeler's instigation in 1959 (after two earlier aborted plans to visit). But Everett's work was not received favorably.[12]

From the evidence adduced by Osnaghi, Freitas, and Freire (2009) it seems clear that Bohr's main objection was that it remained completely unsolved how Everett's scheme made contact with physical reality. Everett's universal wave function struck the Copenhagenists as a purely formal piece of mathematics, because it did not explicitly relate to an experimental context which can be described in classical terms. So the spot where the shoe pinched was in the epistemological part of Bohr's interpretation, the doctrine of the inevitability of the use of classical concepts (Section 3). Indeed, both Everett and Wheeler relativized the inevitability of classical concepts and thought of the quantum formalism as a more general picture of quantum reality (although Wheeler later retraced his steps). This appears to highlight a significant difference between Everett and Bohr: although Bohr was not an instrumentalist, his views were firmly bound up with the experimental practice of physics, and he took the concepts that derive from this practice as primordial. Accordingly, any consideration of the formalism and its meaning should be preceded by the specification of *physical concepts* that ultimately derive from the description of macroscopic experimental situations. This need not be regarded as a Kantian move: it is first and foremost a pragmatic physicist's attitude that serves to safeguard the direct link to concrete physical experience. By contrast, Everett's attitude was completely abstract. Everett did not delve into the epistemological problem of how the mathematical symbols receive their physical meaning and did not analyze physical practice. It is no wonder then that the Copenhagen camp considered Everett as someone who had dabbled at an abstract mathematical structure and naively took it for a picture of the world, without asking himself what the picture-reality relation consisted in. For them, he was someone who had not even understood what the interpretational question was about.

This intermezzo further clarifies Bohr's attitude with respect to the quantum formalism. He certainly took the formalism seriously, was not an instrumentalist and was guided by the mathematical structure of the formalism when he developed his interpretation. But physical concepts, which he took to be indispensable for providing meaning to the formalism, always remained the most important ingredient in his analyses. It remains to be seen, however, whether from a present-day point of view Bohr's objections to Everett provide weighty arguments against an interpretation of Bohr's interpretation as a non-collapse interpretation in the modern sense. As far as I can see, Bohr's doctrine of classical concepts may very well be incorporated in formal non-collapse schemes, even if these make use of the notion of a universal wave function. This, however, is better kept for another study.

Notes

1 At the end of the Como lecture Bohr puts this in the perspective of science leading us ever further away from everyday intuitions: although we make contact with the world through "ordinary perception," to which each and every word in our language is tailored, we have to adapt our theoretical interpretations of these perceptions to "our gradually deepening knowledge of the laws of Nature." Bohr adds that this same point should already have become evident from Einstein's theory of relativity.

2 The remark that the occurrence of the imaginary number i signals the symbolic nature of a formalism appears without further explanation passim in Bohr's publications. It seems therefore not completely impossible that Bohr did fall prey to a confusion between the technical mathematical and the intuitive meaning of "imaginary"—if so, this would reinforce the comment that not too much subtle philosophizing should be imputed to Bohr.

3 Thereby recognizing the *existence* of quantum objects, even if not characterizable in a detached and noncontextual way; note also that Bohr talks about "not *directly* visualizable formalism" (emphasis added), leaving it open that indirect ways of endowing the symbols with physical content may be possible.

4 In 1936 the 2nd International Congress for the Unity of Science took place in Copenhagen at Bohr's honorary residence at Carlsberg (Faye 2008). Neurath, Hempel, Popper, Philipp Frank, and other logical empiricist luminaries were in attendance, whereas Reichenbach, Schlick, and Carnap intended to be present but were finally unable to come. Bohr later complained, however, that the philosophers had not understood his ideas about complementarity.

5 Faye (2014) and Howard (1994) present a list of such "philosophical interpretations" of Bohr.

6 Zinkernagel (2015, 2016) has recently defended a more sophisticated version of a Bohrian ontological "cut," according to which it is true that each physical system can be described quantum mechanically, but not all systems can be treated quantum mechanically at the same time. We shall come back to this "anti-quantum-fundamentalism" shortly.

7 Moreover, Bohr tells us in the same passage that it is arbitrary how we divide the total system into a measured and a measuring part. This would make no sense at all if there were an ontological "cut" between a quantum and a classical realm.

8 Howard (2005) presents a detailed and illuminating survey of the physical developments that drove Bohr to his "wholeness" conclusion, and also discusses how Einstein dealt with these same developments. As Howard makes clear, it is in the early period of the history of quantum theory—1905–27—that we can identify the roots both of Bohr's complementarity, and of Einstein's later resistance to quantum mechanics on the grounds that it does not satisfy a principle of separability. It is sometimes claimed that Bohr only came to embrace "holism" after the 1935 EPR discussion, but this is in conflict with the textual evidence. The doctrine already occurs in the 1927 Como lecture.

9 By "agencies of observation" Bohr just means "measuring devices." Bohr never assigns a special role to *human* agency, apart from the capacity of human observers to decide to build one measuring apparatus rather than another

10 Bohr (1949, 220). Bohr's use of the expressions "knowledge of the position of the slit" and "accuracy of the momentum control" may suggest that there *are* in fact precise positions and momenta, although we do not *know* them; but this is not what he intends, as we have seen before and as he explains on p. 237 of the same text. It is the unsharpness of *definition* and the corresponding "latitude" that are decisive here.

11 This assumes that Bohr did not suppose that there exists a special form of interaction, resulting in a "collapse of the wave function," manifesting itself only in measurements (as opposed to ordinary physical interactions). This non-collapse assumption is eminently plausible, given that Bohr always speaks about interactions between measuring devices and objects in ordinary physical terms, such as "uncontrollable disturbance," "discrete exchange of energy and momentum," leading to the typical "individuality of the phenomenon." Moreover, as we have seen, Bohr in many places stresses the universality of quantum theory, which is hardly compatible with a special role for macroscopic measuring instruments as far as physical evolution is concerned. Finally, Bohr never mentions "collapses" in his published writings. Nevertheless, it has been part of the folklore surrounding the "Copenhagen interpretation" to maintain that Bohr's views are characterized precisely by such "collapses"—this is now generally rejected by Bohr scholars (Howard 2004; Faye 2014). Zinkernagel (2016) has recently drawn attention to a number of unpublished letters in which Bohr mentions the updating of our knowledge when measurement results become available, which seems similar to collapses. But this updating of information can easily be given an epistemic interpretation.

12 This historical episode is described in an informative article by Osnaghi, Freitas, and Freire (2009). This article contains a detailed account of the correspondence between Wheeler and various members of the Copenhagen group, which makes it clear that Bohr himself did not seriously study Everett's work and left it to his collaborators to respond. They did so in ways that were not always consistent with each other, nor with Bohr's own general views. Part of Bohr's lack of interest may have been due to Everett's unforthcoming attitude with respect to "Copenhagen"—which was repaid in kind. However this may be, a basic theme can be distilled from the few remarks by Bohr himself that have been passed down and from the diverse arguments coming from Bohr's collaborators. This theme is well described by Osnaghi, Freitas, and Freire (2009), even though these authors place their findings against the mistaken backdrop of Bohr's supposed instrumentalism.

References

Beller, M. 1999. *Quantum Dialogue: The Making of a Revolution*. Chicago: University of Chicago Press.

Bene, G., and D. Dieks. 2002. "A perspectival version of the modal interpretation of quantum mechanics and the origin of macroscopic behavior." *Foundations of Physics* 32: 645–71.

Bohr, N. 1928. "The quantum postulate and the recent development of atomic theory." *Nature* 121: 580–90. Reprinted in *Collected Works, Vol. 6*.

Bohr, N. 1935. "Can quantum-mechanical description of physical reality be considered complete?" *Physical Review* 48: 696–702.

Bohr, N. 1948. "On the notions of causality and complementarity." *Dialectica* 2: 312–19.

Bohr, N. 1949. "Discussions with Einstein on epistemological problems in atomic physics." In *Albert Einstein: Philosopher-Scientist*, edited by P. A. Schilpp, 201–41. La Salle: Open Court. Reprinted in *Collected Works, Vol. 6*.

Bohr, N. 1985. *Collected Works, Vol. 6: Foundations of Quantum Mechanics I (1926–1932)*, edited by J. Kalckar. Amsterdam: North-Holland.

Bub, J. 1997. *Interpreting the Quantum World*. Cambridge: Cambridge University Press.

Camilleri, K., and M. Schlosshauer. 2015. "Niels Bohr as philosopher of experiment: Does decoherence theory challenge Bohr's doctrine of classical concepts?" *Studies in History and Philosophy of Modern Physics* 49: 73–83.

Cushing, J. T. 1994. *Quantum Mechanics: Historical Contingency and the Copenhagen Hegemony*. Chicago: University of Chicago Press.

Dieks, D. 1989a. "Quantum mechanics without the projection postulate and its realistic interpretation." *Foundations of Physics* 19: 1395–423.

Dieks, D. 1989b. "Resolution of the measurement problem through decoherence of the quantum state." *Physics Letters A* 142: 439–46.

Dieks, D. 2016. "Information and the quantum world." *Entropy* 18: 26. doi:10.3390/e18010026.

Drieschner, M. (ed.) 2014. *Carl Friedrich von Weizsäcker: Major Texts in Physics*. Heidelberg: Springer. A video recording of Teller recounting the story himself can be found on the website http://www.webofstories.com/play/edward.teller/32 (visited January 2016).

Everett, H. 1957. "'Relative state' formulation of quantum mechanics." *Reviews of Modern Physics* 29: 454–62.

Faye, J. 2008. "Niels Bohr and the Vienna circle." In *The Vienna Circle in the Nordic Countries*, edited by J. Mannheim and F. Stadler, 4–10. Berlin: Springer.

Faye, J. 2014. "Copenhagen interpretation of quantum mechanics." In *The Stanford Encyclopedia of Philosophy* (Summer 2014 Edition), edited by E. N. Zalta. Stanford: Metaphysics Research Lab of Stanford University.

Folse, H. 1985. *The Philosophy of Niels Bohr: The Framework of Complementarity*. Amsterdam: North-Holland.

Howard, D. 1994. "What makes a classical concept classical? Toward a reconstruction of Niels Bohr's philosophy of physics." In *Niels Bohr and Contemporary Philosophy*, edited by J. Faye and H. Folse, 201–29. Dordrecht: Kluwer Academic Publishers.

Howard, D. 2004. "Who invented the 'Copenhagen interpretation'? A study in mythology." *Philosophy of Science* 71: 669–82.

Howard, D. 2005. "Revisiting the Einstein-Bohr dialogue." Available at: https://www3.nd.edu/ dhoward1/Revisiting.

Landsman, N. P. 2006. "When champions meet: Rethinking the Bohr-Einstein debate." *Studies in History and Philosophy of Modern Physics* 37: 212–42.

Landsman, N. P. 2007. "Between classical and quantum." In *Handbook of the Philosophy of Science, Vol. 2: Philosophy of Physics*, edited by J. Earman and J. Butterfield, 417–554. Amsterdam: Elsevier.

Lombardi, O., and D. Dieks. 2014. "Modal interpretations of quantum mechanics." In *The Stanford Encyclopedia of Philosophy*, Winter 2016 Edition, edited by E. N. Zalta. https://plato.stanford.edu/archives/win2016/entries/qm-modal/.

Osnaghi, S., F. Freitas, and O. Freire Jr. 2009. "The origin of the Everettian heresy." *Studies in History and Philosophy of Modern Physics* 40: 97–123.

Rosaler, J. 2016. "Interpretation neutrality in the classical domain of quantum theory." *Studies in History and Philosophy of Modern Physics* 53: 54–72.

Schrödinger, E. 1935. "Die gegenwärtige situation in der quantenmechanik." *Die Naturwissenschaften* 23: 807–12.

Von Neumann, J. 1932. *Mathematische Grundlagen der Quantenmechanik*. Berlin: Springer.

Zinkernagel, H. 2015. "Are we living in a quantum world? Bohr and Quantum fundamentalism." In *One Hundred Years of the Bohr Atom: Proceedings from a Conference. Scientia Danica, Series M: Mathematica et physica, vol. 1*, edited by F. Aaserud and H. Kragh, 419–34. Copenhagen: Royal Danish Academy of Sciences and Letters.

Zinkernagel, H. 2016. "Niels Bohr on the wave function and the classical/quantum divide." *Studies in History and Philosophy of Modern Physics* 53: 9–19.

15

Bohrification

From Classical Concepts to Commutative Operator Algebras*

Klaas Landsman

1. Introduction

Despite considerable research (e.g., Joos et al. 2003; Schlosshauer 2007; Landsman 2007; Wallace 2012), it remains unclear how quantum mechanics—accepted as the fundamental theory of the physical world—gives rise to classical physics, or at least to the appearance thereof. The relationship between classical and quantum physics played a fundamental role in the work of Niels Bohr, who contributed two important ideas (or perhaps ideologies):

- The *correspondence principle* (Bohr 1976), which—originally in the context of atomic radiation—suggested an asymptotic relationship between classical and quantum mechanics (see also Mehra and Rechenberg 1982; Hendry 1984; Darrigol 1992)
- The *doctrine of classical concepts*, which Bohr (1949, 209) succinctly summarized as follows:

However far the phenomena transcend the scope of classical physical explanation, the account of all evidence must be expressed in classical terms.... . The argument is simply that by the word *experiment* we refer to a situation where we can tell others what we have done and what we have learned and that, therefore, the account of the experimental arrangements and of the results of the observations must be expressed in unambiguous language with suitable application of the terminology of classical physics.

In brief, the only way to look at the quantum world is through classical glasses. See also Bohr (1985, 1996), as well as Scheibe (1973), Folse (1985), Camilleri and Schlosshauer (2015), and Zinkernagel (2016), for further analysis (which we recommend but shall not review, Bohr's role in this work being that of an inspirator rather than a subject of exegesis—we will similarly not discuss (anti-)realism etc.).

The tension between these ideas was well captured by Landau and Lifshitz (1977, 3): "Thus quantum mechanics occupies a very unusual place among physical theories: it contains classical mechanics as a limiting case, yet at the same time it requires this limiting case for its own formulation." What adds to the difficulties is the need to combine conceptual and mathematical analysis, but precisely at this point the approach we advocate promises a clean separation between the two: the necessary conceptual analysis has largely been performed by Bohr, while von Neumann's theory of operator algebras—as extended to C*-algebras (Gelfand and Naimark 1943) and subsequently to noncommutative geometry (Connes 1994)—provides the required mathematical tools. What has been lacking so far is a relationship between the two, and this is what our "Bohrification" program intends to provide. Indeed, we feel that each author on his own failed to solve the foundational problems of quantum mechanics: Bohr's conceptual ideas must first be interpreted within the formalism of quantum mechanics before they can be applied to the physical world (an intermediate step Bohr himself seems to have considered of little interest, if not wholly superfluous), whereas conversely, von Neumann's brilliant mathematical work needs to be supplemented by sound conceptual moves before it, too, may be related to actual physics. Von Neumann (1932) did write down the most general form of the Born rule, which is an important link between physics and the Hilbert space formalism of quantum mechanics he developed, but this is not nearly enough; what we propose is that Bohr provided these conceptual moves.

In this respect, Bohr and von Neumann historically played asymmetric roles: whereas Bohr hardly attempted to relate his ideas to the mathematical formalism of quantum mechanics (not even to Dirac's), von Neumann (1932) famously did spend a great deal of attention to conceptual issues in quantum mechanics. Although this work was extremely influential, and von Neumann's analysis of hidden variables did set the stage for part of our Bohrification program (albeit in a contrapositive way), his work on entropy or the measurement problem has had little influence on the program proposed here. In any case, von Neumann never advocated the particular use of commutative operator algebras we now see as the basis of the relationship between his mathematics and the real world—seen through the classical glasses suggested by Bohr's doctrine. In fact, although Bohr and von Neumann tend to be jointly bracketed under the Copenhagen Interpretation—a notion to be treated with due reservation (Scheibe 1973; Faye 2002; Howard 2004; Camilleri 2009)—a synthesis of their ideas on quantum mechanics is difficult to find in earlier literature.

The central idea of Bohrification is the following interpretation of Bohr's doctrine: "In so far as their physical relevance is concerned, the noncommutative operator algebras of quantum-mechanical observables are accessible only through commutative algebras."

On this interpretation, the Bohrification program splits into two parts, distinguished by the specific relationship between a given noncommutative operator algebra A and the commutative algebras that supposedly give access to A. Either way, we assume that A is a C*-algebra: this is an associative algebra (over $\mathbb{C}$) equipped with an involution (i.e., a real-linear map $a \to a^*$ such that $a^{**} = a$, $(ab)^* = b^* a^* F$, and $(\lambda a)^* = \bar{\lambda} a^*$ for all

$a, b \in A$ and $\lambda \in \mathbb{C}$), as well as a norm in which A is complete (i.e., a Banach space), such that algebra, involution, and norm are related by the axioms

$$\| ab \| \leq \| a \| \| b \|;$$

(1)

$$\| a^* a \| = \| a \|^2 .$$

(2)

The two main classes of C*-algebras are:

- The space $C_0(X)$ of all continuous functions $f : X \to \mathbb{C}$ that vanish at infinity (i.e., for any $\varepsilon > 0$ the set $\{x \in X \mid |f(x)| \geq \varepsilon\}$ is compact), where X is some locally compact Hausdorff space, with pointwise addition and multiplication, and

$$\| f \|_\infty = \sup\{| f(x) |\};$$

(3)

$$f^*(x) = \overline{f(x)}.$$

(4)

It is of fundamental importance that $C_0(X)$ is *commutative*. Conversely, Gelfand and Naimark (1943) proved that every commutative C*-algebra is isomorphic to $C_0(X)$ for some locally compact Hausdorff space X, which is determined by A up to homeomorphism (X is called the *Gelfand spectrum* of A). Note that $C_0(X)$ has a unit (viz., the function 1_X that is equal to 1 for any x) iff X is compact.

- Norm-closed subalgebras A of the space $B(H)$ of all bounded operators on some Hilbert space H for which $a^* \in A$ iff $a \in A$; this includes the case $A = B(H)$. Here,

$$\| a \| = \sup \{\| a\psi \|, \psi \in H, \| \psi \| = 1\},$$

(5)

the algebraic operations are the natural ones, and the involution is the adjoint (i.e., hermitian conjugate). If $\dim(H) > 1$, $B(H)$ is a *noncommutative* C*-algebra.
An important special case is the C*-algebra $B_0(H)$ of all *compact* operators on H, which fails to have a unit whenever H is infinite-dimensional (whereas $B(H)$ is always unital). In their fundamental paper, Gelfand and Naimark (1943) also proved that every C*-algebra is isomorphic to $A \subseteq B(H)$ for some Hilbert space X.

These two classes are related as follows: in the commutative case $A = C_0(X)$ one may take $H = L^2(X, \mu)$ where the measure μ is supported by all of X, on which $C_0(X)$ acts by multiplication operators, that is, $m_f \psi = f\psi$, where $f \in C_0(X)$ and $\psi \in L^2(X, \mu)$.

C*-algebras were introduced by Gelfand and Naimark (1943), generalizing the rings of operators studied by von Neumann during the period 1929–49, partly in collaboration with Murray (von Neumann 1929, 1931, 1938, 1940, 1949; Murray and von Neumann 1936, 1937, 1943). These rings are now aptly called *von Neumann algebras*, and arise as the special case where a C*-algebra $A \subset B(H)$ satisfies

$$A = A'',$$

(6)

in which for any subset $S \subset B(H)$ we define the *commutant* S' of S by

$$S' = \{a \in B(H) \,|\, ab = ba \,\forall\, b \in S\}, \tag{7}$$

and in terms of which the *bicommutant* of S is defined as $S'' = (S')'$. A *state* on a C*-algebra A is a linear map $\omega : A \to \mathbb{C}$ that is *positive* in that

$$\omega(a^* a) \geq 0 \tag{8}$$

for each $a \in A$, and *normalized* in that, first noting that positivity implies boundedness,

$$\| \omega \| = 1, \tag{9}$$

where $\| \cdot \|$ is the usual norm on the Banach dual A^*. If A has a unit 1_A, then (8) is equivalent to $\| \omega \| = \omega(1_A)$, so that ω is a state iff (8) holds together with

$$\omega(1_A) = 1. \tag{10}$$

The Riesz–Radon representation theorem in measure theory gives a bijective correspondence between states ω on $A = C_0(X)$ and probability measures μ on X, given by

$$\omega(f) = \int_X d\mu \, f, \quad f \in C(X). \tag{11}$$

If $A = B(H)$, then any density operator ρ (i.e., $\rho \geq 0$ and $\mathrm{Tr}(\rho) = 1$) gives a state

$$\omega(a) = \mathrm{Tr}(\rho a), \tag{12}$$

on $B(H)$, but if H is infinite-dimensional there are interesting other states, too, cf. Sections 2 and 3.

A state ω on some C*-algebra A is called *pure* if it has no decomposition

$$\omega = t\omega_1 + (1-t)\omega_2, \tag{13}$$

where $t \in (0,1)$ and $\omega_1 \neq \omega_2$ are states on A (note that the set of all states is convex). The pure states on $A = C_0(X)$ are given by the evaluation maps

$$\omega_x(f) = f(x), \tag{14}$$

and hence correspond to points $x \in X$. For any Hilbert space H, the *normal* pure states on $B(H)$ are the states (called *vector states*) defined in terms of a unit vector $\psi \in H$ by

$$\omega_\psi(b) = \langle \psi, b\psi \rangle. \tag{15}$$

See Kadison and Ringrose (1983, 1986) for a first introduction to operator algebras.

The use of C*-algebra and von Neumann algebras in mathematical physics has been widespread for over half a century now, typically in the context of quantum systems with infinitely many degrees of freedom (Bratteli and Robinson 1981; Haag 1992). More recently, it has become clear that they also form an effective tool when other limiting operations play a role, such as $\hbar \to 0$, where $\hbar$ is Planck's constant (Rieffel 1989, 1994; Landsman 1998). Returning to our main topic of Bohrification, both of these applications of C*-algebras have now found a common conceptual umbrella:

- *Exact Bohrification* studies commutative C*-algebras contained in A, assembled into a poset (i.e., partially ordered set) $\mathcal{C}(A)$ whose partial ordering is given by set-theoretic inclusion (for technical reasons we assume that A has a unit 1_A and that each commutative C*-subalgebra $C \in \mathcal{C}(A)$ of A contains 1_A). Applications to the foundations of quantum mechanics so far include the Born rule (for single experiments), the Kochen–Specker Theorem, and (intuitionistic) quantum logic. The topos-theoretic approach to quantum mechanics (which incorporates the last two themes just mentioned) is heavily based on $\mathcal{C}(A)$, and also, in a purely mathematical setting, $\mathcal{C}(A)$ turns out to be a useful new tool for the classification of C*-algebras.
- *Asymptotic Bohrification* looks at commutative C*-algebras asymptotically included in A by a deformation procedure involving so-called continuous fields of C*-algebras (see appendix). As we shall see, this covers the Born rule (in its frequency interpretation), the measurement problem (which we relate to the classical limit of quantum mechanics), and the closely related issue of spontaneous symmetry breaking. All of this may be bracketed under the general theme of emergence (Landsman 2013).

The conceptual roots of the exact Bohrification program lie in Bohr's doctrine of classical concepts, but the mathematical interpretation of this doctrine as proposed here starts with von Neumann's (1932) proof that (in current parlance) noncontextual linear hidden variables compatible with quantum mechanics do not exist; see Caruana (1995) for historical context. Von Neumann's assumption of linearity of the hidden variable on all operators was criticized by Hermann (1935), whose work was all but ignored, and subsequently by Bell (1966), whose paper became well known. Both Bell (1966) and Kochen and Specker (1967) then obtained a similar result to von Neumann's on the assumption of linearity merely on *commuting* operators. Furthermore, responding to different ideas in von Neumann (1932), the mathematician Mackey (1957) introduced probability measures μ on the closed subspaces of a Hilbert space, based on an additivity assumption on mutually orthogonal subspaces. These formed the basis of his colleague Gleason's (1957) deservedly famous characterization of such measures in terms of density operators.

As we shall explain in Section 2, *commutative* C*-subalgebras of $B(H)$ form an appropriate context for Mackey, Gleason, and Kochen–Specker; this shared background is no accident, since the Kochen–Specker Theorem is a simple corollary of Gleason's. Moreover, Kadison and Singer (1959) studied pure states on *maximal* commutative C*-subalgebras of $B(H)$, launching what became known as the *Kadison–Singer conjecture*, which we also regard as a founding influence on the (exact) Bohrification program.

Two decades later, the reinterpretation of the Kochen–Specker Theorem in the language of topos theory led Isham and Butterfield (1998), Hamilton, Isham, and Butterfield (2000), and Döring and Isham (2010) to a poset similar to $\mathcal{C}(A)$ in the setting of von Neumann algebras (though with the opposite ordering). Finally, $\mathcal{C}(A)$ was introduced by Heunen, Landsman, and Spitters (2009), again in the context of topos theory. It soon became clear that even apart from the latter context $\mathcal{C}(A)$ is an interesting object on its own, especially as a tool for analyzing C*-algebras.

Asymptotic Bohrification conceptually originated in Bohr's correspondence principle, which originally related the statistical results of quantum theory to the classical theory of radiation in the limit of large quantum numbers (Bohr 1976), but was later generalized to the idea that classical physics should arise from quantum physics in some appropriate limit (Bokulich 2008). Mathematically, the correspondence principle eventually gave rise to the field of *semiclassical analysis* (Ivrii 1998; Martinez 2002), and, more importantly for the present chapter, to a new approach to quantization theory developed in the 1970s under the name of *deformation quantization* (Berezin 1975; Bayen et al. 1978). Here the noncommutative algebras characteristic of quantum mechanics arise as deformations of so-called Poisson algebras; the underlying mathematical idea ultimately goes back to Dirac (1930). In Rieffel's (1989, 1994) approach to deformation quantization, the algebras in question are C*-algebras and hence the entire apparatus of operator algebras—and of the closely related field on noncommutative geometry (Connes 1994)—becomes available.

Apart from deformation quantization, a second source of asymptotic Bohrification was the mathematical analysis of the BCS-model of superconductivity initiated by Haag (1962), which, in the general setting of mean-field models of solid state physics, culminated in the work of Bona (1988, 2000), Raggio and Werner (1989), and Duffield and Werner (1992). An amazing feature of this work is that in the macroscopic limit $N \to \infty$ noncommutative algebras of quantum-mechanical observables (which are typically N'th tensor powers of matrix algebras B) converge to some commutative algebra (typically $C(S(B))$, where $S(B)$ is the state space of B), at least as far as macroscopic averages are concerned.

These examples led to a unified operator-algebraic approach to (deformation) quantization, the classical limit of quantum mechanics, and the macroscopic limit of quantum statistical mechanics which deals with both limits $\hbar \to 0$ and $N \to \infty$ in a single mathematical framework (Landsman 1998, 2007). This, in turn, led to a new approach to the measurement problem (Landsman and Reuvers 2013) and to spontaneous symmetry breaking (Landsman 2013), based on the instability of ground states in the two limits at hand discovered by Jona-Lasinio, Martinelli, and Scoppola (1981) and Koma and Tasaki (1994), respectively. Thus asymptotic Bohrification firmly reaches back to physics.

The plan of this chapter is as follows. We first elaborate on the path alluded to above from von Neumann (1932) to the poset $C(A)$ and hence to exact Bohrification. Next, the Born rule provides a nice context for a first acquaintance with the Bohrification program in its two variants (i.e., exact and asymptotic), as it falls under both headings and hence may be understood from two rather different points of view. After an intermezzo on conceptual issues with limits in the context of asymptotic Bohrification, we end with an overview of our strategy for first (re)formulating and then solving the measurement problem.

This paper mainly concentrates on the Bohrification program in physics. For applications to mathematics, see Döring and Harding (2010), Hamhalter (2011), Hamhalter and Turilova (2013), Wolters (2013), Döring (2014), Heunen (2017, b), Lindenhovius (2016), as well as a review by Landsman and Lindenhovius (2017).

2. Gleason, Kochen–Specker, and commutative C*-algebras

Let H be a Hilbert space. Von Neumann (1932) introduced the concept of an *Erwartung* (i.e., *expectation value*) as a linear map

$$\omega' : B(H)_{\text{sa}} \to \mathbb{R} \tag{16}$$

that satisfies

$$\omega'(1_H) = 1; \tag{17}$$

$$\omega'(a^2) \geq 0, \tag{18}$$

for each

$$a \in B(H)_{\text{sa}} = \{a \in B(H) \,|\, a^* = a\}. \tag{19}$$

For general $a \in B(H)$, using the canonical decomposition

$$a = a' + ia''; \tag{20}$$

$$a' = \frac{1}{2}(a + a^*); \tag{21}$$

$$a'' = -\frac{1}{2}i(a - a^*), \tag{22}$$

where a' and a'' are obviously self-adjoint, ω' may be extended to a linear map

$$\omega : B(H) \to \mathbb{C}; \tag{23}$$

$$\omega(a) = \omega'(a') + i\omega'(a''), \tag{24}$$

which is positive in the equivalent sense (8), and of course satisfies (9). Thus ω is a state.

Von Neumann's controversial argument against hidden variables is a simple corollary of his characterization of states, which we review first. As already mentioned in Section 1, if H is finite-dimensional, then any state (23) takes the form (12) for some density operator ρ. In general, density operators define *normal* states, which by definition have the following property: for each orthogonal family (e_i) of projections on H (i.e., $e_i^* = e_i$ and $e_i e_j = \delta_{ij} e_i$),

$$\omega\left(\sum_i e_i\right) = \sum_i \omega(e_i); \tag{25}$$

states that are not normal exist even if H is separable, see below.

Now suppose that some nonzero linear map $\omega' : B(H)_{\text{sa}} \to \mathbb{R}$ is *dispersion-free*, that is,

$$\omega'(a^2) = \omega'(a)^2; \tag{26}$$

without loss of generality we may also assume (17). This implies that ω' is positive and canonically extends to a state (23). If ω' is also normal (which is automatic if H is finite-dimensional), then one has (12), as above, from which (provided dim $(H) > 1$) it is easy to see that ω cannot be dispersion-free on all self-adjoint operators a. In the eyes of von Neumann and most of his contemporaries, this contradiction proved that quantum mechanics did not admit underlying hidden variables (at least of a kind we now call *non-contextual*). Belinfante (1973, 24, 34) recalls that "the authority of von Neumann's overgeneralized claim for nearly two decades stifled any progress in the search for hidden-variable theories ... for decades nobody spoke up against von Neumann's arguments, and his conclusions were quoted by some as the gospel." However, the philosopher Hermann (1935) already pointed out that von Neumann's linearity assumption, though valid for quantum-mechanical expectation values, was not justified for dispersion-free states, if only because addition of non-commuting observables was not physically defined (a point of which von Neumann (1932) was actually well aware, since he attempted to justify such additions through the use of ensembles). See also Seevinck (2012). Unaware of this—Hermann was one of the very few women in academia at the time and also published in an obscure Neokantian journal—Bell (1966) and Kochen and Specker (1967) made a similar point and also remedied it, proving what we now call the Kochen–Specker Theorem (sometimes also named after Bell). It is sad to note that Bell and some of his followers later left the realm of decent academic discourse (and displayed the depth of their own misunderstanding) by calling von Neumann's argument "silly" and "foolish." In fact, von Neumann carefully qualifies his result by stating that it follows "*im Rahmen unserer Bedingungen,*" that is, "*given our assumptions,*" and should be credited rather than ridiculed for being the first author to impose some useful constraints on hidden variable theories, anticipating all later literature on the subject; see also Bub (2011).

In order to explain the link of this development to (exact) Bohrification, we now go back to Gleason (1957), but in addition use later results as reviewed in Maeda (1990) and Hamhalter (2004). We first collect all relevant notions in to a single definition.

Definition 2.1. *Let H be an arbitrary Hilbert space with projection lattice*

$$\mathcal{P}(H) = \{e \in B(H) \mid e^2 = e^* = e\}. \tag{27}$$

1. *A map $P : \mathcal{P}(H) \to [0,1]$ that satisfies $P(1_H) = 1$ is called:*

    (a) *a finitely additive probability measure if*

$$P\left(\sum_{j \in J} e_j\right) = \sum_{j \in J} P(e_j) \tag{28}$$

for any finite collection $(e_j)_{j \in J}$ of mutually orthogonal projections on H (i.e., $e_j H \perp e_k H$, or equivalently, $e_j e_k = 0$, whenever $j \neq k$); equivalently,

$$P(e + f) = P(e) + P(f) \text{ whenever } ef = 0.$$

    (b) *a probability measure if (28) holds for any countable collection $(e_j)_{j \in J}$ of mutually orthogonal projections on H;*

(c) a completely additive probability measure *if* (28) *holds for* arbitrary *collections* $(e_j)_{j \in J}$ *of mutually orthogonal projections on* H.

In (b) *and* (c) *the sum* $\sum_j$ *is defined in the strong operator topology on* $B(H)$.

2. A strong quasi-state on $B(H)$ *is a map* $\omega : B(H) \to \mathbb{C}$ *that is positive (i.e.,* $\omega(a^* a) \geq 0$ *for each* $a \in B(H)$) *and normalized (i.e.,* $\omega(1_H) = 1$), *and otherwise:*

(a) *satisfies* $\omega(a) = \omega(a') + i\omega(a'')$, *where* $a' = \dfrac{1}{2}(a + a^*)$ *and* $a'' = -\dfrac{1}{2}i(a - a^*)$.

(b) *is linear on all commutative unital* C^*-*algebras in* $B(H)$.

Note that a' *and* a'' *are self-adjoint, so that* ω *is fixed by its values on* $B(H)_{sa}$. *Hence we have* $\omega(za) = z\omega(a)$, $z \in \mathbb{C}$, *and* $\omega(a + b) = \omega(a) + \omega(b)$ *whenever* $ab = ba$.

3. A weak quasi-state *on* $B(H)$ *satisfies all properties of a strong one except that linearity is only required on singly generated commutative* C^*-*algebras in* $B(H)$, *that is, those of the form* $C^*(a)$, *where* $a = a^* \in B(H)$; *this is the* C^*-*algebra generated by* a *and* 1_H *(which is the norm-closure of the space of all finite polynomials in* a).
4. A strong (weak) dispersion-free state on $B(H)$ *is a strong (weak) quasi-state that is* pure *on each (singly generated) commutative unital* C^*-*algebra in* $B(H)$.

It was shown by Aarens (1970) that the map $\omega \mapsto \omega_{|\mathcal{P}(H)}$ gives a bijective correspondence between weak quasi-states ω on $B(H)$ and finitely additive probability measures on $\mathcal{P}(H)$.

A probability measure is by definition σ-additive in the usual sense of measure theory; the other two cases are unusual from that perspective. However, if H is separable, then J can be at most countable, so that complete additivity is the same as σ-additivity and hence any probability measure is automatically completely additive. Surprisingly, assuming the *Continuum Hypothesis* (CH) of set theory, it can be shown that this is even the case for arbitrary Hilbert spaces (Eilers and Horst 1975). The fundamental distinction, then, is between *finitely* additive probability measures and probability measures (which by definition are *countably* additive). As we shall see shortly, this reflects the (often neglected) distinction between *arbitrary* and *normal* states on $B(H)$, respectively. Indeed, we now have the following generalization (and bifurcation) of Gleason's Theorem:

Theorem 2.2. *Let H be a Hilbert space of dimension greater than 2.*

1. *Each probability measure P on $\mathcal{P}(H)$ is induced by a normal state on $B(H)$ via*

$$P(e) = \mathrm{Tr}\,(\rho e), \quad e \in \mathcal{P}(H), \tag{29}$$

where ρ is a density operator on H uniquely determined by P. Conversely, each density operator ρ on H defines a probability measure P on $\mathcal{P}(H)$ via (29). Without CH, this is true also for non-separable H if P is assumed to be completely additive.
2. *Each finitely additive probability measure P on $\mathcal{P}(H)$ is induced by a unique state ω on $B(H)$ via*

$$P(e) = \omega(e), \ e \in \mathcal{P}(H). \tag{30}$$

Conversely, each state ω on H defines a probability measure P on $\mathcal{P}(H)$ via (30).

Corollary 2.3. *If* dim $(H) > 2$, *then each weak quasi-state on $B(H)$ is linear and hence is actually a state. In particular, weak and strong quasi-states coincide.*

In this language, the Kochen–Specker Theorem takes the following form:

Theorem 2.4. *If* dim $(H) > 2$, *there are no strong (weak) quasi-states $\omega : B(H) \to \mathbb{C}$ whose restriction to each (singly generated) commutative C^*-subalgebra of $B(H)$ is pure.* For an efficient proof, see Döring (2005). In summary, the (trivial) first step is to show that $\omega(e) \in \{0,1\}$ for any projection e. From this, one shows that ω must be multiplicative on all of $B(H)$ (Hamhalter 1993). If ω is normal, Gleason's Theorem already excludes this. If not, H must be infinite-dimensional, in which case there is a projection e and an operator v such that

$$e = vv^*; \tag{31}$$

$$1_H - e = v^*v; \tag{32}$$

this is sometimes called the *halving lemma*, which is easy to prove. Multiplicativity of ω then implies $\omega(e) = \omega(1_H - e)$, which contradicts the additivity property

$$\omega(e) + \omega(1_H - e) = \omega(1_H) = 1 \tag{33}$$

obtained from Aarens (1970) cited above: if $\omega(e) = 0$ one finds $0 = 1$, whereas $\omega(e) = 1$ implies $2 = 1$. To understand the connection with the original Kochen–Specker Theorem, let some map $\omega' : B(H)_{sa} \to \mathbb{R}$ be dispersion-free, cf. (26), for each $a \in B(H)_{sa}$, and quasi-linear, that is, linear on commuting operators; it was shown by Fine (1974) that these conditions are equivalent to the ones originally imposed by Kochen and Specker (1967), namely,

$$\omega(a) \in \sigma(a); \tag{34}$$

$$\omega(f(a)) = f(\omega(a)), \tag{35}$$

for each Borel function $f : \sigma(a) \to \mathbb{R}$. Canonically extending such a map ω' to a map $\omega : B(H) \to \mathbb{C}$ by complex linearity (cf. Definition 1.2 (a)) and noting that dispersion-freeness implies positivity and hence continuity on each subalgebra $C^*(a)$, we see that dispersion-freeness and quasi-linearity imply that ω is multiplicative on $C^*(a)$, and hence pure (conversely, pure states on $C^*(a)$ are dispersion-free). Finally, although linearity on all commuting self-adjoint operators seems stronger than linearity on each $C^*(a)$, Theorem 2.2 shows that these conditions are in fact equivalent.

In sum, Gleason's Theorem and the Kochen–Specker Theorem give results (of a positive and a negative kind, respectively) on the behaviour of state-like functionals on all of $B(H)$ *given their behaviour on commutative C^*-subalgebras thereof.* This supports the idea of exact Bohrification, which suggests that access to $B(H)$ should precisely be gained through such subalgebras. We now turn to another major influence on this kind of thinking.

3. The Kadison–Singer conjecture

Around the same time as Gleason (1957), and similarly inspired by quantum mechanics through the work of von Neumann (1932) and Mackey (1957), Kadison and Singer (1959) wrote a visionary paper on pure states on *maximal* commutative C*-subalgebras of $B(H)$. Their work arose from the desire to clarify a potential mathematical ambiguity in the Dirac notation commonly used in quantum mechanics. Namely, if $\underline{a} = (a_1,...,a_n)$ is a maximal set of commuting self-adjoint operators on some *finite-dimensional* Hilbert space H, then H has a basis of joint eigenvectors $|\underline{\lambda}\rangle$ of $\underline{a}$, which physicists typically label by the corresponding eigenvalues $\underline{\lambda} = (\lambda_1,...,\lambda_n)$. Equivalently, given a single self-adjoint operator a that is maximal in the sense that its spectrum $\sigma(a)$ is nondegenerate, the eigenvectors of a, denoted by $|\lambda\rangle$, $\lambda \in \sigma(a)$, form a basis of H. The first situation actually reduces to the second, since $a_i = f_i(a)$ for a single a and suitable functions $f_i : \sigma(a) \to \mathbb{R}$.

Algebraically, what happens here is that each $\lambda \in \sigma(a)$ initially defines a pure (and hence multiplicative) state ω'_λ on $C^*(a)$ by $\omega'_\lambda(a) = \lambda$ and of course $\omega'_\lambda(1_H) = 1$, which enforces $\omega'_\lambda(f(a)) = f(\lambda)$ for any polynomial f (this much is true also if dim $(H) = \infty$, in which case the last equation holds for any $f \in C(\sigma(a))$; see also Section 5 below). Since λ defines a unit eigenvector $|\lambda\rangle$ that, given that a is maximal, is unique up to a phase, it also defines a unique extension ω_λ of ω'_λ to a pure state on $B(H)$, namely (in Dirac notation),

$$\omega_\lambda(b) = \langle \lambda|b|\lambda \rangle, \ b \in B(H). \tag{36}$$

Indeed, it is the uniqueness of this extension that makes the labeling $|\lambda\rangle$ unambiguous in the first place. By contrast, if a is not maximal it has an eigenvalue λ having at least two orthogonal eigenvectors, which clearly define different vector states on $B(H)$. Similarly for the commutative C*-algebra $C^*(\underline{a})$ generated by the a_i; the two examples can be united by noting that—assuming maximality—both $C^*(a)$ and $C^*(\underline{a})$ are unitarily equivalent to the algebra $D_n(\mathbb{C})$ of diagonal matrices on $\mathbb{C}^n$ for $n = \dim(H)$ (viz., by changing some given basis of H to a basis of eigenvectors of a or $\underline{a}$).

All of this is true for any maximal commutative C*-algebra $A \subset B(H)$, since $A = C^*(a)$ for some maximal self-adjoint a. We may therefore summarize the discussion so far as follows: *if H is finite-dimensional, then any pure state on any maximal commutative C*-algebra $A \subset B(H)$ has a unique extension to a pure state on $B(H)$, and this is what makes the Dirac notation $|\lambda\rangle$ unambiguous.* This supports the idea of exact Bohrification, since it shows that a state on the noncommutative algebra $B(H)$ of all observables is already determined by its value on an arbitrary maximal commutative C*-subalgebra $A \subset B(H)$.

What about infinite-dimensional Hilbert spaces? The first difference with the finite-dimensional case is that even if H is separable (which is the only case that has been studied so far, sufficient as it is for most applications), maximal commutative C*-subalgebras A of $B(H)$—which are always von Neumann algebras, that is, $A'' = A$—are no longer unique (up to unitary equivalence). Indeed, A is unitarily equivalent to exactly one of the following:

1. $L^\infty(0,1) \subset B(L^2(0,1))$;
2. $\ell^\infty(\mathbb{N}) \subset B(\ell^2(\mathbb{N}))$;
3. $L^\infty(0,1) \oplus \ell^\infty(\kappa) \subset B(L^2(0,1) \oplus \ell^2(\kappa))$,

where either $\kappa = \{1,\dots,n\}$, in which case

$$\ell^2(\kappa) \cong \mathbb{C}^n; \tag{37}$$

$$\ell^\infty(\kappa) \cong D_n(\mathbb{C}), \tag{38}$$

or $\kappa = \mathbb{N}$; the inclusions are given by realizing each commutative algebra by multiplication operators. This classification was stated without proof in Kadison and Singer (1959); the details appeared in Kadison and Ringrose (1986, §9.4), based on von Neumann (1931), who initiated the study of commutative von Neumann algebras. See also Stevens (2016).

The second difference is that if dim $(H) = \infty$, then $B(H)$ admits *singular* pure states that cannot be represented by some unit vector in H; the difficulty of the state extension property in question entirely comes from the singular case. Note that any state ω on $B(H)$ has a unique decomposition as a convex sum $\omega = t\omega_n + (1-t)\omega_s$ where $t \in [0,1]$, ω_n is normal (and hence can be represented by a density operator), and ω_s is singular in that it annihilates all finite-dimensional projections and hence all compact operators on H. A pure state ω must have $t = 0$ or $t = 1$ and hence it is either normal or singular. For example, if $H = L^2(0,1)$ and a is the position operator, that is, $a\psi(x) = x\psi(x)$, having spectrum $\sigma(a) = [0,1]$, then for any $\lambda \in \sigma(a)$ there exists a singular pure state ω on $B(H)$ such that $\omega(a) = \lambda$. More generally, this is true whenever $a \in B(H)_{sa}$ and $\lambda \in \sigma_c(a)$ (i.e., the continuous spectrum of a, defined as the complement of the set of eigenvalues in $\sigma(a)$).

This apparently gives rigorous meaning to the Dirac notation $|\lambda>$ for improper eigenstates, though with an important caveat. In the light of the discussion above, one might rephrase the situation as follows: the position operator a generates a maximal commutative C*-algebra $L^\infty(0,1)$ in $B(L^2(0,1))$), which admits no *normal* pure states whatsoever. However, each $\lambda \in \sigma(a)$ defines a *singular* pure state ω'_λ on $L^\infty(0,1)$, which has at least one pure extension ω_λ to $B(L^2(0,1))$. But is this extension unique? This question also arises for $\ell^\infty(\mathbb{N})$, although this algebra does admit normal pure states (viz., $\omega_k(f) = f(k)$, where $k \in \mathbb{N}$). However, the "overwhelming majority" of pure states on $\ell^\infty(\mathbb{N})$ are singular (corresponding to elements of the so-called Čech–Stone compactification $\beta\mathbb{N}$ of $\mathbb{N}$).

For separable Hilbert spaces, the pure state extension problem has been solved:

Theorem 3.1

- *Any normal pure state on any maximal commutative C*-subalgebra $A \subset B(H)$ has a unique (and hence pure) extension to $B(H)$.*
- *Any pure state on $\ell^\infty(\mathbb{N})$ has a unique extension to $B(\ell^2(\mathbb{N}))$.*
- *No singular pure state on $L^\infty(0,1)$ has a unique extension to $B(L^2(0,1))$, and similarly for $L^\infty(0,1) \oplus \ell^\infty(\kappa)$, for any κ.*

Uniqueness of the state extension property in question—or the lack of it—is preserved under unitary equivalence. The first part of this theorem is even true for non-separable Hilbert spaces; it may be proved along the lines of the finite-dimensional case. If one agrees that only normal states (i.e., density operators) can be realized physically—in that singular states (such as exact eigenstates of position or momentum) are idealizations—this settles the issue and supports exact Bohrification, as explained above. The singular case, however, is of much more mathematical interest and might even be physically relevant, depending on the way one thinks about idealizations (see also Section 6 below).

The second part of the theorem is the pathbreaking solution of the *Kadison–Singer conjecture* due to Marcus, Spielman, and Srivastava (2014a, b), with important earlier contributions by Weaver (2004); see also Tao (2013) and Stevens (2016) for lucid expositions of the proof. Though less well known, the third part, due to Kadison and Singer themselves, whose arguments were later simplified by Anderson (1979) and Stevens (2016), is as remarkable as the second; it shows that Dirac's notation $|\lambda>$ may be ambiguous, or, equivalently, that maximal commutative C*-subalgebras of $B(H)$ that are unitarily equivalent to $L^\infty(0,1)$ (like the one generated by the position operator or the momentum operator) do not suffice to characterize pure states. What to make of this is unclear.

4. The poset $\mathcal{C}(A)$, topos theory, and quantum logic

Returning to the story in Section 2, the next step was to collect all unital commutative C*-subalgebras of $B(H)$, or indeed an arbitrary unital C*-algebra A, into a single mathematical object $\mathcal{C}(A)$; with some goodwill, one might call $\mathcal{C}(A)$ the mathematical home of complementarity (although the construction applies even when A itself is commutative). This decisive idea goes back to Isham and Butterfield (1998) for $A = B(H)$ and, more generally, to Hamilton, Isham, and Butterfield (2000) for arbitrary von Neumann algebras (and hence these authors consider von Neumann subalgebras instead of C*-subalgebras).

Heunen, Landsman, and Spitters (2009) introduced the poset $\mathcal{C}(A)$ whose elements are commutative C*-subalgebras of A that contain the unit 1_A, ordered by (set-theoretic) inclusion (note that Isham et al. use the opposite order, which even apart from their von Neumann algebraic setting gives the theory a totally different flavor as soon as sheaves are defined). At this stage of the Bohrification program it was perhaps not realized what a powerful object $\mathcal{C}(A)$ already is by itself, even though it has "forgotten" the commutative subalgebras (which are merely points in $\mathcal{C}(A)$) and just "remembers" their inclusion relations; this realization only came with papers like Döring and Harding (2010), Hamhalter (2011), Heunen (2014a), Heunen and Lindenhovius (2015), and Lindenhovius (2016). Thus we introduced the "tautological map" $C \mapsto C$, where the C on the left-hand side is an element of $\mathcal{C}(A)$ (now seen as a posetal category, in which C and D are connected by an arrow iff $C \subseteq D$), whereas the C on the right-hand side is C itself, initially seen as a set. Seen as a functor $\underline{A}$ from $\mathcal{C}(A)$ to the category **Sets** of sets (where the arrow $C \to D$, i.e., $C \subseteq D$, is mapped to the inclusion map $C \hookrightarrow D$ of the underlying sets), our tautological map turns out to define a *commutative* internal C*-algebra in the topos $[\mathcal{C}(A), \textbf{Sets}]$ of all functors from $\mathcal{C}(A)$ to **Sets**. See Mac Lane and Moerdijk (1992) for an introduction to topos theory and Banaschewski and Mulvey (2006) for the general theory of commutative C*-algebras and Gelfand duality in toposes: broadly speaking, a topos provides a universe—alternative to set theory—in which to do mathematics, generally based on intuitionistic rather than classical logic (see below). It was this transfiguration of a (generally noncommutative) C*-algebra A into an internal commutative C*-algebra $\underline{A}$, which essentially consists of all (unital) commutative C*-subalgebras of A but may be treated as a single C*-algebra (in a different topos from **Sets**), that was initially meant by the term "Bohrification", which we now use much more liberally.

So far, the main harvest of this construction has been:

- The definition (Heunen, Landsman, and Spitters 2009) and explicit computation (Heunen, Landsman, and Spitters 2012; Wolters 2013) of the (internal) Gelfand spectrum $\underline{\Sigma}(\underline{A})$ of $\underline{A}$, playing the role of a quantum-mechanical phase space (which is lacking in the usual formalism).
- The realization of states on A as probability measures on $\underline{\Sigma}(\underline{A})$ (Heunen, Landsman, and Spitters 2009).
- The reformulation of the Kochen–Specker Theorem to the effect that (at least for C*-algebras like $A = B(H)$ for dim $(H) > 2$) the phase space $\underline{\Sigma}(\underline{A})$ has no points (Isham and Butterfield 1998; Caspers et al. 2009; Heunen, Landsman, and Spitters 2009).
- Intuitionistic logic as the logic of quantum mechanics (Caspers et al. 2009; Döring and Isham 2010; Döring 2012; Heunen et al. 2012; Hermens 2009, 2016; Wolters 2013).

All this requires extensive explanation, especially because these claims are meant in the sense of internal reasoning in a topos (Mac Lane and Moerdijk 1992), but let us restrict ourselves to some brief comments on the last two points (see references for details).

Even in the topos **Sets**, which is the home of classical mathematics (using the Zermelo–Fraenkel axioms), spaces may or may not have points, provided we interpret the notion of a "space" as a *frame*, which is a (necessarily distributive) complete lattice F in which

$$U \wedge VS = V\{U \wedge V, V \in S\}, \tag{39}$$

for arbitrary elements $U \in F$ and subsets $S \subset F$. Frames are motivated by the example $F = \mathcal{O}(X)$, that is, the collection of open sets on some topological space X, where the supremum is given by set-theoretic union. To see if a given frame F is of this form, one looks at the *points* of F, defined as frame maps

$$p : F \to \{0,1\}, \tag{40}$$

where the frame $\{0,1\} \equiv \mathcal{O}(\text{point})$ has order $0 \leq 1$. The set $\mathrm{Pt}(F)$ of points of F is a topological space, with open sets

$$\{p \in \mathrm{Pt}(F) \mid p^{-1}(U) = 1\}, \; U \in F. \tag{41}$$

If $F \cong \mathcal{O}(\mathrm{Pt}(F))$, we say that F is *spatial*. Even in set theory, frames are not necessarily spatial (and if they are, the axiom of choice—which is unavailable in many toposes— is often needed to prove this). All this can be internalized in topos theory, and the Kochen–Specker Theorem states that the Gelfand spectrum of $\underline{\Sigma}(\underline{A})$ of $\underline{A}$ not merely fails to be spatial: it has no points whatsoever! Looking at points of phase space in classical physics as truth-makers (of propositions), this is no surprise: after all, the very point (*sic*) of the Kochen–Specker Theorem is that in quantum mechanics not all propositions can have simultaneous (sharp) truth values.

Frames are also closely related to *Heyting algebras*, defined as distributive lattices H (with top 1 and bottom 0) equipped with a binary map

$$\to : H \times H \to H, \tag{42}$$

playing the role of implication in logic, that satisfies the axiom

$$U \leq (V \to W) \text{ iff } (U \wedge V) \leq W. \tag{43}$$

In a Heyting algebra (unlike a Boolean algebra), negation is a derived notion, defined by

$$\neg U = U \to \bot. \tag{44}$$

Every Boolean algebra is a Heyting algebra, but not vice versa; in fact, a Heyting algebra is Boolean iff $\neg\neg U = U$ for all U, or, equivalently, $(\neg U) \vee U = \top$, which states the law of the excluded middle (famously denied by the Dutch mathematician L. E. J. Brouwer). Thus Heyting algebras formalize intuitionistic propositional logic. A Heyting algebra H is *complete* when it is complete as a lattice, in which case H is a frame. Conversely, a frame is a complete Heyting algebra with implication $V \to W = V\{U \mid U \wedge V \leq W\}$.

The point, then, is that exact Bohrification implies that quantum logic is given by a particular Heyting algebra $H(A)$ eventually defined by the given C*-algebra A. Thus quantum logic turns out to be intuitionistic: it preserves distributivity (i.e., of "and" over "or") but denies the law of the excluded middle. The simplest example is $A = M_n(\mathbb{C})$, which describes an n-level system, for which the relevant Heyting algebra comes down as

$$H(M_n(\mathbb{C})) = \{\mathsf{e} : \mathcal{C}(A) \to \mathcal{P}(A) \mid \mathsf{e}(C) \in \mathcal{P}(C), \mathsf{e}(C) \leq \mathsf{e}(D) \text{ if } C \subseteq D\}, \tag{45}$$

with pointwise order, that is, $e \leq f$ iff $e(C) \leq f(C)$ for each $C \in \mathbb{C}(M_n(\mathbb{C}))$. We see that whereas in the traditional (von Neumann) approach to quantum logic a *single* projection $e \in \mathcal{P}(A)$ defines a proposition, eqn (45) suggests that a proposition consists of a *family* e of projections $e(C)$, one for each classical context C; see also Hermens (2016) for a different interpretation of (45), in which elements of $H(M_n(\mathbb{C}))$ are disjunctions of propositions.

Intuitionistic logic goes flatly against Birkhoff and von Neumann (1936), whose quantum logic has exactly the opposite features in denying distributivity but keeping the law of the excluded middle. It also seems to go against Bohr (1996, 393), who in 1958 claimed that "all departures from common language and ordinary logic are entirely avoided by reserving the word 'phenomenon' solely for reference to unambiguously communicable information, in the account of which the word 'measurement' is used in its plain meaning of standardized comparison." However, the preceding text makes it clear that Bohr refers to certain multivalued logics (Gottwald 2015) rather than to the intuitionistic logic of Brouwer and Heyting, which might even model everyday reasoning better than classical logic does (Moschovakis 2015).

5. The Born rule revisited

In this section we apply exact Bohrification to singly generated commutative C*-algebras in $A = B(H)$, where H is an arbitrary Hilbert space (see Section 2), so let $a = a^* \in B(H)$ and consider the C*-algebra $C^*(a)$ generated by a and 1_H. Another commutative C*-algebra defined by a is $C(\sigma(a))$, that is, the algebra of continuous complex-valued functions on the spectrum $\sigma(a)$ of a (which is a compact subset of $\mathbb{R}$), equipped with pointwise addition, (scalar) multiplication, and complex conjugation (as the involution), and the supremum norm, *cf.* Section 1. One version of the spectral theorem now states that $C(\sigma(a))$ and $C^*(a)$ are isomorphic (as commutative C*-algebras) under a map $f \mapsto f(a)$ that is given by (norm-) continuous extension of the map defined on polynomial functions simply by $p \mapsto p(a)$, that is, if $p(x) = \sum_n c_n x^n$, where $x \in \sigma(a)$ and $c_n \in \mathbb{C}$, then $p(a) = \sum_n c_n a^n$. In particular, the unit function $1_{\sigma(a)} : x \mapsto 1$ maps to the unit operator $1_{\sigma(a)}(a) = 1_H$, while the identity function $\mathrm{id}_{\sigma(a)} : x \mapsto x$ is mapped to $\mathrm{id}_{\sigma(a)}(a) = a$. See, for example, Pedersen (1989), Theorem 4.4.1.

Now recall the Riesz–Radon representation theorem from Section 1. Take a unit vector $\psi \in H$ (or, more generally, a state ω on $B(H)$). Combining this with the spectral theorem above, we obtain a unique probability measure μ_ψ (or μ_ω) on the spectrum $\sigma(a)$ of a such that

$$\langle \psi, f(a)\psi \rangle = \int_{\sigma(a)} d\mu_\psi \, f, \tag{46}$$

for each $f \in C(\sigma(a))$ (or, more generally, $\omega(f(a)) = \int_{\sigma(a)} d\mu_\omega \, f$). This is the Born measure; for example, if $\dim(H) < \infty$, we recover the familiar result:

$$\mu_\psi(\{\lambda\}) = \| e_\lambda \psi \|^2, \tag{47}$$

where e_λ is the spectral projection onto the eigenspace $H_\lambda \subset H$ for the eigenvalue $\lambda \in \sigma(a)$; of course, the right-hand side equals $\langle \psi, e_\lambda \psi \rangle$. If the spectrum is non-degenerate, this expression becomes $|\langle v_\lambda, \psi \rangle|^2$, where v_λ is "the" unit eigenvector of a with eigenvalue λ.

Although the underlying mathematical reasoning is well known (Pedersen 1989, §4.5), the thrust of (exact) Bohrification is that the Born rule—which is the main connection between the mathematical formalism of quantum mechanics and laboratory physics—comes from a simple look at appropriate commutative subalgebras of $B(H)$. Namely, the (vector) state $b \mapsto \langle \psi, b\psi \rangle$ on $B(H)$ defined by $\psi \in H$ (or indeed any state on $B(H)$) is simply restricted to $C^*(a) \subset B(H)$, and is subsequently transferred to $C(\sigma(a))$ through the spectral theorem, where it becomes a probability measure (Landsman 2009).

One might argue that this argument validates Heisenberg's (1958, 53–4) claim that "one may call these uncertainties objective, in that they are simply a consequence of the fact that we describe the experiment in terms of classical physics; they do not depend in detail on the observer. One may call them subjective, in that they reflect our incomplete knowledge of the world." Nonetheless, by itself such mathematical reasoning is insufficient to *derive* the Born rule, but it does show that all one needs to postulate in quantum mechanics in this respect is its *function* rather than its *form* (which, as shown above, is simply given by the formalism).

Asymptotic Bohrification closes also this gap. Following Landsman (2008) we complete (and—in the light of valid critique by Cassinello and Sánchez-Gómez (1996) and Caves and Schack (2005)—correct) a program begun by Finkelstein (1965) and Hartle (1968), as further developed by Farhi, Goldstone, and Gutmann (1989), and Van Wesep (2006).

Referring to the appendix for notation and background, let $I = 1/\dot{\mathbb{N}}$, take some fixed unital C*-algebra B (for which we will just need the $n \times n$ matrices $B = M_n(\mathbb{C})$), and define a bundle $(A_h)_{h \in I}$ of C*-algebras by taking $A_{1/N} = B^N$ (i.e., the N-fold projective tensor product $\otimes^N B$ of B with itself) and $A_0 = C(S(B))$, where $S(B)$ is the state space of B, seen as a compact convex set in the weak*-topology. For example, the state space of $B = M_2(\mathbb{C})$ is affinely homeomorphic to the unit ball in $\mathbb{R}^3$, whose boundary is the familiar Bloch sphere of qubits. The commutative C*-algebra A_0, then, is our classical window on B (Raggio and Werner 1989; Duffield and Werner 1992). To this effect, we now turn this family into a continuous bundle of C*-algebras by stipulating what the continuous sections are (Landsman 2007, 2008). We first define the usual symmetrization operator $S_N : B^N \to B^N$ by linear (and if necessary continuous) extension of

$$S_N(b_1 \otimes \cdots \otimes b_N) = \frac{1}{N!} \sum_{\sigma \in \mathfrak{S}_N} b_{\sigma(1)} \otimes \cdots \otimes b_{\sigma(N)}, \tag{48}$$

where $\mathfrak{S}_N$ is the permutation group (i.e., symmetric group) on N elements, and $b_i \in B$ for all $i = 1, \ldots, N$. For $N \geq M$ we then define $S_{M,N} : B^M \to B^N$ by linear extension of

$$S_{M,N}(a_M) = S_N(a_M \otimes 1_B \otimes \cdots \otimes 1_B), \tag{49}$$

where one has $N - M$ copies of the unit $1_B \in B$ so as to obtain an element of B^N.

For example, the symmetrizer $S_{1,N} : B \to B^N$ is given by

$$S_{1,N}(b) = \frac{1}{N} \sum_{k=1}^{N} 1_B \otimes \cdots \otimes b_{(k)} \otimes 1_B \cdots \otimes 1_B, \tag{50}$$

which is just the "average" of b over all N copies of B. Our main examples of (50) will be *frequency operators*, where $b = e$ for some projection $e \in B$ (i.e., $e^2 = e^* = e$). In particular, if $a = a^* \in B = M_n(\mathbb{C})$ and $\lambda \in \sigma(a)$, we may take the spectral projection e_λ. Applied to states of the kind $v_1 \otimes \cdots \otimes v_n \in \mathbb{C}^N$, where each v_i is an eigenstate of a, so that $a v_i = \lambda_i v_i$ for some $\lambda_i \in \sigma(a)$, the corresponding operator

$$f_N^\lambda = S_{1,N}(e_\lambda) \tag{51}$$

counts the relative frequency of λ in the list $(\lambda_1,\ldots,\lambda_n)$.

We return to the construction of a continuous bundle of C*-algebras. As explained in the appendix, given the fibers A_0 and $A_{1/N}$, $N \in \mathbb{N}$, we may define the topological structure by specifying the continuous cross-sections. To this end, we say that a sequence $(a_N)_{N \in \mathbb{N}}$, with $a_N \in B^N$, is *symmetric* when $a_N = S_{M,N}(a_M)$ for some fixed M and all $N \geq M$, and *approximately symmetric* if $a_N = S_N(a_N)$ for each $N \in \mathbb{N}$ and for any $\varepsilon > 0$ there is a symmetric sequence (a'_N) and some $N(\varepsilon) \in \mathbb{N}$ such that $\| a_N - a'_N \| < \varepsilon$ for all $N \geq N(\varepsilon)$ (norm in A_N). Each section a of the bundle $(A_h)_{h \in 1/\dot{\mathbb{N}}}$ is a sequence $(a_0, a_N)_{N \in \mathbb{N}}$, where $a_0 \in C(S(B))$ and $a_N \in B^N$. The *continuous* sections are those for which $(a_N)_{N \in \mathbb{N}}$ is approximately symmetric and a_0 is given by

$$a_0(\omega) = \lim_{N \to \infty} \omega^N(a_N). \tag{52}$$

Here $\omega \in S(B)$ and $\omega^N \in S(B^N)$ is defined by linear (and continuous) extension of

$$\omega^N(b_1 \otimes \cdots \otimes b_N) = \omega(b_1) \cdots \omega(b_N). \tag{53}$$

This limit exists by definition of an approximately symmetric sequence. The point of this is that since the frequency operator (51), seen as a sequence $(f_N^\lambda)_{N \in \mathbb{N}}$, is evidently symmetric (and hence a fortiori approximately symmetric), it therefore defines a continuous section of our bundle $(A_h)_{h \in 1/\dot{\mathbb{N}}}$ if we complete it with its limit (52), which is given by

$$f_0^\lambda(\omega) = \omega(e_\lambda). \tag{54}$$

This is the Born probability for the outcome $a = \lambda$ in the state ω: if $B = M_n(\mathbb{C})$ and $\omega(b) = \langle \psi, b\psi \rangle$ for some unit vector $\psi \in \mathbb{C}^n$, then $\omega(e_\lambda) = \| e_\lambda \psi \|^2$; cf. (47).

The physical interpretation of this mathematical result is as follows: if the system is prepared in some state ω, the number $f_0^\lambda(\omega)$ is (by construction) equal to the limiting frequency of either a single experiment on a large number of sites or a long run of individual experiments on a single site. A classical perspective on either of those a priori quantum-mechanical situations only arises in the limit $N \to \infty$ (i.e., $1/N \to 0$), in which the Born probability arises as an objective property of the experiment(s). Nothing is implied about single cases; these will be discussed in Section 7 below, preceded by an intermezzo on the complications caused by the fact that in reality the limit $N = \infty$ (or $\hbar = 0$) is never reached. This is arguably why Bohr insists that merely the *results of the observations*—as opposed to the underlying *phenomena*—must be expressed in classical terms.

Our two derivations of the Born rule reflect the two ways one may look at probabilities in physics (as opposed to betting, which context is inappropriate here), namely, either as relative frequencies (which corresponds to the point of view offered by asymptotic Bohrification) or as chances for outcomes of individual random events (which is the perspective given by exact Bohrification); the relationship between these two approaches to probability is as much in need of clarification as the link between exact and asymptotic Bohrification!

6. Intermezzo: Limits and idealizations

Asymptotic Bohrification always involves some classical theory, described by a commutative C*-algebra of observables A_0 (typically of the form $A_0 = C_0(X)$ for some phase space X) and a family of quantum theories indexed by $\hbar \in I \subset [0,1]$, each of which is described by some commutative C*-algebra of observables $A_\hbar$. This includes both the case where $\hbar \in (0,1]$ and the limit of interest is $\hbar \to 0$, and the case where $\hbar = 1/N$, where $N \in \mathbb{N}$ and the limit of interest is $N \to \infty$. We regard the A_0-theory as an idealization that never occurs in physical reality, but which approximates the family of $A_\hbar$-theories in asymptotic regimes described by the parameter $\hbar$ (or N). Thus we follow Norton (2012, abstract):

> [A_0 is] another system whose properties provide an inexact description of the target system [i.e., $A_\hbar$].

Mathematically, A_0 is a limiting case of the theories $A_\hbar$ as $\hbar \to 0$ (or $N \to \infty$), in the sense explained in the appendix. Examples one may have in mind are:

- A_0 is either classical mechanics or thermodynamics (of an infinite system);
- $A_\hbar$ is either quantum mechanics or quantum statistical mechanics (for finite N).

See Landsman (1998) for the classical-quantum link in this language and Landsman (2007) for thermodynamics as a limiting case of quantum statistical mechanics of finite systems. The latter also has another limit, namely, quantum statistical mechanics of infinite systems, described by a different continuous bundle of C*-algebras, which is of no concern here.

Physicists typically call A_0 a *phenomenological theory* and refer to $A_\hbar$ at $\hbar > 0$ as a *fundamental theory*; we sometimes write $A_{>0}$ for the latter family. In the philosophy of science, A_0 is often referred to as a *higher-level theory* or a *reduced theory*, in which case $A_{>0}$

is said to be a *lower-level theory*, or a *reducing theory*, respectively. It is important to realize that A_0 plays a double role: on the one hand, it is an idealization or a limiting case of the family of $A_\hbar$-theories as $\hbar \to 0$, while on the other hand, it is defined and understood by itself; indeed, in our examples the A_0-theories historically predate the $A_{>0}$ theories and were once believed to be real and absolutely correct (if not holy). Furthermore, in many interesting cases (including the ones above) A_0 *taken by itself* has important features that are surprising or even seem out of the question in its role as a limiting theory. For example, classical mechanics and thermodynamics both allow spontaneous symmetry breaking, which according to conventional wisdom (corrected below) is absent in any finite quantum system. Moreover, measurements in classical physics have outcomes, whereas quantum mechanics faces the infamous measurement problem (see Section 7 below).

The fact that the phenomenological theory does not really exist in nature, as opposed to the fundamental theory that gives rise to it—or so we assume—poses a severe problem: *Real physical systems are supposed to be described by $A_{>0}$ rather than by the idealization A_0, yet in nature these systems display the surprising features claimed to be intrinsic to A_0 (and denied to $A_{>0}$).* For example, the reality of ferromagnetism shows that spontaneous symmetry breaking occurs in nature, although finite materials are supposed to be described by the $A_{>0}$-theory that allegedly forbids it. Similarly, measurements have outcomes, although the world is described by the $A_{>0}$-theory (viz., quantum mechanics) that fails to predict this. In sum, *the $A_{>0}$-theory fails to describe key features of the real systems that should fall within its scope, whereas the A_0-theory describes these features despite the fact that, being an idealization, real systems in principle do not fall within its scope.*

To resolve this paradox, we introduce two principles that, if adhered to, should guarantee the link between theory and reality. Earman's (2011, 1065) principle states that "While idealizations are useful and, perhaps, even essential to progress in physics, a sound principle of interpretation would seem to be that no effect can be counted as a genuine physical effect if it disappears when the idealizations are removed." Butterfield's (2011, 1065) principle follows up on this, claiming what should happen instead is that "There is a weaker, yet still vivid, novel and robust behaviour that occurs before we get to the limit, i.e. for finite N. And it is this weaker behaviour which is physically real." Although these principles should, in our view, be uncontroversial, since the link between theory and reality stands or falls with them, it is remarkable how often idealizations violate them. All rigorous theories of spontaneous symmetry breaking in quantum statistical mechanics strictly apply to infinite systems only (Bratteli and Robinson 1981; Liu and Emch 2005), and similarly in quantum field theory (Haag 1992; Ruetsche 2011). The same is true for the "Swiss" approach to the measurement problem based on superselection rules (Hepp 1972; Emch and Whitten-Wolfe 1976), and more recent developments thereof by Landsman (1991, 1995) and Sewell (2005). For a change, we here side with Bell (1975)!

7. Rethinking the measurement problem

To the best of our knowledge, apart from some very general comments, for example, in his Como lecture and in his correspondence (Zinkernagel 2016), the only time Bohr wrote in some technical detail about the quantum-mechanical measurement process was in his papers with Rosenfeld on the measurability of the electromagnetic field (Bohr 1996, 53–66). However, these comments and papers predate the Cat problem raised by Schrödinger (1935), and do not seem to address what we now see as the measurement problem of quantum mechanics. Indeed, one wonders how Bohr would respond to the amazing recent experiments like Hornberger et al. (2012), Kaltenbaek et al. (2015), Palomaki et al. (2013), and Kovachy et al. (2015).

Nonetheless, Bohrification provides a clear lead on the measurement problem. Let us begin with the proper formulation of the problem, on which its solution should evidently be predicated. To set the stage, Maudlin (1995) distinguishes three measurement problems:

1. *The problem of outcomes* states that the following assumptions are contradictory:
   (a) The wave-function of the system is complete;
   (b) The wave-function always evolves linearly (e.g., by the Schrödinger equation);
   (c) Measurements have determinate outcomes.
2. *The problem of statistics* is that 1(a) = 2(a) and 1(b) = 2(b) also contradict:
   (c) Measurement situations which are described by identical initial wave-functions sometimes have different outcomes, and the probability of each possible outcome is given by the Born rule.
3. *The problem of effect* requires that any (physical or philosophical) mechanism producing measurement outcomes should also update the predictions of quantum mechanics for subsequent measurements (typically that these have the same outcome).

Furthermore, Maudlin (1995) also gives a classification of potential solutions:

- Hidden-variable theories abandon 1(a) = 2(a);
- Collapse theories abandon 1(b) = 2(b)
- Multiverse theories abandon 1(c) and reinterpret the word "different" in 2(c).

As quite rightly emphasized by Maudlin, turning superpositions into mixtures does not even address any of these problems, which is why decoherence (or, for that matter, superselection rules), though an interesting prediction of quantum mechanics (Joos et al. 2003; Schlosshauer 2007) by itself fails to solve the measurement problem, despite occasional but persistent claims to the contrary (especially in the literature on experimental physics).

Our own take on the measurement problem is a bit different. First, an apparent problem famously emphasized by Bell (1990) is that "measurement" is a priori undefined within quantum mechanics. Hence the first part of the measurement problem is to define measurement. In our view, this part of the problem *was* solved by Bohr through his doctrine of classical concepts, which effectively says that a physical interaction *defines* a measurement as soon as the measurement device—though ultimately quantum-mechanical in an ontological sense—is *described* or *perceived* classically. Thus the doctrine of classical concepts is an epistemological move (Scheibe 1973; Camilleri and Schlosshauer 2015).

Without such an interpretation of measurement (which places it outside quantum mechanics), there is no measurement problem in the first place, since there are no outcomes. To Bohr, his definition of measurement through the doctrine of classical concepts seems to have been the end of it, but to us, it is just the beginning! Recalling (from Section 6) that the A_0-theory in which measurement outcomes are defined is a limit of the $A_{>0}$-theory responsible for these outcomes, it should be the case that the limits of the underlying (pure) quantum states of the measurement device as $\hbar \to 0$ (or $N \to \infty$, either of which limit would enforce a classical description) are pure states on A_0, since facts in classical physics correspond to pure (i.e., dispersion-free) states, at least approximately. But this is not at all what quantum mechanics predicts! In typical Schrödinger Cat situations the limits in question are *mixed* states that are not even approximately pure (Landsman and Reuvers 2013; Landsman 2013). But in reality sharp outcomes always occur, and so we obtain a violation of Butterfield's principle: outcomes (mathematically represented by pure states in A_0) are not foreshadowed in $A_{>0}$ (whose superpositions induce mixtures in the limit), although at the same time A_0 should arise from $A_{>0}$ as a limit theory.

With Butterfield's principle, the link between theory and reality falls. In other words, it is by no means enough that (in Schrödinger Cat situations) a classical description of the measurement device turns pure quantum superpositions into mixtures (which is trivially the case in the limit): all terms but one eventually have to *disappear* from the state. According to Butterfield's principle this disappearance must (approximately) take place already in quantum mechanics (if only in the limiting regime); a *Deus ex Machina* that suddenly turns classical mixtures into classical pure states (like the ignorance interpretation of classical probability) but did not previously act on the underlying quantum theory violates Earman's principle, since in reality the classical limit is never reached; cf. Section 6.

To see the mathematics of asymptotic Bohrification in a simple model of measurement, consider the continuous bundle of C*-algebras over $I = [0,1]$ that has fibers

$$A_0 = C_0(\mathbb{R}^2); \tag{55}$$

$$A_\hbar = B_0(L^2(\mathbb{R})) \ (\hbar > 0), \tag{56}$$

and whose continuous cross-sections are defined as follows. First, we recall the well-known *coherent states* that for each $(p,q) \in \mathbb{R}^2$ and $\hbar > 0$ are given by

$$\phi_\hbar^{(p,q)}(x) = (\pi\hbar)^{-1/4} e^{-ipq/2\hbar} e^{ipx/\hbar} e^{-(x-q)^2/2\hbar}. \tag{57}$$

These states were originally introduced by Schrödinger as models for wave packets localized in phase space, and they minimize the Heisenberg uncertainty relations. In terms of these, for each $f \in C_0(\mathbb{R}^2)$ and $\hbar > 0$ we define (compact) operators $Q_\hbar(f)$ on $L^2(\mathbb{R})$ by

$$Q_\hbar(f) = \int_{\mathbb{R}^{2n}} \frac{dpdq}{2\pi\hbar} f(p,q) \, | \phi_\hbar^{(p,q)} >< \phi_\hbar^{(p,q)} |, \tag{58}$$

where is the projection onto ϕ. In the literature, $Q_\hbar(f)$ is often called the *Berezin quant-ization* of the classical observable f. Finally, the continuous cross-sections of the bun-dle with fibers (55) - (56) are the maps $\hbar \mapsto Q_\hbar(f)$, with $Q_0(f) = f$; see Landsman (1998) for details. For our present purposes, the main point is that we may now track down quantum states to classical states, all the way in the limit $\hbar \to 0$. Namely, for each unit vector $\psi \in L^2(\mathbb{R})$ the map $f \mapsto \langle \psi, Q_\hbar(f)\psi \rangle$ defines a state on $C_0(\mathbb{R}^2)$, so that by the Riesz–Radon representation theorem we obtain a probability measure μ_ψ on $\mathbb{R}^2$, given by

$$d\mu_\psi(p,q) = \frac{dpdq}{2\pi\hbar}|\langle \phi_\hbar^{(p,q)}, \psi \rangle|^2 . \tag{59}$$

The probability density $\chi_\psi(p,q) = |\langle \phi_\hbar^{(p,q)}, \psi \rangle|^2$ is called the Husimi function of ψ; it gives a phase space portrait of ψ, which is especially useful in studying the limit $\hbar \to 0$. Let us illustrate this formalism for a one-dimensional anharmonic oscillator, that is,

$$H_\hbar = -\frac{\hbar^2}{2m}\frac{d^2}{dx^2} + \tfrac{1}{4}\lambda(x^2 - a^2)^2, \tag{60}$$

where $\lambda > 0$, as in the Higgs mechanism. It is well known that the ground state of this Hamiltonian is unique, despite the fact that the corresponding classical system has two degenerate ground states, given by the phase space points $(p_0 = 0, q_0 = \pm a)$. In particular, classically this model displays spontaneous symmetry breaking, whereas quantum-mechanically it does not, for any value of $\hbar > 0$. This already seems to spell doom for Earman's and Butterfields's principles (and hence for the link between theory and reality).

A quantitative analysis confirms this threat. The ground state wave-function $\psi_\hbar^{(0)}$ is real and positive definite, and has two peaks, above $x = \pm a$, with exponen-tial decay $|\psi_\hbar^{(0)}(x)| \sim \exp(-1/\hbar)$ in the classically forbidden region. In the limit $\hbar \to 0$ the associated probability measure $\mu_\hbar^{(0)}$ converges to the classical mix-ture $\tfrac{1}{2}(\delta(p,q-a) + \delta(p,q+a))$:

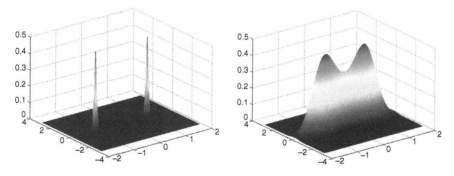

Figure 15.1 Husimi functions of double well ground state for $\psi_{\hbar=0.5}^{(0)}$ (left) and $\psi_{\hbar=0.01}^{(0)}$ (right).

It takes little imagination to see the classical ground states of this systems as pointer states, the (unique) quantum-mechanical ground state then corresponding to a Schrödinger Cat state (Landsman and Reuvers 2013). The limit $\hbar \to 0$, which in the real world (where $\hbar$ is constant), corresponds to $m \to \infty$ or $\lambda \to \infty$, or even to $N \to \infty$, as in the closely related but superior model of Spehner and Haake (2008), fails to predict the empirical fact that in reality the system (typically a particle) is found at one of the two minima (i.e., the cat is either dead or alive), and hence Earman's and Butterfield's principles are indeed violated: the quantum theory ought to have foreshadowed the symmetry breakdown of the limiting classical theory, but it fails to do so however close it is to the classical limit.

Motivated by examples like this—cf. Landsman (2013) for some others—we are led to a formulation of the measurement problem that is a bit different from the ones above:

4. *Problem of classical outcomes.* The following assumptions are contradictory:
   (a) Outcomes of measurements on quantum systems are classical.
   (b) Classical physics is an appropriate limit of quantum physics.
   (c) Quantum states are dynamically unaffected by this limit.

Compared with Maudlin's three measurement problems, the most striking difference with the formulation just given is our merely implicit reference to the linear nature of (unitary) time-evolution in quantum mechanics, through the word "dynamical" in (c). Moreover, all reference to the possible completeness of quantum states has also, implicitly, been bracketed under (c). The reason is that Maudlin's (and everyone else's) assumption 1(b), which forms the basis of all discussions of the measurement problem, is actually a counterfactual of the kind one should avoid in quantum mechanics (Landsman 2017). What this assumption says is that *if* ψ_D *were* the initial state of Schrödinger's Cat, then it *would* evolve (linearly) according to the Schrödinger equation with *given* Hamiltonian h. *If* the initial state *were* ψ_L, then it *would* evolve according to the *same* Hamiltonian h, and *if* it *were* $(\psi_D + \psi_L)/\sqrt{2}$, then it *would* evolve according to h, too.

However, a fundamental lesson from the models in Landsman and Reuvers (2013) is that in typical measurement situations unitary quantum dynamics is extremely unstable in the classical limit $\hbar \to 0$. See Jona-Lasinio, Martinelli, and Scoppola (1981) and Simon (1985) for the original mathematical arguments to this effect (in our view these two papers belong to the most important ever written about the foundations of quantum mechanics, although they were not intended as such), as well as Landsman (2013) for similar results in the limit $N \to \infty$. This implies that at least in the classical regime of quantum mechanics the counterfactual linearity assumption underlying the usual discussions of the measurement problem is meaningless, because very slight modifications to the Hamiltonian (which are harmless in the quantum regime), while keeping the dynamics unitary, suffice to destabilize the wave-function. For example, as $\hbar \to 0$ a tiny asymmetric perturbation to the double well potential is enough to localize the quantum ground state, and this happens in accordance with Butterfield's principle, as the effect already occurs for small $\hbar > 0$. Similarly, ground states of spin chains that are of Schrödinger Cat type for small N localize (in spin configuration space) under tiny asymmetric perturbation of the Hamiltonian as $N \to \infty$, and this already happens for large but finite N (Landsman 2013).

Thus we envisage the solution to the measurement problem (at least in the above reformulation) to be as follows: the task being to find a dynamical mechanism that removes all but one term of a typical post-measurement superposition of quantum states, the key *conceptual* point is that according to Bohr this task only needs to be accomplished near the classical limit, upon which the key *technical* point is that asymmetric perturbations to the Hamiltonian, while having a negligible effect in the quantum regime, destabilize the superposition so as to leave just one term in the classical regime. Thus the dynamical mechanism is effective where it should, and does not operate where it should not. In response to Leggett (2002), the eventual collapse of macroscopic Schrödinger Cat states does not require the demise of quantum theory, as this collapse is guaranteed by the instability under tiny perturbations of quantum dynamics in the classical regime.

In conclusion, of our formulation no. 4 of the measurement problem we reject assumption (c), effectively proposing a (dynamical) collapse theory of a new kind (both technically and conceptually). This also explains 2(c) and solves 3. Assumption 1(a) is somewhat irrelevant, since what is incomplete in practice is not the state but the Hamiltonian (since its tiny perturbations are typically hidden). As explained above, Assumption 1(b) is a dangerous counterfactual, so we do not make it, whereas we wholeheartedly endorse 1(c)!

Having said this, many problems remain to be overcome since the first steps were taken in Landsman and Reuvers (2013), Landsman (2013), and van Heugten and Wolters (2016).

Appendix: The mathematics of asymptotic Bohrification

Asymptotic Bohrification is based on the notion of a *continuous bundle of C*-algebras*. This concept may sound unnecessarily technical, but it provides the mathematical bridge between the classical and the quantum and hence deserves to be more widely known.

Let I be a compact space, which for us is always a subset of the unit interval $[0,1]$ that contains 0 as an accumulation point, so one may have, for example, $I = [0,1]$ itself, or $I = (1/\mathbb{N}) \cup \{0\} \equiv 1/\dot{\mathbb{N}}$, where $\mathbb{N} = \{1, 2, \ldots\}$. In asymptotic Bohrification, I either plays the role of the value set for Planck's constant, in which case one is interested in the limit $\hbar \to 0$, or it indexes $1/N$, where $N \to \infty$ is the number of particles in some system, or the number of experiments (as in our derivation of the Born rule in Section 5), or the principal quantum number in atomic orbit theory (as in Bohr's original correspondence principle). Also in the latter case we generically write $\hbar \in I$, so that $N = 1/\hbar$. A *continuous bundle of C*-algebras* over I, then, consists of a C*-algebra A, a collection of C*-algebras $(A_\hbar)_{\hbar \in I}$, and surjective homomorphisms $\varphi_\hbar : A \to A_\hbar$ for each $\hbar \in I$, such that:

1. The function $\hbar \mapsto \| \varphi_\hbar(a) \|_\hbar$ is in $C(I)$ for each $a \in A$ (in particular at $\hbar = 0$!).
2. Writing $\| \cdot \|_\hbar$ for the norm in $A_\hbar$, the norm of any $a \in A$ is given by

$$\| a \| = \sup_{\hbar \in I} \| \varphi_\hbar(a) \|_\hbar . \tag{61}$$

3. For any $f \in C(I)$ and $a \in A$ there is an element $fa \in A$ such that for each $\hbar \in I$,

$$\varphi_\hbar(fa) = f(\hbar)\varphi_\hbar(a). \tag{62}$$

A *continuous (cross-)section* of such a bundle is a map $\hbar \mapsto a(\hbar) \in A_\hbar$, $\hbar \in I$, for which there is an $a \in A$ such that $a(\hbar) = \varphi_\hbar(a)$ for each $\hbar \in I$. Thus the C*-algebra A may be identified with the space of continuous sections of the bundle: if we do so, the homomorphism $\varphi_\hbar$ is just the evaluation map at $\hbar$. The structure of A as a C*-algebra then corresponds to pointwise operations on sections. The idea is that the family $(A_\hbar)_{\hbar \in I}$ of C*-algebras is glued together by specifying a topology on the disjoint union $\bigsqcup_{\hbar \in [0,1]} A_\hbar$, seen as a fiber bundle over I. However, this topology is in fact given rather indirectly, namely, via the specification of the space of continuous sections.

Another way to look at continuous bundles of C*-algebras is to start from a non-degenerate homomorphism φ from $C(I)$ to the center $Z(M(A))$ of the multiplier algebra $M(A)$ of A; we write fa for $\varphi(f)a$ (in this notation non-degeneracy means that $C(I)A$ is dense in A, which implies that $C(I)A = A$). Given φ, one may define fiber algebras by

$$A_\hbar = A / (C_0(I;\hbar) \cdot A); \tag{63}$$

$$C_0(I;\hbar) = \{f \in C(I) \mid f(\hbar) = 0\}, \tag{64}$$

and $C_0(I;\hbar) \cdot A$ is an ideal in A, making the quotient $A_\hbar$ a C*-algebra. The projections $\varphi_\hbar : A \to A_\hbar$ are then given by the corresponding quotient maps (sending $a \in A$ to its equivalence class in $A_\hbar$). Since in general the function $\hbar \mapsto \| \varphi_\hbar(a) \|_\hbar$ is merely upper semicontinuous, one only obtains a structure equivalent to the one described in the above definition if one explicitly requires the above function to be in $C(I)$, in which case property 2 follows, too. See, for example, Williams (2007, App. C) for the material in this appendix.

Note

*   Dedicated to the memory of Rudolf Haag (1922–2016).
Research on Bohrification has been supported by the Radboud University, the Netherlands Organization for Scientific Research (NWO), and the Templeton World Charity Foundation (TWCF). The author was also very fortunate in having been surrounded by such good (PhD) students and postdocs contributing to the Bohrification program: in alphabetical order these were Martijn Caspers, Ronnie Hermens, Jasper van Heugten, Chris Heunen, Bert Lindenhovius, Robin Reuvers, Bas Spitters, Marco Stevens, and Sander Wolters.

References

Aarens, J. F. 1970. "Quasi-states on C*-algebras." *Transactions of the American Mathematical Society* 149: 601–25.

Anderson, J. 1979. "Extensions, restrictions, and representations of states on C*-algebras." *Transactions of the American Mathematical Society* 249: 303–29.

Banaschewski, B., and C. J. Mulvey, 2006. "A globalisation of the Gelfand duality theorem." *Annals of Pure and Applied Logic* 137: 62–103.

Bayen, F., M. Flato, C. Fronsdal, A. Lichnerowicz, and D. Sternheimer. 1978. "Deformation theory and quantization I, II." *Annals of Physic (N.Y.)* 110: 61–110, 111–51.

Bell, J. S. 1966. "On the problem of hidden variables in quantum mechanics." *Reviews of Modern Physics* 38: 447–52.

Bell, J. S. 1975. "On wave packet reduction in the Coleman-Hepp model." *Helvetica Physica Acta* 48: 93–8.

Bell, J. S. 1990. "Against 'measurement.'" In *Sixty-Two Years of Uncertainty*, edited by A. I. Miller, 17–31. New York: Plenum Press.

Belinfante, F. J. 1973. *A Survey of Hidden-Variable Theories*. Oxford: Pergamon Press.

Berezin, F. A. 1975. "General concept of quantization." *Communications in Mathematical Physics* 40: 153–74.

Birkhoff, G., and J. von Neumann, 1936. "The logic of quantum mechanics." *Annals of Mathematics* 37: 823–43.

Bohr, N. 1949. "Discussion with Einstein on epistemological problems in atomic physics." In *Albert Einstein: Philosopher-Scientist*, edited by P. A. Schlipp, 201–41. La Salle: Open Court.

Bohr, N. 1976. *Collected Works. Vol. 3: The Correspondence Principle (1918–1923)*, edited by J. Rosenfeld and R. Nielsen. Amsterdam: North-Holland.

Bohr, N. 1985. *Collected Works. Vol. 6: Foundations of Quantum Physics I (1926–1932)*, edited by J. Kalckar. Amsterdam: North-Holland.

Bohr, N. 1996. *Collected Works. Vol. 7: Foundations of Quantum Physics II (1933–1958)*, edited by J. Kalckar. Amsterdam: North-Holland.

Bokulich, A. 2008. *Reexamining the Quantum-Classical Relation: Beyond Reductionism and Pluralism*. Cambridge: Cambridge University Press.

Bona, P. 1988. "The dynamics of a class of mean-field theories." *Journal of Mathematical Physics* 29: 2223–35.

Bona, P. 2000. "Extended quantum mechanics." *Acta Physica Slovaca* 50: 1–198.

Bratteli, O., and D. W. Robinson 1981. *Operator Algebras and Quantum Statistical Mechanics. Vol. II: Equilibrium States, Models in Statistical Mechanics*. Berlin: Springer.

Bub, J. 2011. "Is von Neumann's 'no hidden variables' proof silly?" In *Deep Beauty: Mathematical Innovation and the Search for Underlying Intelligibility in the Quantum World*, edited by H. Halvorson, 393–408. Cambridge: Cambridge University Press.

Butterfield, J. 2011. "Less is different: Emergence and reduction reconciled." *Foundations of Physics* 41: 1065–135.

Camilleri, K. 2009. *Heisenberg and the Interpretation of Quantum Mechanics: The Physicist as Philosopher*. Cambridge: Cambridge University Press.

Camilleri, K., and M. Schlosshauer 2015. "Niels Bohr as philosopher of experiment: Does decoherence theory challenge Bohr's doctrine of classical concepts?" *Studies in History and Philosophy of Modern Physics* 49: 73–83.

Caruana, L. 1995. "John von Neumann's 'impossibility proof' in a historical perspective." *Physis* 32: 109–24.

Caspers, M., C. Heunen, N. P. Landsman, and B. Spitters 2009. "Intuitionistic quantum logic of an *n*-level system." *Foundations of Physics* 39: 731–59.

Cassinello, A., and J. L. Sánchez-Gómez 1996. "On the probabilistic postulate of quantum mechanics." *Foundations of Physics* 26: 1357–74.

Caves, C., and R. Schack, 2005. "Properties of the frequency operator do not imply the quantum probability postulate." *Annals of Physics (N.Y.)* 315: 123–46.

Connes, A. 1994. *Noncommutative Geometry*. San Diego: Academic Press.

Darrigol, O. 1992. *From c-Numbers to q-Numbers*. Berkeley: University of California Press.

Dirac, P. A. M. 1930. *The Principles of Quantum Mechanics*. Oxford: Clarendon Press.

Döring, A. 2005. "Kochen–Specker theorem for von Neumann algebras." *International Journal of Theoretical Physics* 44: 139–60.

Döring, A. 2012. "Topos-based logic for quantum systems and Bi-Heyting algebras." In *Logic and Algebraic Structures in Quantum Computing*, edited by J. Chubb, A. Eskandarian, and V. Harizanov, 151–73. Cambridge: Cambridge University Press.

Döring, A. 2014. "Two new complete invariants of von Neumann algebras." arXiv:1411.5558.

Döring, A., and J. Harding, 2010. "Abelian subalgebras and the Jordan structure of a von Neumann algebra." arXiv:1009.4945.

Döring, A., and C. J. Isham. 2010. "'What is a thing?': Topos theory in the foundations of physics." *Lecture Notes in Physics* 813: 753–937.

Duffield, N. G., and R. F. Werner, 1992. "Local dynamics of mean-field quantum systems." *Helvetica Physica Acta* 65: 1016–54.

Earman, J. 2004. "Curie's principle and spontaneous symmetry breaking." *International Studies in the Philosophy of Science* 18: 173–98.

Eilers, M., and E. Horst 1975. "The theorem of Gleason for nonseparable Hilbert spaces." *International Journal of Theoretical Physics* 13: 419–24.

Emch, G. G., and B. Whitten-Wolfe. 1976. "Mechanical quantum measuring process." *Helvetica Physica Acta* 49: 45–55.

Farhi, E., J. Goldstone, and S. Gutmann 1989. "How probability arises in quantum mechanics." *Annals of Physics (N.Y.)* 192: 368–82.

Faye, J. 2002. "Copenhagen interpretation of quantum mechanics." In *The Stanford Encyclopedia of Philosophy (Summer 2002 Edition)*, edited by E. N Zalta. http://plato.stanford.edu/archives/sum2002/entries/qm-copenhagen/.

Fine, A. 1974. "On the completeness of quantum theory." *Synthese* 29: 257–89.

Finkelstein, D. 1965. "The logic of quantum physics." *Transactions of the New York Academy of Science* 25: 621–37.

Folse, H. J. 1985. *The Philosophy of Niels Bohr*. Amsterdam: North-Holland.

Gelfand, I. M., and M. A. Naimark, 1943. "On the imbedding of normed rings into the ring of operators in Hilbert space." *Sbornik: Mathematics* 12: 197–213.

Gleason, A. M. 1957. "Measures on the closed subspaces of a Hilbert space." *Journal of Mathematics and Mechanics* 6: 885–93.

Gottwald, S. 2015. "Many-valued logic." In *The Stanford Encyclopedia of Philosophy (Spring 2015 Edition)*, edited by E. N. Zalta. http://plato.stanford.edu/archives/spr2015/entries/logic-manyvalued/.

Haag, R. 1962. "The mathematical structure of the Bardeen–Cooper–Schrieffer model." *Nuovo Cimento* 25: 287–98.

Haag, R. 1992. *Local Quantum Physics: Fields, Particles, Algebras.* Heidelberg: Springer-Verlag.

Hamhalter, J. 1993. "Pure Jauch-Piron states on von Neumann algebras." *Annales de l'IHP Physique théorique* 58: 173–87.

Hamhalter, J. 2004. *Quantum Measure Theory.* Dordrecht: Kluwer Academic Publishers.

Hamhalter, J. 2011. "Isomorphisms of ordered structures of abelian C*-subalgebras of C*-algebras." *Journal of Mathematical Analysis and Applications* 383: 391–9.

Hamhalter, J., and E. Turilova 2013. "Structure of associative subalgebras of Jordan operator algebras." *Quarterly Journal of Mathematics* 64: 397–408.

Hamilton, J., C. J. Isham, and J. Butterfield, 2000. "Topos perspective on the Kochen–Specker Theorem: III. Von Neumann Algebras as the base category." *International Journal of Theoretical Physics* 39: 1413–36.

Hartle, J. B. 1968. "Quantum mechanics of individual systems." *American Journal of Physics* 36: 704–12.

Heisenberg, W. 1958. *Physics and Philosophy: The Revolution in Modern Science.* London. Allen & Unwin.

Hendry, J. 1984. *The Creation of Quantum Mechanics and the Bohr-Pauli Dialogue.* Dordrecht: Reidel.

Hepp, K. 1972. "Quantum theory of measurement and macroscopic observables." *Helvetica Physica Acta* 45: 237–48.

Hermann, G. 1935. "Die naturphilosphischen Grundlagen der Quantenmechanik." *Abhandlungen der Fries'schen Schule* 6: 75–152.

Hermens, R. 2009. *Quantum Mechanics: From Realism to Intuitionism.* MSc thesis, Radboud University Nijmegen. http://philsci-archive.pitt.edu/5021/.

Hermens, R. 2016. *Philosophy of Quantum Probability: An Empiricist Study of Its Formalism and Logic.* PhD thesis, Rijksuniversiteit Groningen.

Heugten, J. van, and S. Wolters. 2016. "Obituary for a flea." In *Proceedings of the Nagoya Winter Workshop 2015: Reality and Measurement in Algebraic Quantum Theory*, edited by M. Ozawa. arXiv:1610.06093.

Heunen, C. 2014a. "Characterizations of categories of commutative C*-algebras." *Communications in Mathematical Physics* 331: 215–38.

Heunen, C. 2014b. "The many classical faces of quantum structures." arXiv:1412.2177.

Heunen, C., N. P. Landsman, and B. Spitters. 2009. "A topos for algebraic quantum theory." *Communications in Mathematical Physics* 291: 63–110.

Heunen, C., N. P. Landsman, and B. Spitters. 2012. "Bohrification of operator algebras and quantum logic." *Synthese* 186: 719–52.

Heunen, C., N. P. Landsman, B. Spitters, and S. Wolters. 2012. "The Gelfand spectrum of a noncommutative C*-algebra: a topos-theoretic approach." *Journal of the Australian Mathematical Society* 90: 32–59.

Heunen, C., and A. J. Lindenhovius. 2015. "Domains of commutative C*-subalgebras." arXiv:1504.02730.

Hornberger, K., S. Gerlich, P. Haslinger, S. Nimmrichter, and M. Arndt. 2012. "Colloquium: Quantum interference of clusters and molecules." *Reviews of Modern Physics* 84: 157–73.

Howard, D. 2004. "Who invented the Copenhagen interpretation?" *Philosophy of Science* 71: 669–82.

Isham, C. J., and J. Butterfield 1998. "Topos perspective on the Kochen–Specker theorem: I. Quantum states as generalized valuations." *International Journal of Theoretical Physics* 37: 2669–733.

Ivrii, V. 1998. *Microlocal Analysis and Precise Spectral Asymptotics.*
New York: Springer-Verlag.

Jona-Lasinio, G., F. Martinelli, and E. Scoppola 1981. "New approach to the semiclassical limit of quantum mechanics." *Communications in Mathematical Physics* 80: 223–54.

Joos, E., H. D. Zeh, C. Kiefer, D. Giulini, J. Kupsch, and I.-O. Stamatescu. 2003. *Decoherence and the Appearance of a Classical World in Quantum Theory,* second edition. Berlin: Springer-Verlag.

Kadison, R. V., and J. R. Ringrose. 1983. *Fundamentals of the Theory of Operator Algebras. Vol. 1: Elementary Theory.* New York: Academic Press.

Kadison, R. V., and J. R. Ringrose. 1986. *Fundamentals of the Theory of Operator Algebras. Vol. 2: Advanced Theory.* New York: Academic Press.

Kadison, R. V., and I. M. Singer. 1959. "Extensions of pure states." *American Journal of Mathematics* 81: 383–400.

Kaltenbaek, R. et al. 2015. "Macroscopic quantum resonators (MAQRO): 2015 Update." arXiv:1503.02640.

Kochen, S., and E. Specker 1967. "The problem of hidden variables in quantum mechanics." *Journal of Mathematics and Mechanics* 17: 59–87.

Koma, T., and H. Tasaki, 1994. "Symmetry breaking and finite-size effects in quantum many-body systems." *Journal of Statistical Physics* 76: 745–803.

Kovachy, T., et al. 2015. "Quantum superposition at the half-metre scale." *Nature* 528: 530–33.

Landau, L. D., and E. M. Lifshitz. 1977. *Quantum Mechanics: Non-relativistic Theory,* third edition. Oxford: Pergamon Press.

Landsman, N. P. 1991. "Algebraic theory of superselection sectors and the measurement problem in quantum mechanics." *International Journal of Modern Physics* A6: 5349–72.

Landsman, N. P. 1995. "Observation and superselection in quantum mechanics." *Studies in History and Philosophy of Modern Physics* 26: 45–73.

Landsman, N. P. 1998. *Mathematical Topics between Classical and Quantum Mechanics.* New York: Springer-Verlag.

Landsman, N. P. 2007. "Between classical and quantum." In *Handbook of the Philosophy of Science. Vol. 2: Philosophy of Physics, Part A,* J. Butterfield and J. Earman, pp. 417–553. Amsterdam: North-Holland.

Landsman, N. P. 2008. "Macroscopic observables and the Born rule." *Reviews in Mathematical Physics* 20: 1173–90.

Landsman, N. P. 2009. "The Born rule and its interpretation." In *Compendium of Quantum Physics,* edited by D. Greenberger, K. Hentschel, and F. Weinert, 64–70. Dordrecht: Springer.

Landsman, N. P. 2013. "Spontaneous symmetry breaking in quantum systems: Emergence or reduction?" *Studies in History and Philosophy of Modern Physics* 44: 379–94.

Landsman, N. P. 2017. *Foundations of Quantum Theory: From Classical Concepts to Operator Algebras.* Heidelberg: Springer.

Landsman, N. P., and A. J. Lindenhovius. 2017. "Symmetries in exact Bohrification." In *Proceedings of the Nagoya Winter Workshop 2015: Reality and Measurement in Algebraic Quantum Theory,* edited by M. Ozawa. Cham: Springer

Landsman, N. P., and R. Reuvers. 2013. "A flea on Schrödinger's Cat." *Foundations of Physics* 43: 373–407.

Leggett, A. J. 2002. "Testing the limits of quantum mechanics: Motivation, state of play, prospects." *Journal of Physics: Condensed Matter* 14: R415–R451.

Lindenhovius, A. J. 2015. "Classifying finite-dimensional C*-algebras by posets of their commutative C*-subalgebras." *International Journal of Theoretical Physics* 54(12): 4615-4635.

Lindenhovius, A. J. 2016. $C(A)$. PhD thesis, Radboud University Nijmegen.

Liu, C., and G. G. Emch 2005. "Explaining quantum spontaneous symmetry breaking." *Studies in History and Philosophy of Modern Physics* 36: 137-63.

Mac Lane, S., and I. Moerdijk. 1992. *Sheaves in Geometry and Logic: A First Introduction to Topos Theory*. New York: Springer.

Mackey, G. W. 1957. "Quantum mechanics and Hilbert space." *American Mathematical Monthly* 64: 45-57.

Maeda, S. 1990. "Probability measures on projections in von Neumann algebras." *Reviews in Mathematical Physics* 1: 235-90.

Marcus, A., D. A. Spielman, and N. Srivastava. 2014a. "Interlacing families II: Mixed characteristic polynomials and the Kadison–Singer Problem." arXiv:1306.3969.

Marcus, A., D. A. Spielman, and N. Srivastava. 2014b. "Ramanujan graphs and the solution of the Kadison–Singer Problem." arXiv:1408.4421.

Martinez, A. 2002. *An Introduction to Semiclassical and Microlocal Analysis*. New York: Springer-Verlag.

Maudlin, T. 1995. "Three measurement problems." *Topoi* 14: 7-15.

Mehra, J., and H. Rechenberg 1982. *The Historical Development of Quantum Theory. Vol. 1: The Quantum Theory of Planck, Einstein, Bohr, and Sommerfeld: Its Foundation and the Rise of Its Difficulties*. New York: Springer-Verlag.

Moschovakis, J. 2015. "Intuitionistic logic." In *The Stanford Encyclopedia of Philosophy* (Spring 2015 Edition), edited by E. N. Zalta, http://plato.stanford.edu/archives/spr2015/entries/logic-intuitionistic/.

Murray, F. J., and J. von Neumann. 1936. "On rings of operators." *Annals of Mathematics* 37: 116-229.

Murray, F. J., and J. von Neumann. 1937. "On rings of operators II." *Transactions of the American Mathematical Society* 41: 208-48.

Murray, F. J., and J. von Neumann. 1943. "On rings of operators IV." *Annals of Mathematics* 44: 716-808.

von Neumann, J. 1929. "Zur Algebra der Funktionaloperatoren und der theorie der normalen operatoren." *Mathematische Annalen* 102: 370-427.

von Neumann, J. 1931. "Über Funktionen von Funktionaloperatoren." *Annals of Mathematics* 32: 191-226.

von Neumann, J. 1932. *Mathematische Grundlagen der Quantenmechanik*. Berlin: Springer-Verlag. Translation *Mathematical Foundations of Quantum Mechanics*. Princeton: Princeton University Press 1955.

von Neumann, J. 1938. "On infinite direct products." *Compositio Mathematica* 6: 1-77.

von Neumann, J. 1940. "On rings of operators, III." *Annals of Mathematics* 41: 94-161.

von Neumann, J. 1949. "On rings of operators, V. Reduction theory." *Annals of Mathematics* 50: 401-85.

Norton, J. D. 2012. "Approximation and idealization: Why the difference matters." *Philosophy of Science* 79: 207-32.

Palomaki, T. A., et al. 2013. "Entangling mechanical motion with microwave fields." *Science* 342 (6159): 710-13.

Pedersen, G. K. 1989. *Analysis Now*, second edition. New York: Springer-Verlag.

Raggio, G. A., and R. F. Werner. 1989. "Quantum statistical mechanics of general mean field systems." *Helvetica Physica Acta* 62: 980-1003.

Rieffel, M. A. 1989. "Deformation quantization of Heisenberg manifolds."
    Communications in Mathematical Physics 122: 531–62.
Rieffel, M. A. 1994. "Quantization and C^*-algebras." *Contemporary Mathematics*
    167: 66–97.
Ruetsche, L. 2011. *Interpreting Quantum Theories*. Oxford: Oxford University Press.
Scheibe, E. 1973. *The Logical Analysis of Quantum Mechanics*. Oxford: Pergamon Press.
Schlosshauer, M. 2007. *Decoherence and the Quantum-to-Classical Transition*. Berlin: Springer.
Schrödinger, E. 1935. "Die gegenwärtige Situation in der Quantenmechanik." *Die*
    Naturwissenschaften 23: 807–12, 823–8, 844–9.
Seevinck, M. P. 2012. "Challenging the gospel: Grete Hermann on von Neumann's no-
    hidden-variables proof." Slides available at http://mpseevinck.ruhosting.nl/seevinck/
    Aberdeen_Grete_Hermann2.pdf.
Sewell, G. L. 2005. "On the mathematical structure of quantum measurement theory."
    arXiv:math-ph/0505032v2.
Simon, B. 1985. "Semiclassical analysis of low lying eigenvalues. IV. The flea on the
    elephant." *Journal of Functional Analysis* 63: 123–36.
Spehner, D., and F. Haake 2008. "Quantum measurements without macroscopic
    superpositions." *Physical Review A* 77: 052114.
Stevens, M. 2016. *The Kadison–Singer Property*. Heidelberg: Springer.
Tao, T. 2013. "Real stable polynomials and the Kadison-Singer problem." http://terrytao.
    wordpress.com/2013/11/04/real-stable-polynomials-and-the-kadison-singer-problem/.
Van Wesep, R. A. 2006. "Many worlds and the appearance of probability in quantum
    mechanics." *Annals of Physics (N.Y.)* 321: 2438–52.
Wallace, D. 2012. *The Emergent Multiverse: Quantum Theory according to the Everett*
    Interpretation. Oxford: Oxford University Press.
Weaver, N. 2004. "The Kadison-Singer problem in discrepancy theory." *Discrete*
    Mathematics 278: 227–39.
Williams, D. P. 2007. *Crossed Products of C*-Algebras*. Providence: American
    Mathematical Society.
Wolters, S. 2013. *Quantum Toposophy*. PhD thesis, Radboud University Nijmegen.
Zinkernagel, H. 2016. "Niels Bohr on the wave function and the classical/quantum divide."
    Studies in History and Philosophy of Modern Physics 53: 9–19.

16

Why QBism Is Not the Copenhagen Interpretation and What John Bell Might Have Thought of It*

N. David Mermin

Our students learn quantum mechanics the way they learn to ride bicycles (both very valuable accomplishments) without really knowing what they are doing.
                                    J. S. Bell, letter to R. E. Peierls, August 20, 1980

I think we invent concepts, like "particle" or "Professor Peierls," to make the immediate sense of data more intelligible.
                                    J. S. Bell, letter to R. E. Peierls, February 24, 1983

I have the impression as I write this, that a moment ago I heard the bell of the tea trolley. But I am not sure because I was concentrating on what I was writing ... The ideal instantaneous measurements of the textbooks are not precisely realized anywhere anytime, and more or less realized, more or less all the time, more or less everywhere.
                                    J. S. Bell, letter to R. E. Peierls, January 28, 1981[1]

For the past decade and a half Christopher Fuchs and Rüdiger Schack (originally in collaboration with Carlton Caves) have been developing a new way to think about quantum mechanics. Fuchs and Schack have called it QBism (Fuchs and Schack 2013). Their term originally stood for "quantum Bayesianism." But QBism is a way of thinking about science quite generally, not just quantum physics,[2] and it is pertinent even when probabilistic judgments, and therefore "Bayesianism," play no role at all. I nevertheless retain the term "QBism," both to acknowledge the history behind it, and because a secondary meaning remains apt in the broader context: QBism is as big a break with twentieth-century ways of thinking about science as Cubism was with nineteenth-century ways of thinking about art.

* Based on a talk at the conference "Quantum [Un]Speakables II: 50 Years of Bell's Theorem," University of Vienna, June 19, 2014. Dedicated to my friend and Cornell colleague Geoffrey Chester (1928–2014), who for over fifty years enjoyed my more controversial enthusiasms, while always insisting that I keep my feet firmly on the ground. From N. David Mermin *Why Quark Rhymes with Pork, and Other Scientific Diversions*. Reprinted with permission of the author and Cambridge University Press.

QBism maintains that my understanding of the world rests entirely on the experiences that the world has induced in me throughout the course of my life. Nothing beyond my personal experience underlies the picture that I have formed of my own external world.[3] This is a statement of empiricism. But it is empiricism taken more seriously than most scientists are willing to do.

To state that my understanding of the world rests on my experience is not to say that my world exists only within my head, as recent popularizations of QBism have wrongly asserted.[4] Among the ingredients from which I construct my picture of my external world are the impact of that world on my experience, when it responds to the actions that I take on it. When I act on my world, I generally have no control over how it acts back on me.

Nor does QBism maintain that each of us is free to construct our own private worlds. Facile charges of solipsism miss the point. My experience of you leads me to hypothesize that you are a being very much like myself, with your own private experience. This is as firm a belief as any I have. I could not function without it. If asked to assign this hypothesis a probability I would choose $p = 1$.[5] Although I have no direct personal access to your own experience, an important component of my private experience is the impact on me of your efforts to communicate, in speech or writing, your verbal representations of your own experience. Science is a collaborative human effort to find, through our individual actions on the world and our verbal communications with each other, a model for what is common to all of our privately constructed external worlds. Conversations, conferences, research papers, and books are an essential part of the scientific process.

Fuchs (2010) himself may be partly responsible for the silly accusations about solipsism. One of his favorite slogans about QBism is "Quantum mechanics is a single-user theory," sometimes abbreviated to "Me, me, me!" (Fuchs 2014, especially 546–9). This invites the s-word. I hurled it at him myself the first time I came upon such slogans. Although susceptible to misinterpretation, they are important reminders that any application of quantum mechanics must ultimately be understood to be undertaken by a particular person[6] to help her make sense of her own particular experience. They were never intended to mean that there cannot be many different users of quantum mechanics. Nor do they require any particular user to exclude from her own experience what she has heard or read about the private experience of others.

Those who reject QBism—currently a large majority of the physicists who know anything about it—reify the common external world we have all negotiated with each other, removing from the story any reference to the origins of our common world in the private experiences we try to share with each other through language. For all practical purposes reification is a sound strategy. It would be hard to live our daily private or professional scientific lives if we insisted on constantly tracing every aspect of our external world back to its sources in our own private personal experience. My reification of the concepts I invent, to make my immediate sense of data more intelligible, is a useful tool of day-to-day living.

But when subtle conceptual issues are at stake, related to certain notoriously murky scientific concepts like quantum states, then we can no longer refuse to acknowledge that our scientific pictures of the world rest on the private experiences of individual scientists. The most famous investigator Vienna has ever produced, who worked just a short walk from the lecture hall for this conference, put it concisely: "A world constitution that takes no account of the mental apparatus by which we perceive it is an empty abstraction." This was said not by Ludwig Boltzmann, not by Erwin Schrödinger, and not even by Anton Zeilinger. It was said by Sigmund Freud ([1927] 1957, concluding paragraph) just down the hill at *Berggasse* 19. He was writing about religion, but his remark applies equally well to science.

After he returned to Vienna in the early 1960s, Schrödinger ([1954] 1996, 92) made much the same point, somewhat less concisely than Freud: "The scientist subconsciously, almost inadvertently simplifies his problem of understanding Nature by disregarding or cutting out of the picture to be constructed, himself, his own personality, the subject of cognizance" (see also Schrödinger 1958). In expressing these views in the 1960s he rarely mentions quantum mechanics. Only thirty years earlier, in a letter to Sommerfeld on December 11, 1931, does Schrödinger (2011) explicitly tie this view to quantum mechanics, and even then, he allows that it applies to science much more broadly: "Quantum mechanics forbids statements about what really exists—statements about the object. It deals only with the object-subject relation. Even though this holds, after all, for any description of nature, it evidently holds in quantum mechanics in a much more radical sense." We were rather successful excluding the subject from classical physics (but not completely). Quantum physics finally forced (or should have forced) us to think harder about the importance of the object-subject relation.

Niels Bohr (1934, 18), whose views on the meaning of quantum mechanics Schrödinger rejected, also delivered some remarkably QBist-sounding pronouncements, though by "experience" he almost certainly meant the objective readings of large classical instruments and not the personal experience of a particular user of quantum mechanics: "In our description of nature the purpose is not to disclose the real essence of the phenomena but only to track down, so far as it is possible, relations between the manifold aspects of our experience." Thirty years later Bohr ([1961] 1987, 10) was saying pretty much the same thing: "Physics is to be regarded not so much as the study of something a priori given, but as the development of methods or ordering and surveying human experience." Bohr and Schrödinger are not the only dissenting pair who might have found some common ground in QBism.

The fact that each of us has a view of our world that rests entirely on our private personal experience has little bearing on how we actually use our scientific concepts to deal with the world. But it is central to the philosophical concerns of quantum foundational studies. Failing to recognize the foundational importance of personal experience creates illusory puzzles or paradoxes. At their most pernicious, such puzzles motivate unnecessary efforts to reformulate in more complicated ways—or even to change the observational content of—theories which have been entirely successful for all practical purposes.

This talk is not addressed to those who take (often without acknowledging it) an idealistic or Platonic position in their philosophical meditations on the nature of quantum mechanics. They will never be comfortable with QBism. My talk is intended primarily for the growing minority of philosophically minded physicists who, far from rejecting QBism, are starting to maintain that there is nothing very new in it.[7] I am thinking of those who maintain that QBism is nothing more than the Copenhagen interpretation.

I may be partly to blame for this misunderstanding. I have used the above quotations from Bohr in several recent essays about QBism, because QBism provides a context which these quotations finally make unambiguous sense. While they made sense for Bohr too, it was not a QBist kind of sense, and I very much doubt that people gave them a QBist reading. Similarly, my quotation from Freud does not mean that QBism should be identified with psychoanalysis, and the three epigraphs from John Bell at the head of this text should not be taken to mean that I believe QBism had already been put forth by Bell in the early 1980s. The quotations from Bell's letters to Peierls are only to suggest that John Bell, who strenuously and elegantly identified what is incoherent in Copenhagen, might not have dismissed QBism as categorically. There are many important ways in which QBism is profoundly different from Copenhagen, and from any other way of thinking about quantum mechanics that I know of. If you are oblivious to these differences, then you have missed the point of QBism.

The primary reason people wrongly identify QBism with Copenhagen is that QBism, like most varieties of Copenhagen, takes the quantum state of a system to be not an objective property of that system, but a mathematical tool for thinking about the system.[8] In contrast, in many of the major nonstandard interpretations—many worlds, Bohmian mechanics, and spontaneous collapse theories—the quantum state of a system is very much an objective property of that system.[9] Even people who reject all these heresies and claim to hold standard views of quantum mechanics are often careless about reifying quantum states. Some claim, for example, that quantum states were evolving (and even collapsing) in the early universe, long before anybody existed to assign such states. But the models of the early universe to which we assign quantum states are models that we construct to account for contemporary astrophysical data. In the absence of such data, we would not have come up with the models. As Rudolf Peierls (1991, 19–20) remarked, "If there is a part of the Universe, or a period in its history, which is not capable of influencing present-day events directly or indirectly, then indeed there would be no sense in applying quantum mechanics to it."

A fundamental difference between QBism and any flavor of Copenhagen is that QBism explicitly introduces each user of quantum mechanics into the story, together with the world external to that user. Since every user is different, dividing the world differently into external and internal, every application of quantum mechanics to the world must ultimately refer, if only implicitly, to a particular user. But every version of Copenhagen takes a view of the world that makes no reference to the particular user who is trying to make sense of that world.

Fuchs and Schack prefer the term "agent" to "user." "Agent" serves to emphasize that the user takes actions on her world and experiences the consequences of her actions. I prefer the term "user" to emphasize Fuchs's and Schack's equally important point that science is a user's manual. Its purpose is to help each of us make sense of our private experience induced in us by the world outside of us.

It is crucial to note from the beginning that "user" does not mean a generic body of users. It means a particular individual person, who is making use of science to bring coherence to her own private perceptions. I can be a "user." You can be a "user." But we are not jointly a user, because my internal personal experience is inaccessible to you except insofar as I attempt to represent it to you verbally, and vice versa. Science is about the interface between the experience of any particular person and the subset of the world that is external to that particular user.[10] This is unlike anything in any version of Copenhagen.[11] It is central to the QBist understanding of science.

The notion that science is a tool that each of us can apply to our own private body of personal experience is explicitly renounced by the Landau-Lifshitz version of Copenhagen. The opening pages of their *Quantum Mechanics*[12] declare that "it must be most decidedly emphasized that we are here not discussing a process of measurement in which the physicist-observer takes part" (Landau and Lifshitz 1958). They explicitly deny the user any role whatever in the story. To emphasize this they add that "by measurement, in quantum mechanics, we understand any process of interaction between classical and quantum objects, *occurring apart from and independently of any observer*" (my italics). In the second quotation Landau and Lifshitz have, from a QBist point of view, replaced each different member of the set of possible users by one and the same set of "classical objects." Their insistence on eliminating human users from the story, both individually and collectively, leads them to declare that "[it is in principle impossible ... to formulate the basic concepts of quantum mechanics without using classical mechanics." Here they make two big mistakes: they replace the experiences of each user with "classical mechanics," and they confound the diverse experiences of many different users into that single abstract entity.

Bohr seems not as averse as Landau and Lifshitz[13] to letting scientists into the story, but they come in only as proprietors of a single large, *classical* measurement apparatus. All versions of Copenhagen objectify each of the diverse family of users of science into a single common piece of apparatus. Doing this obliterates the fundamental QBist fact that a quantum-mechanical description is always relative to the particular user of quantum mechanics who provides that description. Replacing that user with an apparatus introduces the notoriously ill-defined "shifty split" of the world into quantum and classical, that John Bell so elegantly and correctly deplored.

Bell's split is shifty in two respects. Its character is not fixed. It can be the Landau-Lifshitz split between "classical" and "quantum." But sometimes it is a split between "macroscopic and microscopic." Or between "irreversible" and "reversible." The split is also shifty because its location can freely be moved along the path between whatever poles have been used to characterize it.

There is also a split in QBism, but it is specific to each user. That it shifts from user to user is the full extent to which the split is "shifty." For any particular user there is nothing shifty about it: the split is between that user's directly perceived internal experience, and the external world that that user infers from her experience.

Closely related to its systematic suppression of the user is the central role in Copenhagen of "measurement" and the Copenhagen view of the "outcome" of a measurement. In all versions of Copenhagen a measurement is an interaction between a quantum system and a "measurement apparatus." Depending on the version of Copenhagen, the measurement apparatus could belong to a "classical" domain beyond the scope of quantum mechanics, or it could itself be given a quantum mechanical description. But in any version of Copenhagen the *outcome* of a measurement is some strictly classical information produced by the measurement apparatus as a number on a digital display, or the position of an ordinary pointer, or a number printed on a piece of paper, or a hole punched somewhere along a long tape—something like that. Words like "macroscopic" or "irreversible" are used at this stage to indicate the objective, substantial, non-quantum character of the outcome of a measurement.

In QBism, however, a measurement can be *any* action taken by *any* user on her external world. The outcome of the measurement is the *experience* the world induces back in that particular user, through its response to her action. The QBist view of measurement includes Copenhagen measurements as a special case, in which the action is carried out with the aid of a measurement apparatus and the user's experience consists of her perceiving the display, the pointer, the marks on the paper, or the hole in the tape produced by that apparatus. But a QBist "measurement" is much broader. Users are making measurements more or less all the time more or less everywhere. Every action on her world by every user constitutes a measurement, and her experience of the world's reaction is its outcome. Physics is not limited to the outcomes of "piddling" laboratory tests, as Bell (1990) complained about Copenhagen.

In contrast to the Copenhagen interpretation (or any other interpretation I am aware of), in QBism the outcome of a measurement is special to the user taking the action—a private internal experience of that user. The user can attempt to communicate that experience verbally to other users, who may hear[14] her words. Other users can also observe her action and, under appropriate conditions, experience aspects of the world's reaction closely related to those experienced by the original user. But in QBism the immediate outcome of a measurement is a private experience of the person taking the measurement action, quite unlike the public, objective, classical outcome of a Copenhagen[15] measurement.

Because outcomes of Copenhagen measurements are "classical," they are ipso facto real and objective. Because in QBism an outcome is a personal experience of a user, it is real only for that user, since that user's immediate experience is private, not directly accessible to any other user. Because the private measurement outcome of a user is not a part of the experience of any other user, it is not as such real for other users. Some version of the outcome can enter the experience of other users and become real for them as well, only if the other users have also experienced aspects of the world's response to the user who took the measurement-action, or if that user has sent them reliable verbal or written reports of her own experience.

This is, of course, nothing but the famous story of Wigner and his friend, but in QBism Wigner's Friend is transformed from a paradox to a fundamental parable. Until Wigner manages to share in his friend's experience, it makes sense for him to assign her and her apparatus an entangled state in which her possible reports of her experiences (outcomes) are strictly correlated with the corresponding pointer readings (digital displays, etc.) of the apparatus.

Even versions of Copenhagen that do not prohibit mentioning users would draw the line at allowing a user to apply quantum mechanics to another user's reports of her own internal experience. Other users are either ignored entirely (along with *the* user), or they are implicitly regarded as part of "the classical world." But in QBism each user may assign quantum states in superposition to all of her still unrealized potential experiences, including possible future communications from users she has yet to hear from. Asher Peres's famous Copenhagen mantra "Unperformed experiments have no results" becomes the QBist user's tautology: "Unexperienced experiences are not experienced."

Copenhagen, as expounded by Heisenberg and Peierls, holds that quantum states encapsulate "our knowledge." This has a QBist flavor to it. But it is subject to John Bell's famous objection: Whose knowledge? Knowledge about what?[16] QBism replaces "knowledge" with "belief." Unlike "knowledge," which implies something underlying it that is known, "belief" emphasizes a believer, in this case the user of quantum mechanics. Bell's questions now have simple answers. Whose belief does the quantum state encapsulate? The belief of the person who has made that state assignment. What is the belief about? Her belief is about the implications of her past experience for her subsequent experience.

No version of Copenhagen takes the view that "knowledge" is the state of belief of the particular person who is making use of quantum mechanics to organize her experience. Peierls (2009, 807) may come closest in a little-known November 13, 1980, letter to John Bell: "In my view, a description of the laws of physics consists in giving us a set of correlations between successive observations. By observations I mean ... what our senses can experience. That we have senses and can experience such sensations is an empirical fact, which has not been deduced (and in my opinion cannot be deduced) from current physics." Had Peierls taken care to specify that when he said "we," "us," and "our" he meant each of us, acting and responding as a user of quantum mechanics, this would have been an early statement of QBism. But it seems to me more likely that he was using the first-person plural collectively, to mean all of us together, thereby promulgating the Copenhagen confusion that Bell so vividly condemned.

Copenhagen also comes near QBism in the emphasis Bohr always placed on the outcomes of measurements being stated in "ordinary language." I believe he meant by this that measurement outcomes were necessarily "classical." In QBism the outcome of a measurement is the experience the world induces back in the user who acts on the world. "Classical" for any user is limited to her experience.[17] So measurement outcomes in QBism are necessarily classical, in a way that has nothing to do with language. Ordinary language comes into the QBist story in a more crucial way than it comes into the story told by Bohr. Language is the only means by which different users of quantum mechanics can attempt to compare their own private experiences. Though I cannot myself experience your own experience, I can experience your verbal attempts to represent to me what you experience. It is only in this way that we can arrive at a shared understanding of what is common to all our own experiences of our own external worlds. It is this shared understanding that constitutes the content of science.

A very important difference of QBism, not only from Copenhagen, but from virtually all other ways of looking at science, is the meaning of probability 1 (or 0).[18] In Copenhagen quantum mechanics, an outcome that has probability 1 is enforced by an objective mechanism. This was most succinctly put by Einstein, Podolsky, and Rosen (1935), though they were, notoriously, no fans of Copenhagen. Probability-1 judgments, they held, were backed up by "elements of physical reality."

Bohr (1935) held that the mistake of EPR lay in an "essential ambiguity" in their phrase "without in any way disturbing." For a QBist, their mistake is much simpler than that: probability-1 assignments, like more general probability-p assignments, are personal expressions of a willingness to place or accept bets, constrained only by the requirement[19] that they should not lead to certain loss in any single event. It is wrong to assert that probability assignments must be backed up by objective facts on the ground, even when p = 1. An expectation is assigned probability 1 if it is held as strongly as possible. Probability-1 measures the intensity of a belief: supreme confidence. It does not imply the existence of a deterministic mechanism.

We are all used to the fact that with the advent of quantum mechanics, determinism disappeared from physics. Does it make sense for us to qualify this in a footnote: "Except when quantum mechanics assigns probability 1 to an outcome"? Indeed, the point was made over 250 years ago by David Hume (1748) in his famous critique of induction. Induction is the principle that if something happens over and over and over again, we can take its occurrence to be a deterministic law of nature. What basis do we have for believing in induction? Only that it has worked over and over and over again.

That probability-1 assignments are personal judgments, like any other probability assignments, is essential to the coherence of QBism. It has the virtue of undermining the temptation to infer any kind of "nonlocality" in quantum mechanics from the violation of Bell inequalities (Fuchs, Mermin, and Schack 2014). Though it is alien to the normal scientific view of probability, it is no stranger or unacceptable than Hume's views of induction.[20] What is indisputable is that the QBist position on probability-1 bears no relation to any version of Copenhagen. Even Peierls, who gets closer to QBism than any of the other Copenhagenists, takes probability 1 to be backed up by underlying indisputable objective facts.

Since this is a meeting in celebration of John Bell, I conclude with a few more comments on the quotations from Bell's little-known[21] correspondence with Peierls at the head of my text.

The first quotation suggests a riddle: Why is quantum mechanics like a bicycle? Answer: Because while it is possible to learn how to use either without knowing what you are doing, it is impossible to make sense of either without taking account of what people actually do with them.

The second quotation indicates Bell's willingness to consider concepts, as fundamental as "particle" or the person to whom he is writing his letter, as "inventions" that help him to make better sense of the data that constitute his experience.

The third reveals a willingness to regard measurements as particular responses of particular people to particular experiences induced in them by their external world.

These are all QBist views. Does this mean that John Bell was a QBist? No, of course not—no more than Niels Bohr or Erwin Schrödinger or Rudolf Peierls or Sigmund Freud were QBists. Nobody before Fuchs and Schack has pursued this point of view to its superficially shocking,[22] but logically unavoidable and, ultimately, entirely reasonable conclusions. However, what Bell wrote to Peierls, and the way in which he criticized Copenhagen, lead me to doubt that Bell would have rejected QBism as glibly and superficially as most of his contemporary admirers have done.

John Bell and Rudolf Peierls are two of my scientific heroes, both for their remarkable, often iconoclastic ideas, and for the exceptional elegance and precision with which they put them forth. Yet in their earlier correspondence, and in their two short papers in *Physics World* at the end of Bell's life, they disagree about almost everything in quantum foundations. Peierls disliked the term "Copenhagen interpretation" because it wrongly suggested that there were other viable ways of understanding quantum mechanics. Bell clearly felt that Copenhagen was inadequate and downright incoherent. I like to think that they too, like Bohr and Schrödinger, might have found common ground in QBism.

Notes

I am grateful to Chris Fuchs and Rüdiger Schack for their patient willingness to continue our arguments about QBism, in spite of my inability to get their point for many years. And I thank them both for their comments on earlier versions of this text.

1 Peierls (2009). I have the impression (confirmed at the conference) that all three of these quotations are unfamiliar even to those who, like me, have devoured almost everything John Bell ever wrote about quantum foundations.

2 When the QBist view of science is used to solve classical puzzles I have suggested calling it CBism (Mermin 2014).

3 For "my," "me," "I," you can read appropriate versions of "each of us"; the singular personal pronoun is less awkward. But unadorned "our," "us," and "we" are dangerously ambiguous. In QBism the first-person plural always means each of us individually; it never means all of us collectively, unless this is spelled out. Part of the ninety-year confusion at the foundations of quantum mechanics can be attributed to the unacknowledged ambiguity of the first-person plural pronouns and the carelessness with which they are almost always used.

4 See von Baeyer (2013) and Chalmers (2014). I believe that in both cases these gross distortions were the fault of overly intrusive copy editors and headline writers, who did not understand the manuscripts they were trying to improve.

5 I have more to say about $p = 1$ below.

6 Generally named Alice.

7 I count this as progress. The four stages of acceptance of a radical new idea are: (1) it's nonsense; (2) it's well known; (3) it's trivial; (4) I thought of it first. I'm encouraged to find that stage (2) is now well under way.

8 Heisenberg and Peierls are quite clear about this. Bohr may well have believed it but never spelled it out as explicitly. Landau and Lifshitz, however, are so determined to eliminate any trace of humanity from the story that I suspect their flavor of Copenhagen might reject the view of quantum states as mathematical tools.

9 In consistent histories, which has a Copenhagen tinge, its quantum state can be a true property of a system, but only relative to a "framework."

10 See in this regard my remarks above about the dangers of the first-person plural.

11 And unlike any other way of thinking about quantum mechanics.

12 Translated into English by John Bell, who was therefore intimately acquainted with it.

13 But Peierls (1991) identifies their positions, referring to "the view of Landau and Lifshitz (and therefore of Bohr)" in his *Physics World* article. He disagrees with all of them, saying that it is incorrect to require the apparatus to obey classical physics.

14 As John Bell may have heard the bell of the tea trolley. Hearing something, of course, is a personal experience.

15 I shall stop adding the phrase "or any other interpretation," but in many cases the reader should supply it.

16 Bell used the word "information," not "knowledge," but his objection has the same force with either term.

17 Indeed, the term "classical" has no fundamental role to play in the QBist understanding of quantum mechanics. It can be replaced by "experience."

18 A good example to keep in mind is my abovementioned assignment of probability-1 to my belief that you have personal experiences of your own that have for you the same immediate character that my experiences have for me.

19 Known as Dutch-book coherence. See Fuchs and Schack (2013).

20 I would have expected philosophers of science, with an interest in quantum mechanics, to have had some instructive things to say about this connection, but I'm still waiting.

21 I have had no success finding any of them with Google. For example, there is no point in googling "Bell bicycle." " 'John S. Bell' bicycle" does no better. Even " 'John S. Bell' bicycle quantum" fails to produce anything useful, because there is a brand of bicycle called "Quantum," and Quantum bicycles have bells.

22 Ninety years after the formulation of quantum mechanics, a resolution of the endless disagreements on the meaning of the theory has to be shocking, to account for why it was not discovered long, long ago.

References

Bell, J. S. 1990. "Against 'measurement.'" *Physics World* 3: 33–40.

Bohr, N. 1934. *Atomic Theory and the Description of Nature.* Cambridge: Cambridge University Press.

Bohr, N. 1935. "Can quantum-mechanical description of reality be considered complete?" *Physical Review* 48: 696–702.

Bohr, N. [1961] 1987. *Essays 1958–1962 on Atomic Physics and Human Knowledge.* New York: Wiley; reprinted Woodbridge, CT: Ox Bow Press.

Chalmers, M. 2014. " QBism: Is quantum uncertainty all in the mind?" *New Scientist.* May 10, 2014, 32–5.

Einstein, A., B. Podolsky, and N. Rosen. 1935. "Can quantum-mechanical description of reality be considered complete?" *Physical Review* 47: 777–80.

Freud, S. [1927] 1957. *The Future of an Illusion,* translated by W. D. Robson-Scott. Garden City, NY: Anchor Books.

Fuchs, C. A. 2010. "Quantum Bayesianism at the perimeter." arXiv:1003.5182 [quant=ph]. Doi: https://arxiv.org/pdf/1003.5182v1.pdf.

Fuchs, C. A. 2014 "My struggles with the block universe." arXchiv 1405.2390 [quant-ph]. Doi: https://arxiv.org/pdf/1405.2390v2.pdf.

Fuchs, C. A., N. D. Mermin, and R. Schack. 2014. "An introduction to Qbism with an application to the locality of quantum mechanics." *American Journal of Physics* 82: 749–54.

Fuchs, C. A., and R. Schack. 2013. "Quantum-bayesian coherence." *Review of Modern Physics* 85: 1693–714.

Hume, D. 1748. *An Enquiry concerning Human Understanding.*

Landau, L. D., and E. M. Lifshitz. 1958. *Quantum Mechanics: Non-relativistic Theory,* translated by J. B. Sykes and J. S. Bell. Amsterdam: Butterworth Heinemann.

Mermin, N. D. 2014. "Physics: QBism puts the scientist back into science." *Nature* 507: 421–3.

Peierls, R. E. 1991. "In defence of 'measurement.'" *Physics World,* January, 19–20.

Peierls, R. E. 2009. *Selected Correspondence of Rudolf Peierls, Vol 2,* edited by S. Lee. Singapore: World Scientific.

Schrödinger, E. 1958. *Mind and Matter.* Cambridge: Cambridge University Press.

Schrödinger, E. [1954 and 1951] 1996. *Nature and the Greeks* and *Science and Humanism.* Cambridge; Cambridge University Press.

Schrödinger, E. 2011. *Schrödingers Briefwechsel zur Wellenmechanik und zum Katzenparadoxon.* Berlin: Springer.

von Baeyer, H. C. 2013. "Quantum weirdness? It's all in your mind." *Scientific American* 308: 46–51.

Index

Afshar, Shahriar/Afshar experiment 11, 67, 80–6
Anderson, J. 347
anti-realism/instrumentalism 14, 25, 92, 122, 123, 126, 150, 155, 157, 159, 161, 162, 163, 164, 174 n.9, 236, 237, 253, 264, 265, 284, 303, 305, 308, 310, 311, 329, 330, 331

Bacciagaluppi, Guido 68
Bächtold, Manuel 31, 37
Bacon, Francis 207, 208
Bataille, G. 188
Belinfante, F. J. 342
Bell, John S. 60, 121, 134, 142, 144, 148–9, 161, 303, 339, 342, 354–5, 367, 370–6 n.16
Beller, Mara 37, 41 n.17, 69, 71, 290
Bergson, Henri 185
Birkhoff, G. 350
Bitbol, Michel 167
Bohm, David 23, 39 n.1, 40 n.5, 60, 134, 147, 174 n.5, 216–19, 289, 303, 370
Boltzmann, Ludwig 369
Bona, P. 340
Born, Max/Born rule 4, 8, 22, 24, 49, 123, 137, 172, 197, 210, 261, 279, 283, 300 n.9, 336, 339–40, 350–5, 359
Bose, S. 9
Brandes, Georg 185
Brandom, Robert 156, 169–72, 174 n.7, 175 n.11
Brigandt, Ingo 36
Brock, Steen 40 n.11
Brouwer, Luitzen E. J. 350
Bub, Jeffery 139–40, 150 n.3, 226–7, 230
Bunge, Mario 125
Butterfield, Jeremy 339, 347, 354, 356, 357
Büttinker, M. 219

Camilleri, Kristian 6, 13, 130, 151, 235, 237, 241, 243, 247–8, 285 nn.10,11,13, 314
Carnap, Rudolph 22, 92, 155, 173, 309–10, 330 n.4
Cassinello, A. 351
Cassirer, Ernst 26, 28, 40 n.15, 47, 64
causality 7, 11, 22, 31, 38, 40, 49, 51–9 passim, 64, 68–77, 85, 136–47, 169, 173, 180–4, 238–46, 257–67, 273–5, 293, 300, 305, 309, 317–18
Caves, Carlton 351, 367
Chevalley, Catherine 20, 25, 26, 39 n.3, 40 n.11, 243
Chiao, R. Y. 218
Chisholm, Roderick 92
Christiansen, Voetmann 40 n.11
classical concepts 1, 3–4, 11–14, 20–1, 28–30, 34–9, 48, 51–3, 59–64, 75–7, 103–12, 115–22, 129, 134–7, 157–68, 223–50 passim, 268–9, 275–7, 283, 285, 290, 294, 304–5, 309–22, 329–30, 335, 339, 355–6
Clifton, Rob 139, 285 n.13
Coffa, Alberto 155, 173
cognitive faculties 3, 38, 50–61, 74–6, 115–21, 129, 310
Cohen, Hermann 64, 72
collapse interpretations 1, 3, 4, 7–9, 14, 63, 108, 124–5, 143, 147–8, 192, 228, 254–5, 273, 278–85, 289, 296–303, 326–31, 355–9
Collingwood, R. G. 20, 27, 38
Como paper 6, 8, 11, 28, 56, 67–70, 73, 85, 94, 145, 183, 189, 197, 212, 265–6, 303–8, 312, 317–18, 321, 330–1, 354
complementarity 2, 4–7, 9–11, 13–14, 19–28, 39–40, 48, 55–9, 67–86 passim, 91–2, 100–1, 113, 117, 134, 138, 140, 143, 149, 155–7, 161, 171, 173, 180, 183–9, 207–20 passim, 245–6, 254, 264–5, 270–7, 283, 290–6 passim,

299, 304, 312, 314, 316–23 *passim*, 329–31, 345
Compton, A. H. 260
Condon, E. U. 210, 211, 214, 215–16, 218
contextualism 118–21, 149, 151 n.23, 322
Copenhagen interpretation 5–9, 15, 21, 39, 47, 50, 124, 162–4, 185, 200, 225–9, 254, 285, 289, 331, 336, 367–78
correspondence principle 13–14, 49–50, 63, 122, 133, 138–9, 172, 189–90, 196, 253–63, 267, 276–7, 281, 309–10, 335, 340, 359
Crull, Elise 290
Cuffaro, Michael 25, 243

Daneri, A. 226
Darwinian theory 12, 115–21 *passim*, 129
de Broglie, Louis 56, 69, 144, 166, 210, 289, 304–5, 321
decoherence interpretation 10, 11, 13–14, 34, 60, 224–5, 229–31, 235–7, 246–50, 255, 277–85, 289, 296–7, 300, 315, 324–7, 355
Delbrück, Max 184
Derrida, J. 25, 188
Deutsch, David 289
Dirac, P. A. M. 5, 7, 47, 179, 191–2, 196–7, 200, 291, 295, 300 n.4, 336, 340, 345–7
Dorato, Mauro 123
Döring, A. 339, 348
Duffield, N. G. 340
Dummett, Michael 170, 174 n.8
Dyson, Freeman 191, 193–6

Earman, John 354, 356, 357
Ehrenfest, Paul 68, 290, 299
Einstein, Albert 5, 11, 14, 37, 68–9, 77–9, 81, 85, 92–3, 100, 120, 134, 135, 137–43 *passim*, 149, 150 n.13, 155–6, 182–3, 186, 197, 200, 239–40, 256, 260, 264, 276, 290–9 *passim*, 312–13, 316, 320–3, 330 n.1, 331 n.8, 373
Englert, B.-G. 79–85 *passim*, 196
entanglement 13, 29, 55, 133, 142, 147–9, 208, 224, 229, 243, 247, 250, 262, 265, 269, 274–82, 297, 308, 323–8, 372
epistemological lesson 2, 8, 20, 25–39 *passim*, 47, 51, 62–4, 92, 103–6, 113, 184–5, 223–31 *passim*, 235–8, 243,

248–9, 253–9 *passim*, 267–76 *passim*, 315–16, 321–9 *passim*, 355
Everett, Hugh 14, 33, 60, 62, 156–66, 173 n.2, 226, 248, 289–90, 295–9, 300 nn.8,9,12, 326, 329–30, 331 n.12
experiment 3–4, 6, 11–15, 21, 28–39, 41 n.19, 50–62, 67–8, 70–86, 108–12, 115, 118–21, 125–9, 135–50, 158–72, 179–200 *passim*, 207–9, 212–20, 223–31 *passim*, 237–51 *passim*, 254, 258–77 *passim*, 279, 284, 285 nn.10,11, 290–4, 299, 307, 310–21, 326–9, 335, 351–5, 373

Farhi, E. 351
Favrholdt, David 26, 40 n.10
Faye, Jan 22, 25, 26, 40 nn.8,10,14, 136, 150 n.7, 173 nn.1,2, 174 nn.7,8,10, 239, 243, 245, 330 n.5
Fermi, Enrico 47, 213
Feyerabend, Paul K. 19, 26, 200 n.2
Feynman, Richard 193, 218–20
Fine, Arthur 37, 41 n.17, 290, 344
Finkelstein, D. 351
Fock, Vladimir 23–4, 54
Folse, Henry J. 25–7, 40 n.8, 74–5, 200 nn.5,6, 239, 245
Fowler, Ralph H. 213
Frank, Philipp 22, 24, 330 n.4
Freire, O. 226
Freitas, F. 226
Freud, Sigmund 369, 370, 374
Fuchs, Christopher 301 n.13, 367, 368, 370, 374, 375

Gamow, George 210, 211, 212, 214–15, 216
Gefand, I. M. 337
Gleason, A. M. 339, 342–4, 345
Goldstone, J. 351
Goodman, Nelson, 92
Greaves, Hilary 289
Groenwald, Thomas H. 225
Gurney, R. W. 210, 211, 214, 215–16
Gutmann, S. 351

Haag, Rudolf 340
Haake, F. 357
Halvorson, Hans 139, 285 n.14
Hamhalter, J. 342, 348

Hamilton, J. 339, 347
Harding, J. 348
Harper, W. 255
Hartle, J. B. 351
Heidegger, Martin 12, 25, 185, 188,
   199–200
Heisenberg, Werner 5–9, 22, 24, 33, 39
   nn.2,4, 47, 49–50, 53–4, 58, 68, 70, 76,
   80, 85, 109, 118, 122–3, 135,
   137–47 *passim*, 151 nn.16,23,
   189–200 *passim*, 207–19 *passim*, 224,
   227–31 *passim*, 240, 254, 260–1, 266,
   272, 285, 293, 300, 304, 311, 316, 351,
   356, 373
Hempel, Carl 330 n.4
Henderson, J. R. 9
Hermann, Grete 22, 47, 54, 293, 300
   nn.6,11, 339, 342
Herz, Heinrich 26, 40 n.11
Heunen, Chris 339, 347, 348
Heyting, Arend 349–50
Higgs boson 12, 179–81, 191, 193, 196–7,
   200, 357
Høffding, Harald 2, 25–6, 38, 40 nn.10,11,
   47, 91, 93
Honner, John 47, 174, 239, 244
Hooker, Clifford A. 26, 47, 71, 160, 208
Hornberger, K. 355
Howard, Don 6, 20, 24, 26, 27, 133, 146–7,
   285 nn.5,6,9,14, 290–4 *passim*, 315,
   317, 323–4, 327, 330 n.5, 331 n.8
Hume, David 185, 374
Hund, Friedrich 13, 210–15, 220
Husserl, Edmund 25, 185, 188

individuality; *see also* wholeness 9, 56, 110,
   183, 189, 242, 243–4, 246, 247, 254,
   262, 265–6, 271, 272, 273, 317–18,
   322, 323–4, 331 n.11
indivisibility/non-separability 12, 54, 79,
   133, 142, 145–8, 171, 183, 188–9,
   229, 242, 254, 260, 262–82 *passim*,
   317–18, 347
interaction 2–12 *passim*, 29–31, 35, 49, 54–
   8, 61, 64, 74, 79, 86, 102, 108–12, 120,
   125, 129, 133–7, 141–8, 150 nn.1,14,
   157–8, 162, 165–7, 180–1, 188, 190,
   193–5, 198, 208, 212–13, 223–4, 227–
   30, 236, 239, 241–3, 247, 250, 258–9,

265–84, 291–301, 316–19, 322–31,
   355, 371
Isham, C. J. 339, 347

Jacobsen, Anja 22, 23
Jakiel, J. 219
James, William 2, 47, 91–2, 113, 184–5
Jammer, Max 70, 226, 293, 300 n.2
Jona-Lasinio, G. 340
Joos, Erich 230
Jordan, Pascual 21, 197
Jørgensen, Jørgen 22
Jung, Karl 184

Kadison, R. V. 339, 345–7
Kaiser, David 47, 53, 71, 168, 208
Kaltenbaek, R. 355
Kant, Immanuel/kantianism 2–3, 11, 20,
   22, 24–6, 36, 38, 40 nn.7,11, 47–55,
   58–63, 70–6, 85, 92–7, 106, 113,
   115–16, 130, 147, 150 n.8, 167–78,
   173, 180, 182, 185, 188, 240, 244, 250,
   300 n.6, 311, 329, 342
Kastner, R. E. 83
Katsumori, Makoto 20, 25, 34, 155
Kauark-Leite, Patricia 237, 240
Kochen-Specker theorem 14, 200 n.3,
   339–44, 348–9
Kovachy, T. 355
Kramers, Hendrik A. 259, 260, 261, 317
Krips, Henry 40 n.8
Kroes, Peter 30, 32, 37
Kuhn, Thomas S. 24, 113 n.1, 122, 200 n.2

Lakatos, Imre 11, 19, 63, 200 n.2
Landau, Lev 134, 142, 147–9, 151 n.22, 201,
   336, 371, 375 n.8, 376 n.13
Landsman, Klaas 314, 339, 347, 354, 358
language 3, 12, 21–3, 25–6, 29–38, 40 n.5,
   50–3, 70–1, 75–6, 104–13, 115–18,
   121–4, 126–9, 135–7, 140, 144, 146–7,
   155, 158–9, 167–8, 175 n.12, 181,
   198, 208, 210, 223, 239–42, 254, 256,
   263–4, 267, 270, 272, 275, 283, 296,
   307–11, 314–15, 318–21, 326–8, 330
   n.1, 335, 339, 344, 350, 353, 368, 373
Laudisa, Frederico 123
Leggett, A. J. 358
Lewis, C. I. 12, 92–113 *passim*, 113 nn.2,3

Lifshitz, E. 134, 142, 147–9, 336, 371, 375
    n.8, 376 n.13
Lindenhovius, Bert 348
logical positivism/empiricism 11, 14, 21–3,
    25, 47, 290, 309–10, 311–12
Loinger, A. 226
Lorentz, H. A. 140
Ludwig, Günther 164, 226, 238
Lyotard, J.-F. 188

Mackey, G. W. 339, 345
MacKinnon, Edward 25, 40 n.8, 41 n.17,
    69–70, 126, 150, 175
Maeda, S. 342
many worlds interpretation 10, 370
Marcus, A. 347
Martinello, F. 340
mathematical symbolism/formalism 3–4,
    7–14, 21, 29–32, 37, 49–53, 56–60,
    62, 64, 67–70, 75, 80, 84–5, 92–3, 104,
    108–13, 115, 118, 121–8, 134–5, 138,
    143, 157–60, 163–4, 168–71, 179, 187,
    190, 193–6, 207, 210–13, 219, 223,
    228–30, 236–7, 249, 254–5, 260–5,
    268–77 *passim*, 283–4, 300 nn.4,9,
    303–30 *passim*, 336, 348–51, 357
Maudlin, Tim 123–4, 355, 358
measurement problem 12, 14, 28, 48,
    59–61, 124–5, 133, 142–7 *passim*, 156,
    164, 171, 226, 254–5, 283–4, 319–20,
    326, 336, 339–40, 354–9
Merzbacher, Eugen 210, 211, 215
Metzinger, Jan 185–6
Miller, A. I. 185–6
mind 2–6, 9–10, 12, 26, 54, 74, 91–4, 98–9,
    102–7, 111, 135, 164, 184, 188, 200,
    228, 282, 297
Morse, P. M. 216, 218
Murdoch, Dugald 26, 40 n.8,14, 47, 56,
    69, 73, 118, 144, 150, 155, 160, 174
    n.5,8, 245
Murray, F. J. 337
Myer-Abich, Klaus-Michel 23

Naimark, M. A. 337
naturalism 2, 5, 11–12, 115, 118, 120,
    124, 129
Neurath, Otto 22, 330 n.4
Nietzsche, Friedrich 185, 188

Nordheim, L. 13, 209–15, 220
Norton, John D. 353

objectivity/objective description 5, 7, 13,
    58, 74, 75, 92, 98, 101, 103, 109, 115,
    127, 155, 159, 161, 172, 175 n.11, 238,
    242–6, 273, 294, 315
obscurity 11, 14, 19–28 *passim*, 50, 134,
    156, 209, 316
Olkhovsky, V. S. 219
Oppenheimer, Robert J. 216
Osnaghi, Stefano 226

Pais, Abraham 40 n.6
Palomaki, T. A. 355
particle picture 8, 9
Pauli, Wolfgang 9, 22, 39 n.4, 49, 54, 68,
    184, 227, 293
Peierls, Rudolf 201 n.9, 367, 370, 373–5
    n.1,8, 376 n.13
Peirce, C. S. 3, 91
Petersen, Aage 23, 24, 33, 40 n.5, 62, 158,
    162, 163, 166, 169, 173 n.2, 225
pictorial description/visualization 5, 7,
    8, 9, 191
pictorial representation 7, 9, 123, 191,
    239–40, 306, 308
Plotnitsky, Arkady 19, 25, 28, 40 n.13
Podolsky, Boris 14, 37, 182, 264, 291, 299,
    312, 321, 323, 373
Popper, Karl 4, 19, 92, 113–14 n.3, 125,
    330 n.4
pragmatism 2, 3, 5, 7, 12, 15, 26, 38, 47, 91–
    113, 115, 122–3, 155, 170, 185
probability 5, 7, 8, 121, 122, 123, 155, 157,
    158–61, 162, 163, 173 n.3, 182, 189,
    191, 192, 210, 214, 215, 216, 218, 228,
    230, 257, 258, 260, 262, 281, 291, 296,
    301 n.12, 326, 335–6 *passim*, 368,
    373–4, 376 n.18
Prosperi, G. M. 226

QBism 11, 15, 367–75 nn.2,3, 376 n.17
quantum/classical cut/distinction/divide
    13, 133, 146, 149, 151, 223–31 *passim*,
    235–8, 311, 321, 331 n.7
quantum electrodynamics/QED 12, 28, 37,
    40 n.13, 179, 180, 190–6
quantum revolution 2, 9, 98, 110, 113

quantum tunneling 13, 207–20
Quine, W. V. O. 92
Qureshi, T. 83

Raggio, G. A. 340
realism 1, 13–14, 22, 25, 27, 74, 92, 115,
  129–30, 150 nn.3,7, 180–1, 183–4,
  187, 188, 236, 265–6, 308, 310, 336
Recami, E. 219
Reck, A. J. 104, 107, 108
Reichenbach, Hans 53, 64, 309, 330 n.4
Reitzner, D. 83
relativity 5, 68, 92, 93, 100–1, 116, 121,
  135–41, 146, 150 n.10, 180, 189–91,
  198, 239, 240, 264–5, 268, 271, 282,
  289, 306–7, 330
representation 3, 4, 7–10, 12, 14, 15, 47,
  52, 55, 56, 71, 73, 74, 115, 118, 122–6,
  128, 129, 130, 136, 139, 161, 169, 170,
  172, 174 n.10, 180, 181, 182, 187–8,
  190–1, 193, 195–6, 225, 240, 244, 254,
  264, 296, 300 n.9, 307, 321, 323, 324,
  325, 329, 350, 357, 368
Reuvers, Robin 358
Rieffle, M. A. 340
Rorty, Richard 174 n.10
Rosaler, J. 325
Röseberg, Ulrich 22
Rosen, Nathan 14, 37, 182, 264, 291, 299,
  312, 321, 323, 373
Rosenfeld, Léon 22, 24, 28, 39 nn.1,4, 40
  n.5,14, 62, 68, 116–17, 164–6, 168,
  194, 195, 225, 226, 227, 355
Rutherford, Ernst 214

Sánchez– Gómez, J. L. 351
Sauders, Simon 289
Schack, Rüdiger 351, 367, 370, 374, 375
Scheibe, Erhard 26, 240
Schlick, Moritz 330 n.4
Schlosshauer, Maximillian 151, 235, 237,
  241, 243, 247–8, 285 nn.10,11,13, 314
Schopenhauer, Arthur 55
Schrödinger, Erwin 4, 32, 58, 73, 122–3,
  155, 182, 197, 200, 210, 216, 217, 229,
  243, 256, 265, 293, 300 nn.4,5, 306,
  323, 328, 355, 356, 369, 374, 375
Schweber, S. S. 195
Schwinger, Julian 193

Scoppola, E. 340
Sellars, Wilfred 92, 169–70
Shimony, Abner 19
Simon, A. W. 260
Singer, I. M. 339, 345–7
Slater, John C. 259, 317
Sommerfeld, Arnold 263
space-time coordination/description 6,
  11, 31, 35, 38, 52, 56, 58, 64 n.1, 68,
  74, 76, 77, 85, 86, 100, 109, 111, 140,
  147, 150 n.11, 165, 190, 207, 239–40,
  243, 245–6, 261–6, 267, 271, 272, 293,
  304–6, 313, 319
spectator epistemology 2, 10, 93
Spehner, D. 357
Spielman, D. A. 347
Spitters, Bas 339, 347
Srivastava, N. 347
Stapp, Henry 200 n.6
Steinle, Friedrich 35
Stern-Gerlach experiment 31, 36
Steuernagel, O. 84
Stevens, Marco 347
Strauss, Martin 24
Strawson, Peter F. 136
subject, subjective 51, 59, 74, 75, 76, 107,
  136, 172, 299 n.1, 301 n.13, 351
symbolic formalism/representation 4, 7,
  8–9, 37, 52, 123, 125, 128, 157, 264,
  265, 268, 271, 276, 277, 303, 304–9,
  312, 321, 322, 328, 329

Tasaki, H. 340
Teller, Edward 311
't Hooft, Gerard 193
Tomonaga, S.-I. 193

unambiguous communication 3, 5, 34,
  55, 75, 101, 103, 127, 244, 245, 246,
  247, 270

Valentini, Antony 68
van Frassen, Bas 160, 161
van Heugten, Jasper 359
Van Wesep, R. A. 351
Veltman, M. 193
Vienna Circle 22
von Helmholz, Herman 26, 40 n.11
von Humboldt, Wilhelm 26

von Neumann, J. 7, 9, 14, 47, 54, 165–6, 226, 227, 228, 254, 323, 324, 336, 337, 338–42, 345, 346–7, 350
von Weizsäcker, C. F. 22, 41 n.19, 47, 58, 70, 225–6, 227, 238, 244, 293, 311

Wallace, David 289
Washburn, S. 219
wave function (state function psi function) 4, 7–8, 9, 15, 37, 58, 108, 118, 122, 123–5, 135, 143, 147, 148, 150, 161, 165, 173 n.2, 191, 192, 210, 213, 291, 295, 296, 297, 300 n.4, 307, 308, 323, 324, 329, 330, 331 n.11, 355, 358
wave-particle dualism/complementarity 69–75 *passim*, 207–20 *passim*
Weaver, N. 347
Weinberg, Steven 162, 174 n.4
Werner, R. F. 340
Weyl, Herman 199

Wheeler, John A. 62, 84, 136, 150 n.6, 156–8, 162, 166–7, 289, 329, 331 n.12
Whitehead, Alfred N. 185, 188, 200 n.6
wholeness/holism 12, 55, 111, 133–51, 167, 171, 174 n.10, 180, 187, 242, 247, 318, 323, 331 n.8
Wigner, Eugene 9, 164, 218, 227, 297, 372
Winful, H. G. 219
Wittgenstein's philosophy of language 12, 126–9, 155, 167, 173
Wolters, Sander 359
Wootters, William K. 79, 80, 248

Zeh, Heinz-Dieter 229, 230, 289
Zeilinger, Anton 369
Zinkernagel, Henrik 134, 141–2, 147–9, 315, 316, 330 n.6, 331 n.11
Zurek, Wojciech H. 79, 80, 162, 229, 230, 248

Lightning Source UK Ltd.
Milton Keynes UK
UKHW020606170519

342845UK00003B/32/P